The Spacetime Origin
Of the Universe

With Visible Dark Matter & Energy

Third Edition

I0041090

by
Vladimir B. Ginzburg

Edited by Ellen Orner & Nobuyuki Kanai
Cover by Eugene B. Ginzburg

Helicola Press
Division of IRMC, Inc.
612 Driftwood Drive
Pittsburgh, Pennsylvania, 15238

The Spacetime Origin of the Universe
With Visible Dark Matter & Energy

Third Edition

Published by
Helicola Press
Division of IRMC, Inc.
612 Driftwood Drive
Pittsburgh, Pennsylvania
USA
www.helicola.com

Printed in the United States of America

ISBN: 978-0-9671432-6-2 (Perfect bound)
ISBN: 978-0-9671432-7-9 (Hardcover)

Current printing (last digit)
10 9 8 7 6 5 4 3 2

My special thanks go to the members of my family: to my gorgeous, friendly and energetic grandson Alex, to my beautiful granddaughter Alexandra, who since the age of 10 became the youngest scientifically-minded person capable of understanding the basic principle of my theory; to my daughter Ellen, with whom I brainstormed some ideas related to my theory, and who edited my three previous books on this subject and also Chapter 1 of this book; to my son Gene, who provided many digital illustrations for my books, and also designed the covers of all my books including this one; to my brother Paul who patiently followed my research in this field and provided me with his valuable comments and corrections, and, finally, to my wife Tanya, whose moral support, advice, and assistance in editing this book were invaluable.

I also dedicate this book to my late wonderful parents and very wise grandparents.

Dedications

This book is dedicated to the German-Swiss physicist Albert Einstein who, in 1914, presented gravity as a geometric property of spacetime. Besides explaining for the first time the physical meaning of gravity, Einstein opened the door to the search for other physical phenomena that could also be explained as properties of spacetime. Einstein was a dreamer; I will not be surprised if in one of his dreams he may have envisioned the spacetime as the origin of the Universe.

This book is also dedicated to Walter Russell, the American architect and artist who envisioned the spacetime origin of the Universe in his book:

> *That the man calls matter, or substance, has no existence whatsoever. So-called matter is but waves of the motion of light, electrically divided into opposed pairs, then electrically conditioned and patterned into what we call various substances of matter. Briefly put, matter is the motion of light, and motion is not substance. It only appears to be. Take motion away and there would not be even the appearance of substance.*
>
> Walter Russell
> The author of *The Secret of Light.*

ACKNOWLEDGEMENTS

In the first Chapter of this book I gave credits to many professional and amateur scientists of all times whose ideas are related to the Universal Spacetime Theory. I am very grateful to three scientists who provided their valuable contributions to this book. The American theoretical physicist Richard Gauthier outlined briefly his transluminal energy quantum (TEQ) models of photon, electron, cosmic quantum and dark matter. The American-Australian molecular biologist Horace R. Drew, the co-author of the ground-breaking book *Understanding DNA – The Molecule and How It Works*, wrote an article on the helical theory of DNA. The third article on the vortices in biology and medicine was written by Dr. James Oschman, a recognized American expert in energy medicine – who is also a physiologist, cellular biologist. The article was written jointly with a biophysicist and his naturalist colleague, Nora Oschman.

I am also grateful to several scientists for making their valuable and stimulating comments on my idea of multiple-level Universe. Among them are: Professor Carlo Rovelli (Department of Physics and Astronomy of the University of Pittsburgh), Professor Gregory M. Townsend (Department of Physics of the University of Akron), Professor Clifford Taubes (Department of Mathematics, Harvard University).

Several American physicists made valuable comments on my model. Among them are: Professor Rudolph Hwa (Department of Physics, University of Oregon), Dr. Blair M. Smith (Innovative Nuclear Space Power and Propulsion Institute, University of Florida), Professor Edward F. Redish (Department of Physics, University of Maryland), Professor David F. Measday (Department of Physics and Astronomy, University of British Columbia, Canada), Dr. Eric Carlson (Department of Physics, Wake Forest University), and Professor Warren Siegel (Department of Physics and Astronomy, Stony Brook University).

Thanks to the recommendations made by Dr. Akhlesh Lakhtakia (Department of Engineering and Mechanics, the Pennsylvania State University), I was able to publish the first three papers describing the earliest versions of my theory in *Speculations in Science and Technology* in 1996-1998. Several my papers on this subject were published with the assistance provided by Dr. Harold Fox (Editor of *The Journal of New Energy*). Natural Philosophy Alliance (NPA) provided me with an opportunity to present various parts of my theory during its several annual conferences and published them in their proceedings. I also contributed the latest article related my theory to the 2015 *Material Science & Technology Conference* held in Columbus, Ohio.

I appreciate very much my interesting discussions with late Marvin Solit, Director of Foundations for New Directions, and his ideas about possible space orientations of toryces. I am very grateful to Dr. Cynthia Kolb Whitney (Editor of the Journal *Galilean Electrodynamics*) for reading my book *Prime Elements of Ordinary Matter, Dark Matter & Dark Energy* and giving me very valuable suggestions. Mr. Nobuyuki Kanai from Osaka, Japan followed my research on spirals for the last several years and provided me with many useful corrections of my writings, including this book. I will always remember very stimulating discussions with the late Distinguished Professor Eli Gorelik of the University of Pittsburgh, Pennsylvania about a crystal structure of a nucleon core. It was my great pleasure to have very fruitful discussions with Dr. Vladimir M. Segal regarding to the theoretical and experimental discoveries of materials with unique properties.

I value greatly my two extensive meetings with Dr. Arlie Oswald Petters, the Benjamin Powell Professor and Professor of Mathematics, Physics and Business Administration at Duke University held at the Duke University in October 2015. After an open-minded review of a basic concept of my theory, Dr. Petters offered me several practical recommendations on how to introduce it to the academic world. Finally, let me express my gratitude to Dr. John Kern II, Chair, Dr. Anna Haensch and Ms. Larisa Shtrahman, Instructor of Mathematics (all from Department of Mathematics & Computer Science of Duquesne University, Pittsburgh, PA), for providing me with an opportunity to conduct a seminar on mathematical aspects of my theory in their department on February 15, 2017. I also appreciate an opportunity to present my theory at the Materials Science & Technology (MS&T) 2017 Conference to be held in Pittsburgh, PA.

Vladimir B. Ginzburg

Contents

THE ESSENCE OF
THE UNIVERSAL SPACETIME THEORY

The Universal Spacetime Theory (UST) is the main subject of this book. It attempts to answer two most intriguing and puzzling questions related to the science and philosophy:

1. How was the Universe created out of nothing?
2. Is it possible to develop a unified theory of both micro- and macro-worlds applied to the ordinary matter, dark matter and dark energy?

Below is a brief summary of how the UST proposes to answer to these questions.

Main principle - What we perceive as matter, field, gravity, mass, charge and energy, etc. are merely various metamorphoses of toroidal spiral spacetimes called *toryces* which degrees of freedom are limited by three *toryx spacetime postulates*.

Quantum vacuum – The micro-toryces are created from *quantum vacuum* - a chaotic assembly of short-lived *virtual spacetimes* without any certain forms and limitations of their degrees of freedom, and with their existence governed by the rules of uncertainty and probability.

Elementary matter particles – The micro-toryces exist in four topologically-polarized states and the polarized micro-toryces are unified to form four kinds of elementary matter particles called *electrons*, *positrons ethertrons* and *singulatrons*, the building blocks of nucleons and atoms.

Elementary radiation particles – The micro-toryces exist in quantum spacetime states and emit the helical spiral spacetimes called *helyces* which degrees of freedom are defined by four *helyx spacetime postulates*. The micro-helyces exist in four topologically-polarized states and they are unified to form elementary radiation particles responsible for the communications between the elementary matter particles. Some of them propagate with superluminal velocities.

Spacetime fields – The spacetime fields are associated with celestial bodies comprising many elementary particles, nucleons and atoms. These fields are made up of the *macro-toryces* which degrees of freedom are restricted by the same three toryx space-time postulates applied to the micro-toryces. The difference between the spacetime properties of the micro- and macro-toryces is only in the equation defining the radius of the toryx eye. The macro-toryces also exist in four topologically-polarized states and they are unified to form four kinds of spacetime fields responsible for the interaction between the celestial bodies.

Gravitons – The gravitons are made up of the *macro-helyces* which degrees of freedom are restricted by the same four space-time postulates applied to the micro-helyces. They also exist in the same four topologically-polarized states. After their unifications, the polarized macro-helyces form gravitons responsible for the communications between celestial bodies.

Math of the UST – The development of the UST was greatly stimulated by several theories. Among them are the vortex theory, the general theory of relativity, the quantum mechanics, the wave mechanics, the standard model and the string theory. Unlike some of these theories employing very sophisticated math, the UST uses mainly elementary math taught in a high school, with some modifications required to comply with the toryx and helyx spacetime postulates.

Basic Concept of the Toryx

As shown in Figures 01 and 02, toryx is the spiral spacetime entity made up of two parts:

- Circular *leading string* with the radius r_1 propagating along its circular path with the velocity V_1 and
- Toroidal *trailing string* with the radius r_2 propagating with velocity of light c along its toroidal spiral path synchronously with the leading string

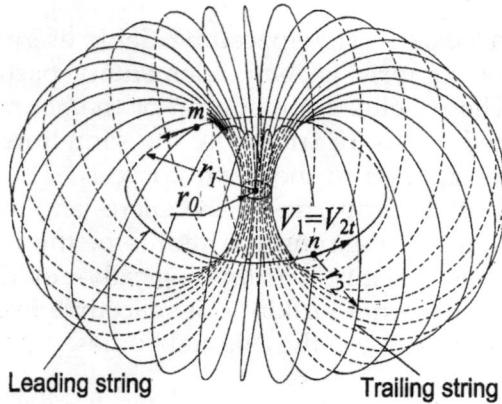

Figure 01. Isometric view of a toryx.

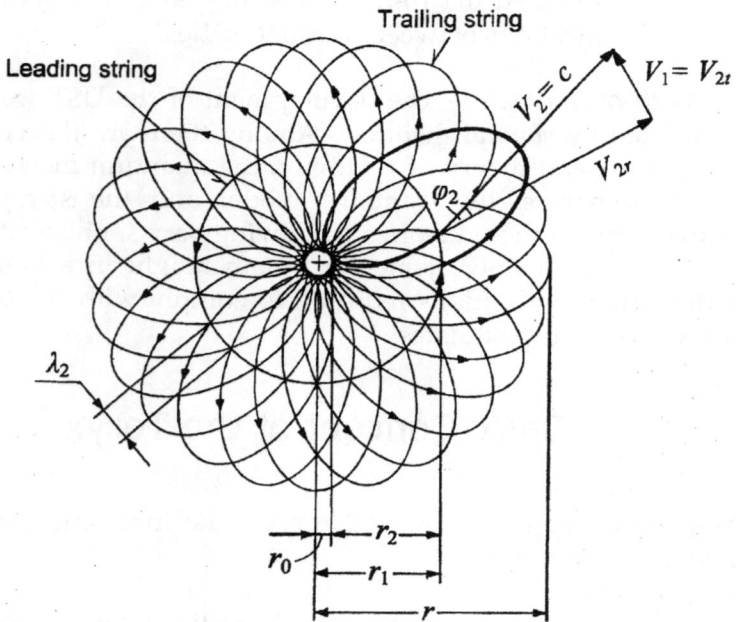

Figure 02. Top view of a toryx.

The toryx spacetime properties are based on three fundamental equations called the *toryx spacetime postulates*. These postulates limit the toryx degrees of freedom, allowing one to establish the relationships between all its spacetime parameters.

Toryx Spacetime Postulates

1. The length of one winding of the trailing string L_2 is equal to the length of one winding of the leading string L_1:

$$L_2 = L_1 = 2\pi r_1 \quad (+\infty < r_1 < -\infty)$$

2. The toryx eye radius r_0 is equal to a real positive constant:

$$r_0 = r_1 - r_2 = const. \quad (+\infty < r_1 < -\infty)$$

3. The spiral velocity of the trailing string V_2 is constant and equals to the velocity of light in vacuum c at each point of its spiral path.

$$V_2 = \sqrt{V_{2t}^2 + V_{2r}^2} = c = const. \quad (+\infty < r_1 < -\infty)$$

In spite of their apparent simplicity, the toryx spacetime postulates yield very complex spacetime transformations of toryces. There are two principal causes for these transformations. Firstly, these transformations occur because all three toryx spacetime postulates allow the radius of leading string r_1 to be reduced from positive to negative infinity. Consequently, at certain values of r_1, this radius and also the radius of trailing string r_2 and the wavelength of trailing string λ_2 reverse their signs, making the toryx either partially or completely topologically-inverted inside out. Secondly, the third toryx spacetime postulate sets no limits on the values of either translational velocity V_{2t} or rotational velocity V_{2r} of trailing string to be superluminal, while its spiral velocity V_2 remains constant and equal to the velocity of light.

Thanks to these transformations, the toryces are capable to exist in four polarized states, creating the conditions for the formation of elementary matter particles by the unification of properly-matched polarized toryces. To provide a mathematical description of the toryx spacetime properties, it is necessary to modify several aspects of elementary math, including the definitions of zero, number line and elementary trigonometric functions. The modified elementary math is called the *spiral spacetime math*.

Generic Properties of the Toryces

The toryces are the smallest entities in the Universe that contribute to the self-preservation of the Universe thanks to their numerous properties found in many larger entities of the Universe.

Motion – The toryx is the smallest entity in the Universe with its components in the state of motion. All entities of the Universe are in the state of motion. Take the motion away and the Universe, as we know it, will cease to exist.

Propagation with the velocity of light – The toryx is the smallest entity in the Universe in which its trailing string propagates with velocity of light. Take the velocity of light away and the Universe, as we know it, will cease to exist.

Limitation of degrees of spacetime freedom – The toryx is the smallest entity in the Universe with limited degrees of spacetime freedom. The degrees of spacetime freedom are limited in all entities of the Universe. Take the limitation of degrees of freedom away and the Universe, as we know it, will cease to exist.

The planetary motion – The toryx is the smallest entity in the Universe in which its leading strings follow the Spacetime Law of Planetary Motion for which the classical law of planetary motion is a particular case. Take the Spacetime Law of Planetary Motion away and the Universe, as we know it, will cease to exist.

Spirality – The toryx is the smallest entity in the Universe in which its trailing string propagates along a spiral path. Spirality pervades the entire universe, including the paths of all celestial bodies. Take the spirality away and the Universe, as we know it, will cease to exist.

Polarization – The toryx is the smallest entity in the Universe that exists in the topologically-polarized states. Polarization is prevalent in the Universe. The atoms are made up of the electrically polarized electrons and nuclei. Similarly, the chemical compositions are made up of the chemically-polarized acids and bases. Take the polarization away and the Universe, as we know it, will cease to exist.

Expansion & contraction – The toryx is the smallest entity in the Universe that is capable to expand and contract. This capability can be found in all entities of the Universe. Take the expansion and contraction away and the Universe, as we know it, will cease to exist.

Quantization – The toryx is the smallest entity in the Universe that exists in quantum spacetime (energy) states. This property extends to all atomic electrons. According to the UST, this capability is also applied to other elementary particles, and possibly to the planets and their moons. Take the quantization away and the Universe, as we know it, will cease to exist.

Absorption & release of spacetime – The toryx is the smallest entity in the Universe that is capable to absorb and release the spacetime (energy). All entities of the Universe possess this capability. Take the absorption and release of spacetime (energy) away and the Universe, as we know it, will cease to exist.

Unification & coexistence – The toryx is the smallest entity in the Universe that can be unified and coexist with an oppositely-polarized toryx, forming an elementary matter particle. In the universe, electrons and protons are unified and co-exist in atoms. The living creatures with opposite sexes are unified to produce their offspring. Take the unification and coexistence away and the Universe, as we know it, will cease to exist.

Radiation – The toryx is the smallest entity in the Universe that is capable to emit radiation in the form of a topologically-polarized helical spacetime called the *helyx*. A pair of the unified polarized helyces forms an elementary radiation particle. The Universe is filled with radiation. Take the radiation away and the Universe, as we know it, will cease to exist.

Spacetime properties – The toryces are the smallest entities in the Universe that possess clearly-defined spacetime properties, such as length, wavelength, radius, time, velocity and frequency. The spacetime properties can be found in all entities of the Universe. Take the spacetime properties away and the Universe, as we know it, will cease to exist.

Physical properties – The toryces are the smallest entities in the Universe that possess physical properties, such as mass, charge, energy, density, elasticity, angular momentum, and magnetic moment. The toryx physical properties are defined by their spacetime properties. The physical properties can be found in all entities of the universe.

Metamorphoses of the Toryces

As the relative radius of the toryx leading string $b_1 = r_1/r_0$ decreases from positive to negative infinity, the steepness angle of the toryx trailing string φ_2 increases from 0 to 360°.

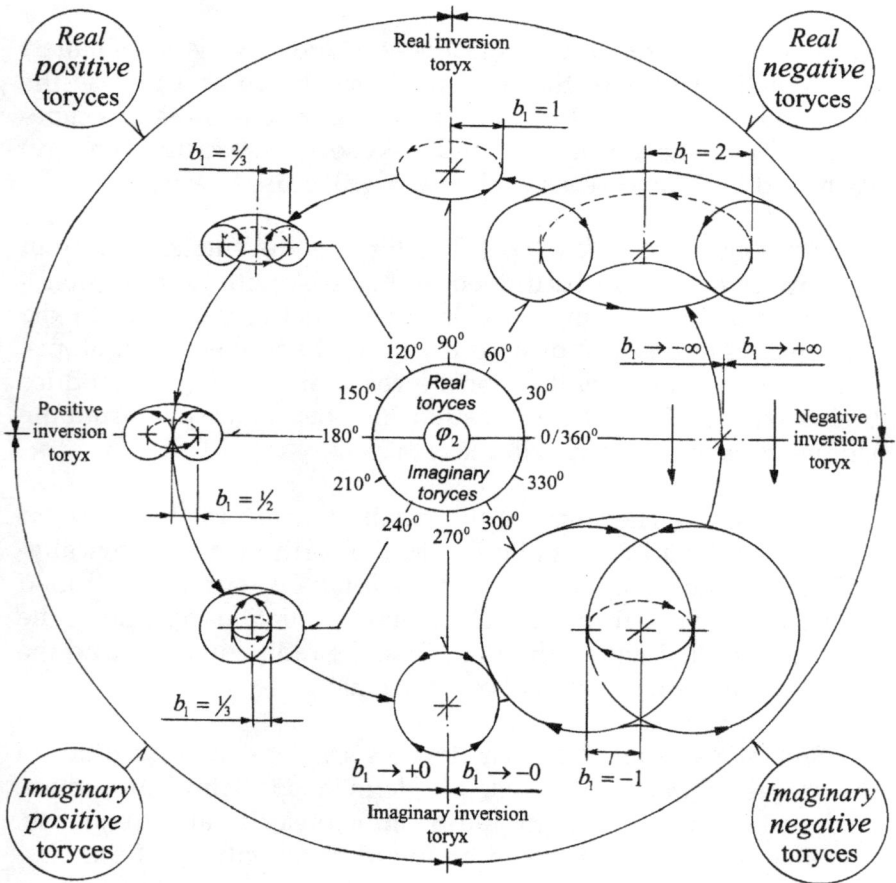

Figure 03. Metamorphoses of the toryx leading and trailing strings as a function of the steepness angle of the trailing string φ_2.

Consequently, the following changes in toryces occur:

- The sign of the radius of trailing string reverses at $\varphi_2 = 90^0$.
- The reality of the wavelength of trailing string changes from real to imaginary at $\varphi_2 = 180^0$.
- The sign of the radius of leading string reverses at $\varphi_2 = 270^0$.
- All three above changes occur at $\varphi_2 = 0^0$ and 360^0.

This produces four kinds of topologically inverted toryces shown in Figure 03: *real negative toryces, real positive toryces, imaginary positive toryces* and *imaginary negative toryces*.

Excitation and Oscillation of the Toryx

Toryces exist in excitation and oscillation quantum states as shown in Figure 04. During the excitation of a toryx its eye radius r_0 remains constant while other parameters change in quantum steps as a function of the *toryx quantization parameter z* given by the equations:

$$z = r_1/r_0 = 2(n\Lambda)^m$$

where:
  $m = 0, 1, 2,...$, toryx exponential excitation quantum state
  $n = 0, 1, 2,...$, toryx linear excitation quantum state
  $\Lambda = 137$, *spacetime quantization constant*.

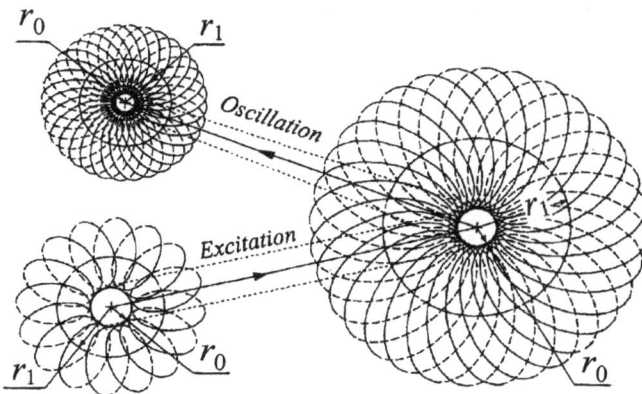

Figure 04. Excitation and oscillation of a toryx.

The exponential excitation quantum state m of the toryces depends on the *spacetime levels*: *L1* (dark energy), *L2* (ordinary matter), *L3* (dark matter), etc. The linear excitation quantum states n exist within each spacetime level.

During its oscillation the entire toryx, including its eye radius r_0, reduces in size and its all spacetime parameters change in the quantum steps defined by the *toryx oscillation quantum states q*.

Elementary Matter Particles

Four kinds of elementary matter particles are formed by the unification of the topologically-polarized toryces: *electrons, positrons, ethertrons* and *singulatrons* as shown in Figure 05. In each particle, one of its constituent toryces absorbs spacetime, while the other one releases it.

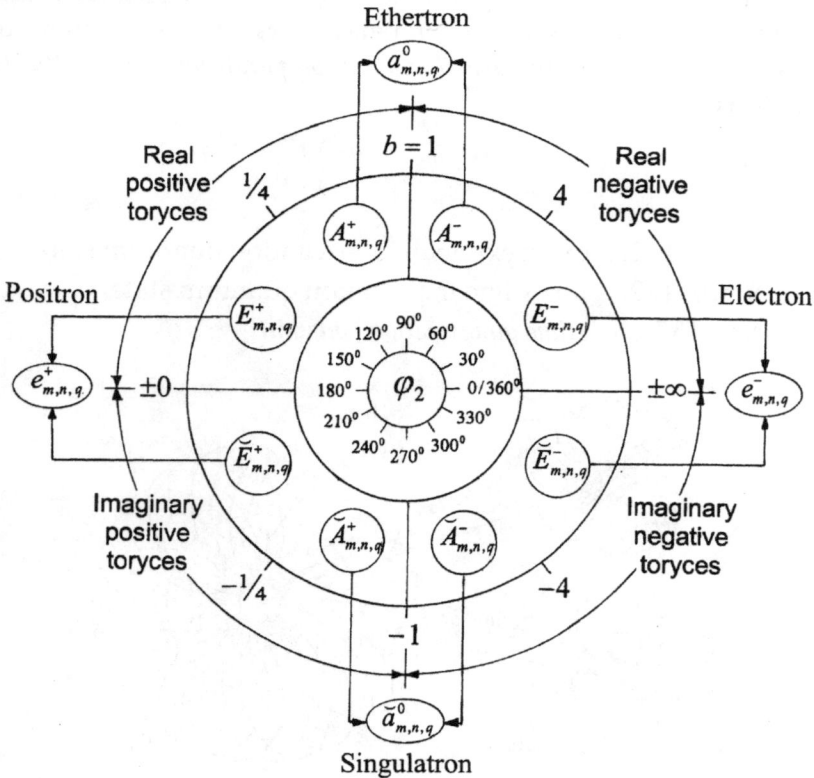

Figure 05. Formation of four elementary matter particles.

Stable elementary matter particles are formed only if they follow the proposed *Spacetime Conservation Law* according to which the sum of polarization parameters of their constituent toryces must be infinitely small.

Helyx Structure

Excited and oscillated toryces emit spiral spacetimes called the *helyces*. The helyx has two levels. As shown in Figure 06, the first level is in the form of a double-helical spiral forming the helyx *leading string* A_1.

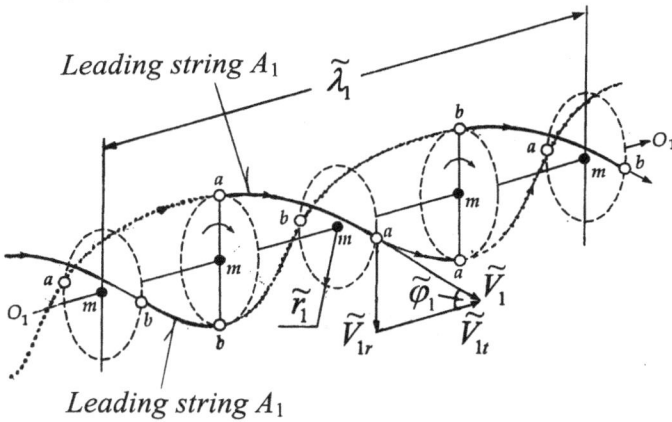

Figure 06. Structure of the helyx leading string.

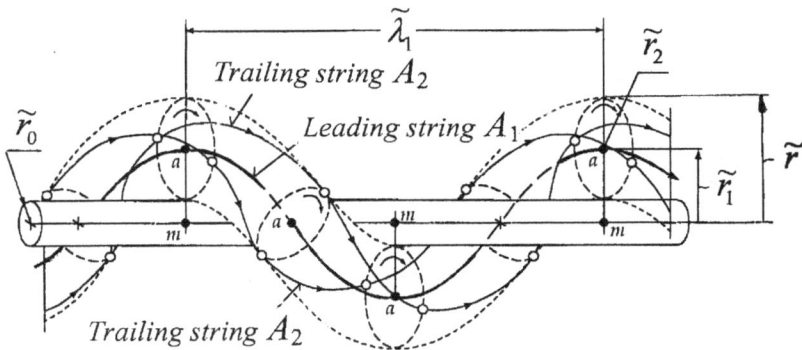

Figure 07. One of two double-helical leading strings A_1 accompanied by two branches of the trailing string A_2.

The second level of the helyx is made up of two double helices forming the helyx *trailing string* as shown in Figure 07. Four branches of the helyx trailing string A_2 are wrapped around two branches of the helyx leading string A_1.

Metamorphoses of the Helyces

Similarly to the toryx, the degrees of freedom of helyces are also limited, allowing them to form four kinds of topologically-polarized helyces shown in Figure 08.

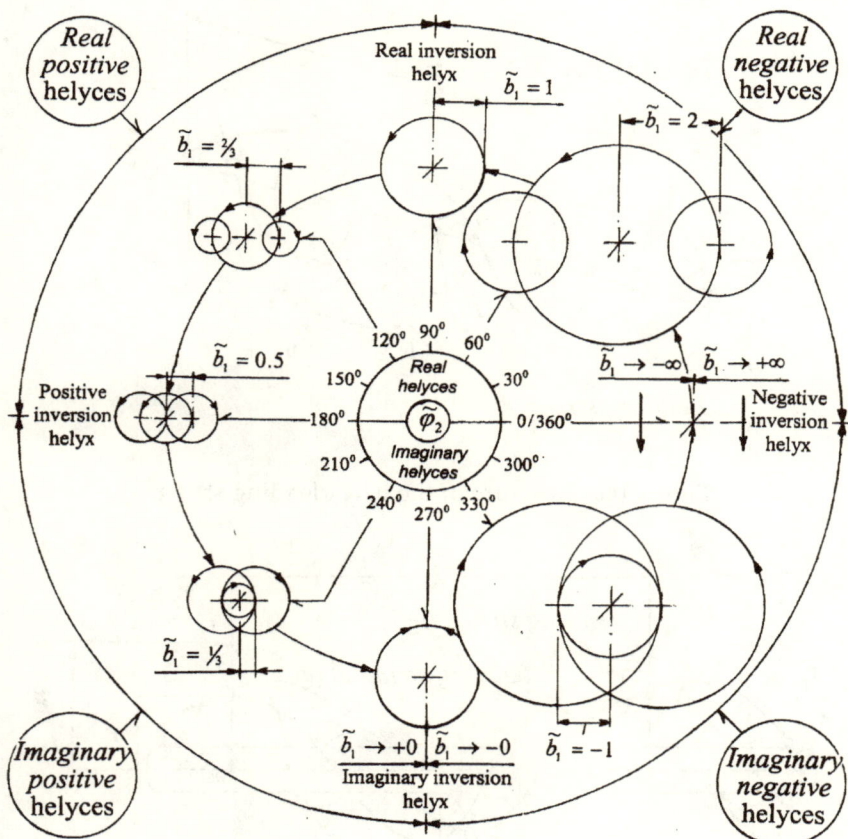

Figure 08. Transformations of cross-sections of helyces as a function of the apex angle of the trailing string φ_2.

Elementary Radiation Particles

Four kinds of elementary radiation particles are formed by the unification of the topologically-polarized helyces: *electons, positons, ethertons* and *singulatons* as shown in Figure 09.

In each particle, one of its constituent helyces absorbs space-time, while the other one releases it. The names of the elementary radiation particles are similar to the names of their parental elementary matter particles responsible for the creation of helyces.

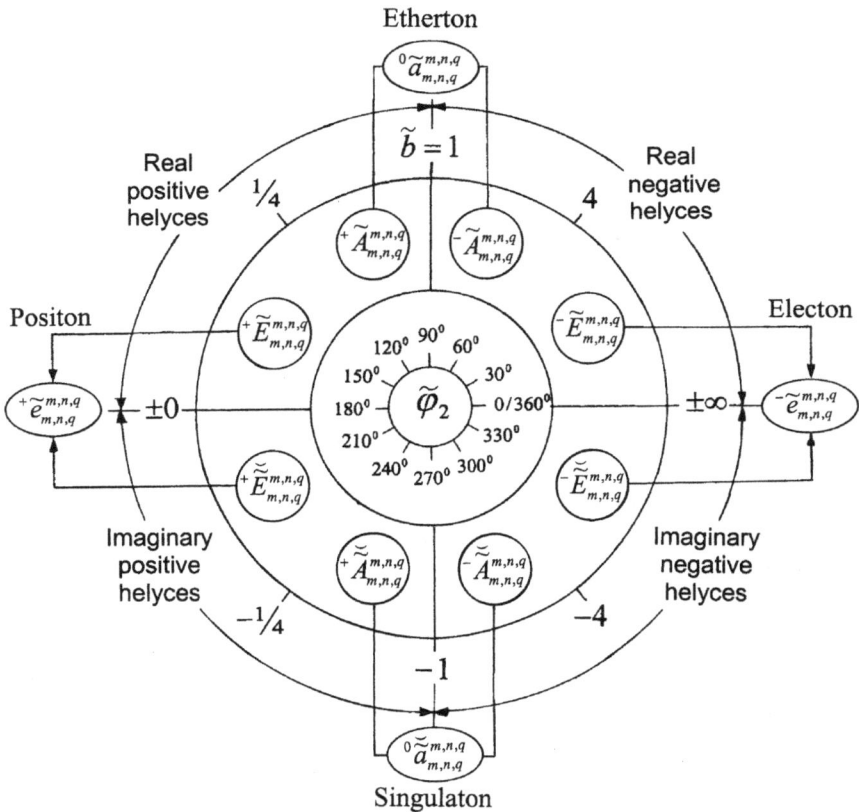

Figure 09 Formation of four elementary radiation particles.

Spacetime Levels of the Universe

According to the UST, the universe exists at several spacetime quantum levels L. The properties of the first three spacetime levels are shown in the Table 1.

Table 1. Properties of three spacetime quantum levels of the Universe.

Spacetime levels L	Frequencies & velocities of radiations emitted by			
	Ethertron	Singulatron	Electron	Positron
$L1$	$m = 0$ $10^{17} - 10^{19}$ Hz luminal	$m = 0$ $\approx 10^{20}$ Hz superluminal	$m = 1$ $10^{16} - 10^{17}$ Hz Luminal	$m = 1$ $10^{16} - 10^{17}$ Hz luminal
$L2$ (ordinary Matter)	$m = 1$ $10^{16} - 10^{17}$ Hz luminal	$m = 1$ $\approx 10^{22}$ Hz superluminal	$m = 2$ $10^{13} - 10^{15}$ Hz Luminal	$m = 2$ $10^{13} - 10^{15}$ Hz luminal
$L3$	$m = 2$ $10^{13} - 10^{15}$ Hz luminal	$m = 2$ $\approx 10^{25}$ Hz superluminal	$m = 3$ $10^{11} - 10^{13}$ Hz Luminal	$m = 3$ $10^{11} - 10^{13}$ Hz luminal

At each spacetime level, the properties of elementary matter particles depend on the exponential excitation quantum states m of their constituent toryces. Each spacetime level of the Universe is identified by the frequency ranges and velocities of the elementary radiation particles emitted by their parental elementary matter particles. The radiations emitted by the electrons, positrons and ethertrons propagate with the velocity of light, while the radiations emitted by the singulatrons propagate with superluminal velocities.

The calculated frequencies of electons emitted by electrons are within the following frequency ranges:

- For the spacetime level $L1$ at which the exponential excitation quantum state $m = 1$, the frequencies of electons are within the frequency range of <u>cosmic X-ray background (CXB) radiations</u>.
- For the spacetime level $L2$ at which the exponential excitation quantum state $m = 2$, the frequencies of electons are within the frequency ranges of <u>infrared, visible and ultraviolet radiations</u>.

- For the spacetime level *L3* at which the exponential excitation quantum state $m = 3$, the frequencies of electons are within the frequency ranges of infrared and microwave frequency ranges known as <u>cosmic microwave background (CMB) radiation</u>.

Table 2. Relative properties of the hydrogen atoms $\downarrow H_L^0$
in the ground state $n = 1$ of several spacetime levels L.

Spacetime levels L	Hydrogen atoms	Electron orbital radius ratio	Orbital magnetic moment ratio	Mass ratio	Atomic density ratio
L1	$\downarrow H_{L1}^0$	1/137.25	1/11.7	1/106.8	24200
L2 (ordinary matter)	$\downarrow H_{L2}^0$	1.0	1.0	1.0	1.0
L3	$\downarrow H_{L3}^0$	137.0	11.7	136.7	1/18800

Table 2 shows relative properties of hydrogen atoms in the ground state $n = 1$ in the spacetime levels *L1* and *L3* in respect to the properties in the spacetime level *L2* (ordinary matter):

- In the spacetime level *L1*, the nucleons are about 107 times lighter, the orbital radii of their atomic electrons are about 137 times smaller and, consequently, the atomic densities are about 24200 times greater than in the spacetime level *L2* (ordinary matter).
- In the spacetime level *L3*, the nucleons are about 137 times heavier, the orbital radii of their atomic electrons are about 137 times larger and, consequently, the atomic densities are about 18800 times smaller than in the spacetime level *L2* (ordinary matter).

Since the atomic density of the hydrogen atoms decreases with the increase in the spacetime level, the UST proposes that the observed expansion of the universe could be caused by the expansion of spacetimes from lower to higher levels *L*.

Similar dependences of atomic properties on the spacetime levels are expected for all atoms. This opens a path for the creation of new materials with unique properties, including lighter materials with greater strength.

Summary of Predictions of the UST

The UST confirms the following already known information:

- Orbital radii of atomic electrons of hydrogen atom
- Orbital velocities of the atomic electrons of hydrogen atom
- Mass, charge and magnetic moment of proton, neutron and muon
- Mass and charge of tau particle
- Decay of a neutron into a proton, electron and electron neutrino.

The UST makes the following new predictions:

- All elementary particles exist in the excited and oscillated quantum states, and also in the spacetime quantum levels.
- Each proton contains an excited positron.
- Each neutron contains a harmonic positron and electron
- There is a stable neutron.
- Leptons are the oscillated excited electrons.
- Neutrinos are the radiation particles emitted by the oscillated excited electrons.
- Electrons, positrons and ethertrons emit radiation particles propagating with velocity of light, while singulatrons emit radiation particles propagating with superluminal velocities.

Table 3. Calculated properties of singulatons emitted by excited singulatrons.

Singulatrons		Emitted singulatons			
$k - j$	Sign	Relative radius*	Relative velocity**	Frequency Hz	Energy GeV
2 − 1	-	- 0.002437	+ 411.3934	1.39×10^{25}	57.4756
	+	+ 0.002437	- 409.3931	1.39×10^{25}	57.6156
3 - 2	-	- 0.001461	+ 685.3210	2.32×10^{25}	95.8393
	+	+ 0.001461	- 683.3208	2.32×10^{25}	95.9794
4 - 3	-	- 0.001044	+ 959.2487	3.25×10^{25}	134.2031
	+	+ 0.001044	- 957.2484	3.25×10^{25}	134.3431

*) in respect to the helyx eye radius, **) in respect to velocity of light.

Table 3 shows the calculated energies of singulatons emitted by excited singulatrons forming nucleon cores when they are transferred from the higher excitation quantum states $n = k$ to the lower quantum states $n = j$.

Notably, the calculated energies of singulatons emitted by excited singulatrons shown in Table 3 are comparable with the measured energies:

- When $k = 4$ and $j = 3$, the calculated energies are approximately 6.8% greater than the measured energy of a new particle discovered during the proton-proton collision experiments conducted at the CERN Large Hadron Collider (LHC) in 2012.
- When $k = 3$ and $j = 2$, the calculated energies are approximately 5.2% greater than the measured energy of the Z particle.

Macro-Trons & Macro-Tons

The UST considers a celestial body as an assembly of micro-toryces. This assembly forms the *macro-toryces* intimately associated with each celestial body.

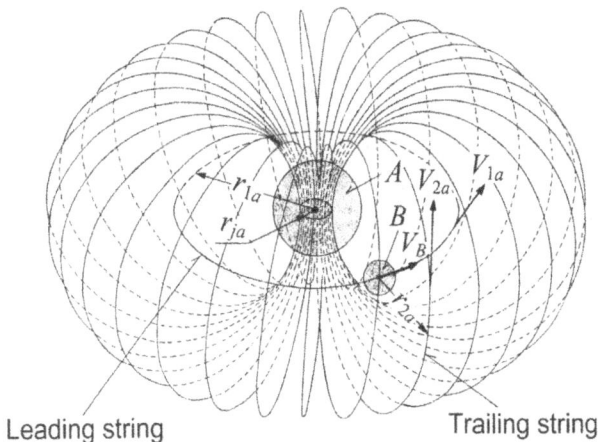

Leading string   Trailing string

Figure 10. Central body A with a macro-toryx encompassing the satellite body B.

Fig. 10 shows a macro-toryx associated with the body A which leading string passes through the center of gravity of the satellite body B. Spacetime properties of the macro-trons are described by the same equations, except for the toryx eye radius r_0 that is dependent on the mass of each celestial body.

Similarly to the four kinds of micro-toryces shown in Figure 03, there are four kinds of topologically-polarized *macro-toryces* that exist in excitation quantum states.

- Real negative macro-toryces
- Real positive macro-toryces
- Imaginary positive macro-toryces
- Imaginary negative macro-toryces.

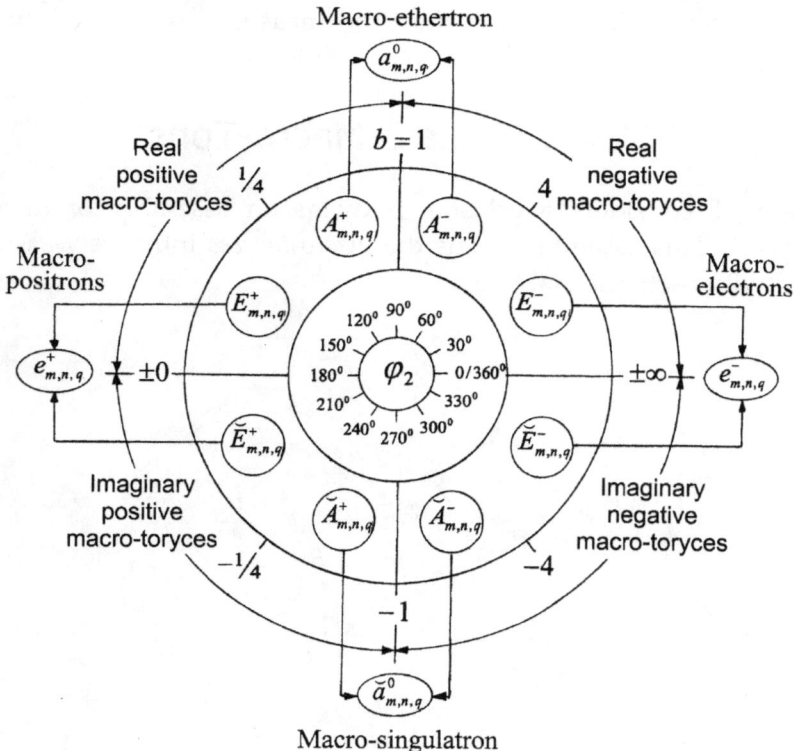

Macro-ethertron

$a^0_{m,n,q}$

$b = 1$

Real positive macro-toryces   ¼

Real negative macro-toryces   4

$A^+_{m,n,q}$ $A^-_{m,n,q}$

Macro-positrons

Macro-electrons

$E^+_{m,n,q}$ $E^-_{m,n,q}$

$120°$ $90°$ $60°$
$150°$ $30°$
$e^+_{m,n,q}$ ± 0 $180°$ φ_2 $0/360°$ $\pm\infty$ $e^-_{m,n,q}$
$210°$ $330°$
$240°$ $270°$ $300°$

$\breve{E}^+_{m,n,q}$ $\breve{E}^-_{m,n,q}$

Imaginary positive macro-toryces $-¼$

Imaginary negative macro-toryces -4

$\breve{A}^+_{m,n,q}$ $\breve{A}^-_{m,n,q}$

-1

$\breve{a}^0_{m,n,q}$

Macro-singulatron

Figure 11. Formation of four macro-trons from polarized macro-toryces.

Similarly to the four kinds of micro-trons shown in Figure 05, there are four kinds of *macro-trons* formed by the unification of the topologically-polarized macro-toryces as shown in Fig. 11.

- Macro-electrons
- Macro-positrons
- Macro-ethertrons
- Macro-singulatrons.

Depending on the relationship between the outer radius of the star body r_b and the radius of the real inversion macro-toryx r_j associated with a star, the stars are divided into two groups, *outverted stars* and *inverted stars*, as shown in Figure 12.

Outverted star   Inverted star

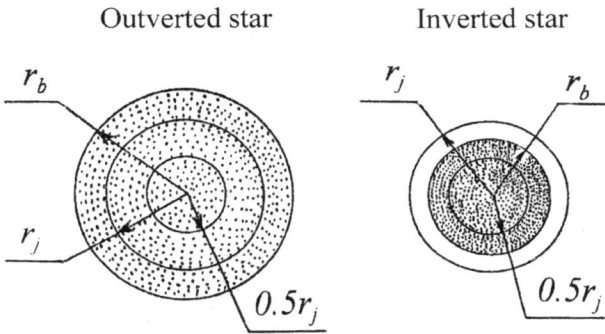

Figure 12. Two kinds of stars.

Outverted stars ($r_b > r_j$) – In the outverted stars, the outer radius r_b of the star body is greater than the radius r_j of the real inversion macro-toryx associated with this star. Our Sun is a typical example of the outverted star.

Inverted stars ($r_b < r_j$) – In the inverted stars, the outer radius r_b of the star body is smaller than the radius r_j of the real inversion macro-toryx associated with this star, These stars are extremely dense, and they are the most likely candidates for *black holes*.

Gravitons

The gravitons are the radiation particles of the macro-world. They are composed of the reality-polarized and charge-polarized matched macro-helyces. The gravitons are emitted after the orbit of either a planet or a satellite collapses to a lower quantum energy state. The names of gravitons are similar to the names of their parental macro-trons responsible for the creation of the macro-helyces.

The gravitons emitted by the excited macro-trons are called *macro-tons*, while the gravitons emitted by the oscillated macro-trons are called *macro-tins* as shown in Table 4. Similarly to the micro-trons, the macro-electrons, macro-positrons and macro-ethertrons emit radiation particles propagating with luminal velocity, while the macro-singulatrons emit radiation particles propagating with superluminal velocity.

Table 4. Types of gravitons.

Parental macro-trons	Gravitons	
	Macro-tons (emitted by excited macro-trons)	Macro-tins (emitted by oscillated macro-trons)
Macro-electron	Macro-electon	Macro-electin
Macro-positron	Macro-positon	Macro-positin
Macro-ethertron	Macro-etherton	Macro-ethertin
Macro-singulatron	Macro-singulaton	Macro-singulatin

The UST explains gravity as a result of violation of the derived Spacetime Law of Planetary Motion. Consequently, the UST derives the Spacetime Law of Gravitation for which Newton's universal law of gravitation is a particular case.

The rest is commentary, as described in this book.

PART 1

Introduction to the Universal Spacetime Theory

Part I

1

IDEAS RELATED TO
THE UNIVERSAL SPACETIME THEORY

As many theories of physics, the Universal Spacetime Theory (UST) is based on several earlier theories developed since the beginning of science. Although it is very difficult for me to evaluate the effect of all of them on the way of my thinking, still the vortex theory was certainly the most influential among them. Introduced at the dawn of science almost 2600 years ago, it had passed through several phases of gaining strength by absorbing the discoveries made by the Greek civilization, the Copernicus Revolution, the age of electromagnetism, the atomic age, and the information age. Each time this theory managed to engage the attention of a new generation of brilliant scientists, enchanting them by the deep physical meaning of its basic concept. Still, although they employed the latest advances in science, none of them was able to produce a mathematical tool to make the vortex theory usable in practice. Consequently, this wonderful theory repeatedly faded from view, awaiting another chance to return.

It appears that its lucky chance finally arrived by the end of the 20th century with the discovery of special spiral strings called the *toryx* and the *helyx*. I provided below brief descriptions of the indirect contributions made by both professional and amateur scientists to the development of the *Universal Spacetime Theory* (UST).

1.1 The Age of the Ancient Greek Civilization

Anaximander of Miletus (611-547 BC) – The Greek natural philosopher Anaximander of Miletus proposed the idea of vortex (δίνη or δῖνος) that closely resembled the mythological "cosmic whirl." At any given time, he professed, there exists an infinite number of worlds that have been separated from the infinite, τόάπειρον, which is eternal and ageless, and also the source and reservoir of all things. After our world was created, a rotary motion in a vortex caused the heavy materials to concentrate at the center, while masses of fire surrounded by a mass of air went to the periphery and later formed the heavenly bodies. The created worlds, however, do not exist forever, and, when perished, they are absorbed back into the infinite.

Anaximander logically argued that the source from which the world begins is not identical with any of the ordinary stuff known to us; it must have the capability of giving rise to the wide variety of things of contrary qualities. According to Anaximander, the things appeared from the boundless source, or *apeiron*, that functions as a storehouse of the world's qualities and provides a mechanism for releasing and absorbing these qualities. This is a never-ending cyclic process. During each cycle, the world's qualities are separated from the apeiron by the rotary motion of a vortex, and as soon as their contrary qualities become manifest, they, in turn, become reabsorbed into the apeiron.

Pythagoras of Samos (c.560-480 BC) – In ancient Greece, there was probably no other philosopher whose name was associated with so much praise and controversy. He founded a religious and philosophical Secret Society devoted to exploring the mysteries of numbers. Through this society, Pythagoras promoted a "rounded" education, introducing *the fourfold way* of study that included geometry, arithmetic, music and astronomy. One of his greatest achievements was the number theory. He and his followers, the Pythagoreans, looked for a direct connection between the whole numbers and geometrical shapes.

Pythagoras came to the conclusion that the numbers themselves expressed not only harmonies and shapes, but the essences of the things having those harmonies and shapes. Moreover, the numbers had to be the generators of everything else. Pythagorean music theory was based upon observations of a monochord, one of the most commonly used instruments of those times. Pythagoras saw a great similarity between the ratios in geometrical figures and the ratios that divided the strings of the monochord, so making music was merely another way of using proportions between numbers. He eventually applied this idea to everything in nature.

The concept of harmony led Pythagoras to a more detailed examination of the relationships between the sides of right triangles that he summarized in his famous Pythagorean Theorem:

In any right triangle, the sum of the squares of the two sides A and B is equal to the square of the hypotenuse C, so:

$$A^2 + B^2 = C^2$$

Although the Egyptians and the Babylonians discovered many right triangles to which this rule was applied, Pythagoras stated that the rule, applicable to the right triangles known before his time, was also true for "any" right triangle. By making this theoretical generalization of practical geometry, Pythagoras made another important step towards the conversion of practical geometry into science. Ironically, Pythagoras's obsession with the idea of harmony forced him to apply his own theorem only to rational numbers. After finding that some applications of his theorem produced irrational numbers, such as the square root of 2, he ruthlessly tried to suppress this discovery, an unwise decision that contributed to his tragic death and the demise of his society.

Anaxagoras of Glazomenae (500-c.428 BC) - In his book *The Nature of Things*, the Greek philosopher Anaxagoras introduced his version of the creation of the universe by the vortex motion. Like Anaximander, he made an ambitious attempt to employ vortex motion to explain the creation of the universe without reference to supernatural forces. But, unlike Anaximander, who did not explain the cause of the vortex motion, he proposed that the primordial soup was set in rotation by an all-pervading intelligence, or Mind.

To explain the multiplicity of things in the universe as well physical and biological change in those things, he proposed that the primordial soup was made of an infinite variety of substances. Consequently, as the different things separated from the primordial soup, they acquired various properties. The main argument of Anaxagoras was that there is something in everything. For instance, the water is perceived to be water because most of its parts are water, but in addition to water it contains a part of every other thing in the universe. According to Anaxagoras, besides the creation of the things with various properties, there was another important role of the vortex motion of the primordial soup. The vortex motion supplied heat through friction that ignited the sun and the stars. His system enabled Anaxagoras and his followers to describe all existing objects.

Empedocles of Acragas (c.490-c.430 BC) - the Greek philosopher Empedocles created vortices so that anyone could see what they looked like and understand how they worked. To explain the formation of celestial bodies by vortex motion, Empedocles dropped tea leaves in the center of the urn (Fig. 1.1.1), while he continued to stir the liquid. It was clearly visible that the leaves migrated towards the center and accumulated there as long as they were heavy enough to withstand the updraft. This effect came to be known to us as the "tea cup phenomenon." Empedocles thought that the Earth hovered in the updraft of the primordial vortex.

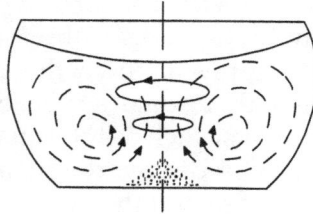

Figure 1.1.1. An illustration of the tea cup phenomenon.
Adapted from H.J. Lugt (1995).

Democritus of Abdera (c.470-c.400 BC) - The Greek philosopher Democritus of Abdera is well-known for his pioneering work on atomic theory, but he also made an important contribution to vortex theory. According to his vortex theory, the primordial soup was made of solid bodily atoms and empty space. The atoms came in various shapes and sizes, and their small dimensions made them invisible. The atoms were indestructible, homogeneous in substance, contained no void and no interstices. They were in perpetual motion in the infinitely extended void, probably moving equally in all directions. When a group of atoms became isolated, a vortex was produced by random collisions of atoms. Consequently, atoms of irregular shape became entangled and moved towards the center. Democritus considered the vortex motion so fundamental that he interpreted it simply as a general law of nature. This was probably the first attempt in the history of science to formulate a "unified theory of physics."

Archimedes of Syracuse (c.287-212 BC) – The Greek mathematician, physicist and inventor Archimedes of Syracuse made an extraordinary contribution to the development of a mathematical concept of spirals and its practical application. He discovered one of the simplest forms of spirals; it is still associated with his name.

In his work *On the Measurement of the Circle* he used the so-called *method of exhaustion* to calculate a more accurate value of π (*pi*). He put regular polygons inside and outside of a circle (Fig. 1.1.2). He then figured out their perimeters. The perimeter of the circle itself had to lie somewhere in between. After increasing the number of sides of the polygon to 96, Archimedes was able to calculate a very accurate value of π (*pi*) for his time, approximately between 3.141 and 3.143.

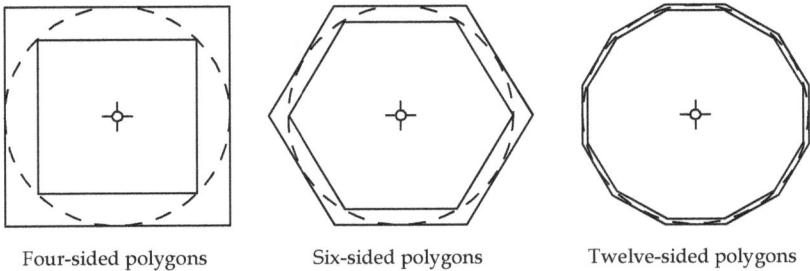

| Four-sided polygons | Six-sided polygons | Twelve-sided polygons |

Figure 1.1.2. Archimedes' method for determining the value for π (pi).

Archimedes realized that the most accurate result would be obtained when the number of sides was infinitely large. Another trend became obvious to him. As he continued to increase the number of sides of the polygon, the length of each side become shorter and shorter, approaching zero. Here, Archimedes saw both zero and infinity face-to-face; he also saw their close relationship, as the appearance of one of them led to the appearance of the other.

Perhaps this encounter with zero and infinity encouraged Archimedes to devise a scheme for naming the numbers that went beyond the limits of the language of his time. A myriad was the largest number used by the ancient Greeks. It was equivalent to our ten thousand. His scheme included any number up to $10^{80,000,000,000,000,000}$ in contemporary notations. This number survived as the largest number known to humanity for more than two millennia. A new term, *googol*, was coined by an American mathematician in 1955.

1.2 The Age of Classical Mechanics

Vortex theory was revived in the beginning of 16th century. Its revival was clearly associated with the advent of the Copernican revolution.

Johannes Kepler (1571-1630) - The German astronomer and physicist Johannes Kepler is well-known for his three laws of planetary motion that he derived from the precise astronomic observations made by the Dutch astronomer Tycho Brache (1546-1601). The most relevant to our theory is the Kepler's third law of planetary motion that states:

> The cubes of the mean distances of the planets from the Sun R
> are proportional to the squares of their periods of revolution T, so:

$$R^3 = kT^2$$

where k is constant.

While working on the development of his laws, Kepler was preoccupied with finding natural causes for the planets' motion according to these laws. He identified three causes. The first cause, he thought, was a "magnetic force," emanated from the center of the Sun. Only many years later, well after Kepler's time, was it discovered that, unlike gravitational forces, magnetic forces acting between celestial bodies are too weak to explain celestial mechanics. The second cause was a natural desire of the planets to have a "sphere, or a layer of space within which the planet must always be found." To avoid wasting of space, he proposed that the spheres of adjacent planets must touch each other through the Platonic solids. Finally, to explain more accurately the physical meaning of his third Law, Kepler imagined that the radii of the orbits of various planets must have ratios to one another, much like sound frequencies of an imaginary celestial organ that plays its eternal harmonies throughout the world.

René Descartes (1596-1650) – The French philosopher and mathematician René Descartes is mostly known for his discovery of analytical geometry. It is much less known that he also developed a general theory of the universe. According to his idea, known as the *theory of vortices,* the planets floated in an ethereal fluid composed of fine globular particles whirling in the form of vortices about the Sun. He envisioned that the stars and the circling planets were simply the visible parts of a great celestial whirlpool or vortices (Fig. 1.2.1).

He realized that every natural body in the Universe could be at rest in respect to the local matter that belonged to the vortices, and yet moving with respect to distant bodies. Therefore, the Earth did not move freely through space, but was carried around the Sun in a vortex of matter without changing its place with respect to the surrounding substance. In this way, it could be said to be "stationary." The phenomenon of whirlpools in rivers provided Descartes with an analogy to explain the mechanism of planetary motion. He envisioned the matter in the sky, where the planets were turning around the Sun, as a vortex with the Sun at its center. The parts of the vortex matter that were nearer to the Sun moved faster than those that were farther away.

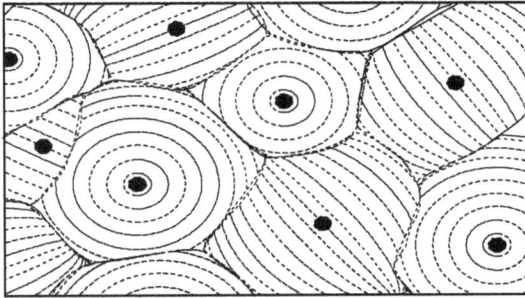

Figure 1.2.1. Congruence of celestial vortices according to Descartes. Adapted from E.J. Aiton (1972).

By using this "self-evident" analogy, Descartes was able to describe qualitatively the main observed characteristics of the planetary system, including the faster velocities of the planets located closer to the Sun and the noncircular shapes of planetary orbits. This paradigm, however, turned out to be the principal weakness in his theory. Since his fundamental cosmology was nearly all qualitative, without any quantitative and mathematical support, he came to fear that he produced nothing more than a beautiful "romance with nature."

Christiaan Huygens (1629-95) – Dutch physicist and astronomer Christiaan Huygens made several outstanding contributions to the development of science. He developed further Descartes' attempts to explain gravity by assuming that a vortex of particles of subtle matter was circling the Earth with great velocity. In his theory, Huygens replaced the cylindrically symmetrical vortices, proposed by Descartes, with a multilaterally moving vortex, in which these particles circled the Earth in all directions.

He was the first scientist who proposed an equation for the calculation of a centrifugal force. He discovered that the centrifugal force F_c, applied to a body with the mass m that orbits another body located at the distance r with the orbital velocity V, is equal to:

$$F_c = -\frac{mV^2}{r}$$

In 1679 Huygens introduced his wave theory of light. According to this theory, light propagates through ether. In Huygens' model, ether is made of small, uniform, elastic particles compressed very close together. Light propagates with very great, but finite, velocity by creating hemispherical waves around the ether particles that the light touches. Huygens' Principle stipulates that spherical waves created by single particles are too weak to transmit light, but light is transmitted when spherical waves of many particles overlap, forming a *wave front*. For his theory to work, Huygens needed ether as a medium in which the waves can propagate. He thought of ether as an "ethernal matter," which vibrates in response to a source of light, allowing the light signal to propagate.

Isaac Newton (1642-1727) – The English physicist and mathematician made a great contribution to the invention of calculus and to the development of theories of mechanics, optics and gravitation. Relevant to our story is Newton's law of universal gravitation, which states:

> Every particle in the universe attracts every other particle with a force F_g that is directly proportional to the product of their masses m_1 and m_2 and inversely proportional to the square of the distance r between them:
>
> $$F_g = \frac{m_1 m_2 G}{r^2}$$

where G is Newtonian constant of gravitation.

Notably, Newton's theory of gravity offered no means of identifying any mediator of gravitational interaction. His theory assumed that gravitation acts instantaneously, regardless of distance. There was a clear distinction between the explanation of gravity offered by Newton's law of universal gravitation, and that proposed by vortex theory. A dispute between Newton and his followers (Newtonians) and the followers of Descartes (Cartesians) became known as the "battle of the century." Among the most prominent Cartesians were the Dutch astronomer and physicist Christiaan Huygens and the German mathematician and philosopher Gottfried Leibniz (1646-1716).

The main goal of the vortex theory was to understand the physics of the process that made celestial bodies move along their orbits in space. From that point of view, this theory was very successful, because it explained many observed phenomena. One could readily visualize the forces acting on small debris in the vortices of a river and be easily persuaded that the vortices serve as a medium for transferring these forces. The problem, however, was that this explanation was largely qualitative. When it came to a precise comparison of its predictions against astronomical observations, vortex theory failed miserably.

The opposite was true with Newton's law of universal gravitation. Its predictions worked very well, but nobody could comprehend the physical meaning of the term *attraction force*, nor explain what created this force or how it was transmitted across the vast distances of empty space from one celestial body to another. This lack of physical meaning in the theory of attraction force was of great concern to many scholars, including Newton. Therefore, for some time he couldn't decide whether or not to disregard the vortex theory.

Eventually, Newton put aside the problem of defining the physical meaning of the forces and concentrated on obtaining mathematical equations for determining the mechanics of celestial bodies in the presence of attraction forces caused by gravity and repulsion centrifugal forces caused by inertia. He looked at the circular motion of a celestial body having the mass m_2 as a state of equilibrium between two equal and opposing forces applied to this body: the centripetal gravitational force F_g generated by a central body with the mass m_1 and the centrifugal force F_c (Fig. 1.2.2). As Huygens discovered earlier, the centrifugal force F_c, applied to a body with the mass m_2 that orbits with the orbital velocity V another body located at the distance r, is equal to:.

$$F_c = -\frac{m_2 V^2}{r}$$

Consequently, by equating the centrifugal force F_c and gravitational force F_g to each other ($F_c = F_g$), Newton obtain the equation expressing orbital velocity V as a function of orbital radius r.

$$V = \sqrt{\frac{m_1 G}{r}}$$

The good thing about the above equation was that it inevitably led back to Kepler's third law of planetary system. This fact was used by Newton as confirmation of his theory. As we all know, his equations worked, but the price he paid, the loss of physical meaning, was too high even for his loyal supporters.

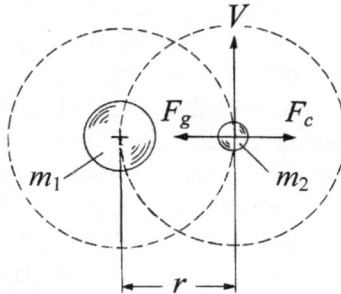

Figure 1.2.2. Celestial planetary system with two bodies.

Emanuel Swedenborg (1688-1772) – The Swedish physicist Emanuel Swedenborg made a serious attempt to challenge Newton's mechanics. He embraced this subject wholeheartedly by following his own path. Instead of applying this theory to the macro-world of celestial bodies, as Descartes, Huygens and Leibniz had done, he concentrated his efforts on applying it to the micro-world of atoms and elementary magnets. He wanted to discover the basic roots of the spiral motion in the universe. In 1734 he published a voluminous book called *Principia Rerum Naturalium*, or simply, *Principia*. According to the concept that Swedenborg presented to the world, physical reality developed from an elementary particle that is a mathematical point. In the elementary particle, "there is an internal condition tending to a spiral motion." Through this internal condition implanted in the elementary particle, a series of material particles developed, eventually producing the Universe in its present state.

A significant part of his *Principia* was dedicated to explaining the causes and mechanisms of magnetic forces. After stating that "in every particle there is a force tending to spiral gyration," Swedenborg then clearly demonstrated that two magnets with concordant spiral gyrations would attract each other while the magnets with opposite spiral gyrations will repulse each other. In spite of the success of his book, Swedenborg nevertheless knew that he was far from reaching his ultimate goal of challenging Newton's *Principia*. He soon realized that his concept of spiral motion as the origin of matter did not capture the hearts of European scientists, who were still enchanted by the simplicity and practical validity of Newton's laws.

Rudjier Bośković (1711-1787) - In 1758 a challenge to both vortex theory advanced by Leibniz and to Newtonian mechanics came from the Croatian physicist, astronomer and mathematician Rudjier Bośković. He formulated his principal concept in natural philosophy in his book *Philosophia Naturalis Theoria*. According to Bośković's natural philosophy, the prime elements of matter are real, homogeneous, simple, indivisible, non-extended geometrical points that possess inertia and mutual interaction.

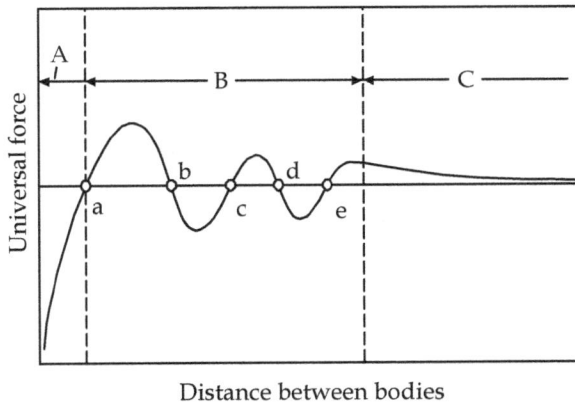

Figure 1.2.3. Graphic presentation of Bośković's Universal Force Law.

Based on that premise, Bośković challenged the validity of Newton's law of universal gravitation for the case when the distances between two elements were so small that they would penetrate each other. This, however, contradicted the main premise of indivisibility of prime elements. Bośković resolved the contradiction (Fig. 1.2.3) by assuming that Newton's law remained valid only when the distance between the bodies is very large (zone C). As this distance decreases (zone B), the force ceases to follow Newton's law and becomes alternatively either attractive or repulsive, depending upon the distance by which the bodies are separated. Then at even shorter distance (zone A), the force becomes purely repulsive. As the distance diminishes to zero, repulsion grows infinite, thus preventing direct contact between particles. Bošković called the proposed relationship between the forces acting on two particles the *universal force law*.

Bošković further speculated that the points of equilibrium between attraction and repulsion forces *a, b, c, d, e...* played a very important role. These points, which he called "boundaries," correspond to either stable or unstable equilibrium between the bodies. By evaluating the behavior of these boundaries and the area of the force-versus-distance curve contained between these boundaries, Bošković was able to explain various physical and chemical properties of matter, including emission of light. This was certainly the first attempt to explain such a great variety of natural phenomena by using a *unified theory*.

1.3 The Age of Electromagnetism

Reincarnation of the vortex theory in the 18th century was triggered by the advent of the age of electricity and magnetism.

Benjamin Franklin (1706-1790) – Benjamin Franklin became the first American to win an international reputation in pure science and the first man of science to gain fame for work done wholly in electricity. His most important achievements included the formulation of the theory of general electrical action that explained the process of production and transfer of electrostatic charges. It had long been common knowledge that by conducting rather simple experiments it is possible to demonstrate that there are two kinds of electric charges, which Franklin called *positive* and *negative*. Positive charges appear when one rubs a glass rod with silk, whereas negative charges arise after rubbing a rubber rod with fur. If one brings a positively charged glass rod near a negatively charged rubber rod, suspended by a nonmetallic thread, then the rubber rod will move toward the glass rod. Conversely, two rods having the electric charges of the same sign will repulse each other. Therefore, it is possible to make a very important conclusion known today as a *basic law of electrostatics*, stating that like electric charges repel one another and unlike charges attract one another.

Knowing the law of electrostatics, Franklin formulated the *principle of conservation of electric charge.* According to this principle, the rubbing of one body against the other does not produce electric charges but rather *transfers* them from one body to another. Additionally, Franklin demonstrated that a transfer of electric charges from one body to another can also be accomplished by induction. Consider a neutrally charged metallic ball (Fig. 1.3.1). This ball charges electrically when a charged rod is placed near it, without any physical contact. In this case half of the ball will have a positive charge and the other half negative.

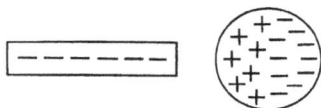

Figure 1.3.1. Charging of a metallic ball by induction.

Franklin explained that electric charge appears because of either excess or deficiency of the so-called "electric fluid." His idea of single electric fluid successfully competed against the idea of two electric fluids advanced by several prominent scientists of that time. Later he astonished the world by discovering a device that was capable of accumulating an electric charge. Franklin spent much of his time studying the nature of lightning. By conducting several ingenious experiments, Franklin proved convincingly that the lightning discharge is an electrical phenomenon.

Charles Coulomb (1736-1806) –To discover a quantitative basis for Franklin's laws of electrostatics, the French physicist Charles Coulomb conducted a series of experiments. Using the torsion balance, a device of his own invention, he measured the twist in a suspended fiber. He then used this technique in his experiments with magnets. Based on the results of these experiments, Coulomb established the law of electric force F_e between two charged particles that states the following:

The electric force between two charged particles F_e is directly proportional to the product of the charges e_1 and e_2 of the particles and is inversely proportional to the square of the distance r between these particles, i.e.:

$$F_e = \frac{e_1 e_2}{4\pi\varepsilon_0 r^2}$$

where ε_0 is the electric constant.

Coulomb obtained a similar equation in respect to magnetic charges, but he claimed that there were no direct relationships between electricity and magnetism. It was obvious to him that his law of electric force F_e was very similar to Newton's gravitational force F_g acting between two bodies with gravitational masses m_1 and m_2 separated by the distance r. To emphasize the similarity between these two laws, Coulomb used the term "electrical mass." But, unlike Newton's law of universal gravitation, his law applied to both attraction and repulsion forces.

André-Marie Ampère (1775-1836) –The discovery of electromagnetism by the Dutch physicist Hans Oersted (1777-1851) in 1820 became a turning point in the scientific career of the French physicist and mathematician André-Marie Ampère. Jointly with the French physicist Augustin Fresnel (1788-1827), Ampère proposed the theory of electromagnetic molecule to explain why a current-carrying helical coil behaved like a magnet.

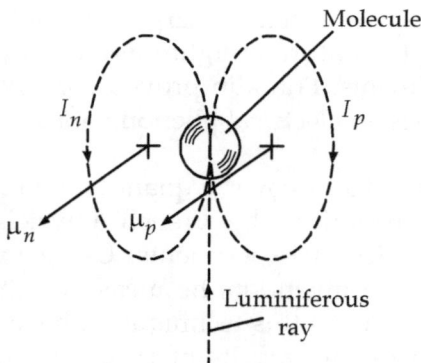

Figure 1.3.2.
Ampère's electrodynamic molecule.

Their idea was that myriads of small circular electric currents existed within the molecules of iron. Once these molecules were aligned, the resultant current would act precisely as the concentric currents in the helical coil. Both positive and negative fluids pour out of the top, flow around the molecule and reenter at the bottom (Fig. 1.3.2). Since the positive and negative fluids I_p and I_n are circling around the molecule in opposite directions, their resultant magnetic moments μ_p and μ_n are pointed in the same direction, thus doubling the magnetic moment of the molecule. Ampère and Fresnel proposed that the molecular electric currents in iron were induced by the luminiferous ether that pervades space and penetrates matter.

After describing the noumena of the electromagnetism, the next step for Ampère was to apply mathematics to the phenomena. He successfully completed this task in 1820 by deducing his famous relationship that became known as *Ampère's law*. This law establishes a relationship between the magnetic induction B along a closed path around a wire and the current I in the wire. If the magnetic path around the wire with radius R is circular, then the magnetic induction B at any distance r from the center of the wire is equal to (Fig. 1.3.3):

$$B = \frac{\mu_0 I}{2\pi r}$$

where μ_0 is the magnetic permeability of free space.

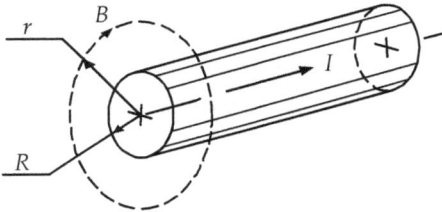

Figure 1.3.3.
Magnetic induction B around a wire with current I.

Michael Faraday (1792-1867) – The British chemist Michael Faraday made a great contribution to science by his discoveries in electromagnetism. His inventions laid out the foundation for the construction of electric motors, generators and transformers based on this phenomenon. Faraday began his discoveries in electromagnetism by making the same very logical proposition that several other scientists advanced at the time.

If magnetism is produced by electricity, as was clearly demonstrated by Ampère, then the reverse phenomenon should also exist and electricity must be produced by magnetism. To prove this proposition, Faraday conducted several experiments. In one of his most interesting experiments, Faraday used what became known as Faraday's induction ring, a simple device that resembles a modern transformer (Fig. 1.3.4). The ring was made of soft iron and contained two coils wound around it. Faraday connected one coil first to a galvanometer, and the other coil to the battery through a switch. After connecting the battery, he noticed that the needle of the galvanometer moved and then came back to zero. The needle moved again but in the opposite direction when Faraday disconnected the battery.

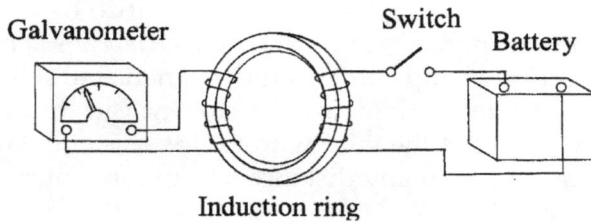

Figure 1.3.4. Faraday's experiments with electromagnetism.

It was clear now that the electric current appeared only when the magnetic field was changing. In his presentation to the Royal Institution, Faraday summarized his historic discovery in a single statement:

Whenever a magnetic force increases, it produces electricity;
the faster it increases or decreases, the more electricity it produces.

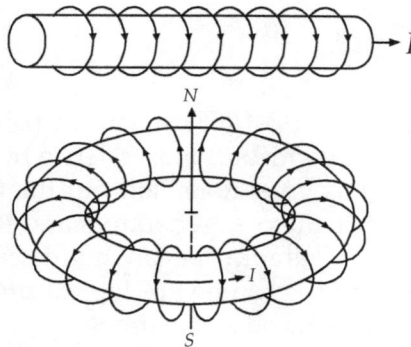

Figure 1.3.5. Faraday's lines of force in a straight wire and in the loop, both carrying electric current.

To illustrate the electromagnetic effects, Faraday employed *lines of force*. A straight current-carrying wire was surrounded by circular lines of force (Fig. 1.3.5). After bending this wire into a loop, the lines of force would become distorted, so a greater concentration of the lines of force would be created inside the loop. This concentration of the lines of force, according to Faraday, produces magnetic forces with the north and south poles located at opposite sides of the loop.

Faraday's philosophical way of thinking inevitably led him to the conviction that all known physical phenomena could be explained by one unified theory. He wondered if besides the electric nature of magnetism the nature of gravity is also electrical. He was very familiar with the idea that gravity may be explicable as the result of slight deviations of celestial bodies from the state of electrical neutrality. Father Giovan Battista (1716-1781), known in the secular world as Francesco Beccaria, originally proposed this fascinating idea in the mid-eighteenth century. Since that time, several scientists, including the German natural philosopher Franz Aepinus (1724-1802) and the Italian physicist Ottaviano-Fabrizio Mossotti (1791-1863), brought to light the idea that the electrostatic forces of attraction between unlike charges slightly outbalance the repulsion forces between like charges.

As was usual in application to any other phenomenon, Faraday began his investigations of the relationship between gravity and electricity by conducting experiments. This time, unfortunately, all of his experiments failed. Insufficient resolution of his instruments, Faraday thought, did not permit him to detect the infinitely small deviation from the state of electrical neutrality that was needed to produce gravity. Nevertheless, he was strongly convinced that, in his words,

Universal gravitation is mere residual phenomenon
of electrical attraction and repulsion.

Wilhelm Weber (1804-1891) – Nowadays, we describe *relativity* as the dependence of the physical properties of objects on their velocity. The German physicist Wilhelm Weber was among the first scientists who described this phenomenon. His long scientific life was both interesting and contentious. In 1846 he formulated his controversial *law of electric force*. According to this law, the electric force F_e between two moving electric charges e_1 and e_2 separated by the distance r is equal to:

$$F_e = \frac{e_1 e_2}{4\pi\varepsilon_0 r^2}\left(1 - \frac{V^2 + 2ra}{2c^2}\right)$$

where
   V = radial velocity of electric charges in respect to each other
   a = radial acceleration of electric charges in respect to each
      other
   c = velocity of light.

The dominant term in the above equation is Coulomb's force $e_1e_2/4\pi\varepsilon_0r^2$. The remaining terms (in brackets) modify this force, making it either attractive or repulsive, when the charges are in motion relative to each other. Weber was the first scientist to propose that the *interaction between particles is velocity-dependent*. Another interesting consequence of Weber's law was that two particles of the same sign of charge can revolve around each other on stable orbits. If this is true, Weber speculated, then his law might lead to an understanding of the laws of light and heat radiation. Not everyone accepted Weber's theory. His strongest opponent was the German physicist Hermann von Helmholtz (1821-1894), who believed that Weber's law of electric force violated the principle of conservation of energy.

In 1882 Weber developed further the ideas of the German astrophysicist Johann Zöllner (1834-1882). According to Zöllner, all matter is compounded of electrically charged particles, held together in various stable configurations by the action of the Weberian law of force. Even gravitation was interpreted by adopting in essence the earlier hypotheses of Aepinus and Mossotti that the attraction electrostatic forces between unlike charges slightly outbalance the repulsion forces between like charges.

William Thomson (1824-1907) - The British physicist William Thomson (Lord Kelvin) is mostly known for his works in thermodynamics. He also turned out to be one of the most ardent supporters of the vortex theory. Thomson argued that from any galvanic current there extends a moving spiral that coils about the line of magnetic forces, which passes through the center of the axis of the current (Fig. 1.3.6). He proposed that the current itself consists of the trapping of a segment of this spiral in ponderable matter (matter that has weight). When extending this idea to light waves, he believed that transverse vibrations of the particles of the moving spirals caused the light waves to propagate. He pictured that the plane of polarization of the waves is rotated depending upon the direction of the magnetic force.

Thomson hoped to explain all electromagnetic phenomena with this representation. For instance, the twisting motion of the spiraling helices could produce magnetic effects. These helices, when located in matter, would become currents. In his vortex theory Thomson treated the luminiferous ether as a medium for the propagation of waves; light waves were viewed as oscillations of ether. Present everywhere, even in empty space, ether was massless and at the same time an elastic solid medium.

The last requirement was imperative to account for the major properties of light. Furthermore, in order for reflection and refraction of light to occur, the ether had to be absolutely incompressible.

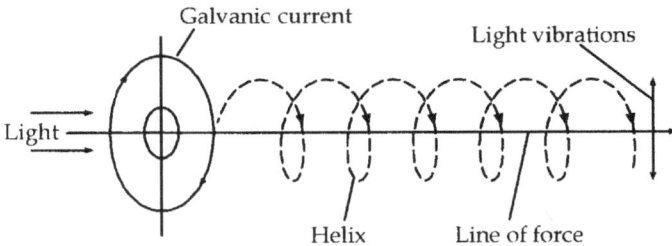

Figure 1.3.6. William Thomsoson's presentation of the electromagnetic wave. Adapted from C.C. Gillispie (1971).

Thomson made his most interesting discoveries in the vortex theory in collaboration with the Scottish physicist Peter Tait (1831-1901). Thomson renewed his interest in spirals after reading a 1858 paper by Helmholtz. In his work, Helmholtz demonstrated that the motions of linear fluid vortices exhibit striking patterns: two adjacent figures appear to repel one another in a complex manner. This effect is the result solely of pressures produced in the medium; once established by some unknown means in a non-viscid medium, the vortices cannot be destroyed by any mechanical process and, thus, are made eternal.

Tait suggested to Thomson that the same concept could be applied to explain the structure of light and atoms, and proposed the idea of a demonstration. He also designed the apparatus for the demonstration. It consisted of a couple of boxes, two pieces of cloth, and a few bottles of chemicals. A sturdy cloth covered one side of each box, while the opposite side had a circular hole. Within each box, Tait mixed the vapors of an acid with ammonia to produce thick clouds of smoke. When he struck firm blows on the cloth, numerous circular rings shot out of the hole. The members of the Royal Society of Edinburgh were amused by the behavior of the circular rings, which filled the entire room with smoke rings bouncing off each other. The rings shook violently from the collisions, making the impression that they were made of rubber. It was impossible to cut these rings with a knife; they simply moved away from the blade.

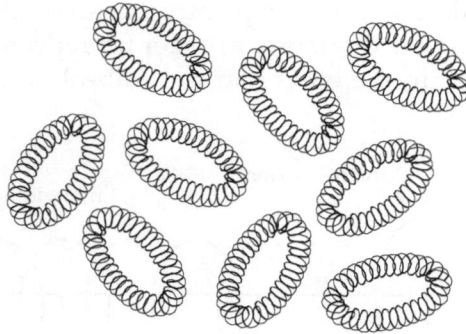

Figure 1.3.7. Tait's smoke rings as toroidal vortices.

When looking closely at the smoke rings (Fig. 1.3.7), one could clearly recognize the neatly wound toroidal spirals. Thomson and Tait used this demonstration to point out that these vortex rings of smoke behaved just like atoms, and that all the properties of atoms stemmed from vortex spin.

James Clerk Maxwell (1831-1879) - The British physicist James Clerk Maxwell is best known for the development of the theory of electromagnetism. Very few of us, however, know that during the many years of his scientific life he was a great supporter of the vortex theory. He worked closely with Faraday, Thomson and Tait. In 1862, Maxwell devised a model to illustrate Faraday's law of electromagnetic induction. He proposed an analogy between lines of force in the electromagnetic field on the one hand, and vortex filaments in a liquid with finite fixed boundaries on the other. The following year he employed vortex theory to explain the rotation of the plane of polarization of light, traveling along the force in magnetic field. He envisioned a beam of linearly polarized light as consisting of two circularly polarized components rotating in opposite directions. The vortices, created by the magnetic field, speeded up the rotation of one component and slowed down that of the other. The outcome was a rotation of the plane of polarization.

In 1864 Maxwell suddenly abandoned the vortex theory. This became evident after he published his famous paper *The Dynamic Theory of the Electromagnetic Field* in that year. In our time this paper is considered to be the cornerstone of the theory of electromagnetism. The essential point of this theory was that the electromagnetic field is transverse vibrations, supported by electromagnetic medium and propagated at the velocity of light. In his theory, Maxwell did not use either the vortices or, in fact, any other physical model, allowing the mathematical equations alone to do the job. The publication of this paper brought Maxwell neither fame nor glory. Moreover, because of this publication, he alienated two of his closest friends in the entire scientific community, Faraday and Thomson.

In fairness to Maxwell we shall remember that before publishing this paper, he did his best to explain electromagnetism by using the vortex theory, to which he himself became attached wholeheartedly. He tried to model the electric current in the electromagnetic wave by employing some sensible physical analogies. This current was badly needed because, according to Faraday's law of electromagnetic induction, the magnetic field in the electromagnetic wave can only be created if there is a change in the electric current. However, to introduce the electric current, which he called the *displacement current*, the electromagnetic wave must contain a device similar to a capacitor that is periodically charged and discharged. Certainly, there was no way of finding such a capacitor in the colorless and weightless electromagnetic wave.

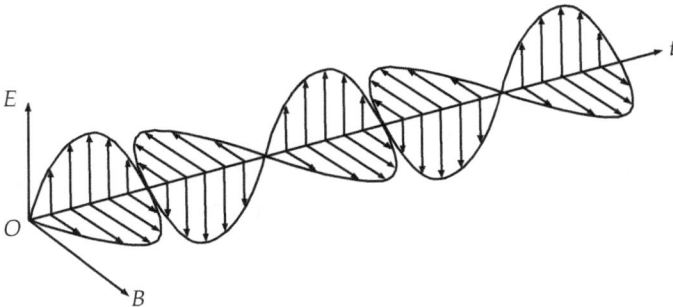

Figure 1.3.8. Maxwell's presentation of the electromagnetic wave.

Frustrated, Maxwell eventually dropped all his attempts to model the electromagnetic wave by employing any physical analogy, and presented his theory with a set of mathematical equations. The good news was that this purely mathematical theory worked. It explained how a sinusoidal electric field E and a sinusoidal magnetic field B, oscillating in two mutually perpendicular planes, were generating each other and forming an electromagnetic wave (Fig. 1.3.8). It also stipulated that the velocity of the propagation of the electromagnetic wave was equal to the velocity of light.

Here again, Maxwell had a serious problem. Unable to derive the above equation from his vortex model, he derived it by making some *ad hoc* assumptions regarding the elastic structure of the medium. Despite strong opposition to his theory, Maxwell published his *Treatise* in 1873. Its publication, however, did not bring him any personal satisfaction: he failed again to explain his equations from a physical point of view and had to drop any sensible reference to any kind of a physical model.

He recognized his failure and was afraid that some young scientists would see in his *Treatise* a sign that there was no longer a need for discovering a physical meaning of the mathematical equations that describe physical phenomena as long as these equations predicted accurately the results of the experiments. Maxwell was concerned that this viewpoint would eventually exclude the use of physics in the models that educated laymen could understand. Therefore, when Maxwell started on a new edition of his *Treatise* in 1879, his goal was to restore Faraday's traditions and to rewrite his theory by employing a clear physical model. Unfortunately, he was not able to meet his goal. The great scientist died of cancer on 5 November 1879.

1.4 Atomic & Nuclear Age

The atomic and nuclear age is associated with experimental discoveries of particles of matter and radiation on atomic and subatomic levels. Two new theories, quantum mechanics and the theory of relativity, provided a strong theoretical base for understanding the structures of elementary particles and the forces acting between them.

Max Planck (1858-1947) – The German physicist Max Planck originated the quantum theory, making the year 1900 a transitional point between classical and modern physics. Planck's path to quanta, his greatest discovery, began with his studies of one of the problems of thermodynamics known as the black body radiation. The ideal black body absorbs all radiation that incidents on it.

The best approximation of a black body is a container with a small hole in it, into which radiation shines and is trapped inside (Fig. 1.4.1). As light enters the container through the hole, a part of it is reflected and another part is absorbed on each reflection from the interior walls. After many reflections, essentially all of the incident energy is absorbed.

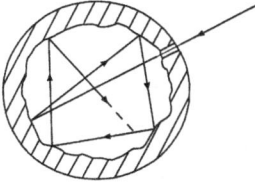

Figure 1.4.1.
An approximation of a black body radiation.

To derive the equation that worked for all wavelengths, Planck had to assume that radiation is emitted or received in energy packets called *quanta* with the quanta energy E equal to a product of the *Planck constant h* and the wave frequency f:

$$E = hf$$

This was the beginning of the quantum mechanics that was further developed by several physicists in the 20th century.

Albert Einstein (1879-1955) - In 1905 the German-Swiss-American theoretical physicist Albert Einstein successfully used Planck's new theory in his prediction of photoelectric effect. Simply, photoelectric effect is the release of electrons from a substance under the influence of light or other electromagnetic radiation. In his 1905 paper, Einstein demonstrated that the particles of light would have to be such that the ratio of energy E to frequency f must be equal to the Planck constant h. Thus, according to Einstein, the concept of quanta is applicable not only to black body thermal radiation, but also to light.

During that same year Einstein introduced his special theory of relativity. Before Einstein published this theory, people calculated very simply the relationships between two objects moving with constant velocities. Suppose, for instance, that you are traveling in a train that moves with velocity V_1 in respect to a train station, and you are walking towards the front of your carriage with velocity V_2 in respect to the carriage floor. What is your velocity in respect to the train station? A logical answer is $V_1 + V_2$. And this is also a correct answer according to the Galilean law of addition for velocities.

Einstein showed that the law of addition for velocities becomes completely different when one of the objects is a photon moving with the velocity of light. Coming from his principle of relativity was one of the most intriguing propositions of his time:

In empty space light travels with the definite velocity c
that does not depend on the motion of its source.

The above proposition obviously clashes with the Galilean law of addition for velocities. For instance, if a source of light travels in respect to a stationary object with the velocity V_1, then, according to Galilei, the light coming from this source will travel in respect to the stationary object with velocity $V_1 + c$, and not c as followed from the above proposition. Einstein resolved this contradiction by employing the Lorentz-Fitzgerald Transformation derived earlier by the Dutch theoretical physicist Hendrik Lorentz (1853-1928) and the Irish physicist George FitzGerald (1851-1901) to prove the existence of ether. According to these equations, the following changes occur with a moving object:

- The observed length of an object decreases with increase of velocity in comparison with its length at rest.
- The observed mass of an object increases with increase of velocity in comparison with its rest mass.
- According to a stationary observer, a moving clock runs slower than an identical stationary clock. This effect is known as *time dilatation*.

Einstein, however, using this known relativistic equation, provided a new physical meaning for it. The outburst of Einstein's new ideas in 1905 did not end with the publication of the special theory of relativity. In his fifth paper, published in November of that year, Einstein showed by calculation that if a body gives an amount E of energy in the form of light, then its mass decreases by an amount E/c^2. Thus, mass and energy are wholly equivalent. In 1907 Einstein presented this idea in a long and mainly expository paper that contained his famous equation, establishing the relation between energy E and mass m:

$$E = mc^2$$

In 1915 Einstein introduced his general theory of relativity in which he applied the concept of *curved spacetime* to explain gravity. The idea of curved space was introduced earlier by several great scientists, including the Russian mathematician Nikolai Lobachevsky (1793-1856), the German mathematician Karl Friedrich Gauss (1777-1855), the Hungarian mathematician János Bolyai (1755-1856) and the German mathematician Georg Riemann (1826-1866).

The term "spacetime" was originally proposed in 1907 by the Russian-German mathematician Hermann Minkowski (1864-1909). He used this term to describe a coordinate system of the four-dimensional space. It was an important step in breaking away from the Newtonian concept of absolute time and space. When time and space are considered independent of each other, the location of an object in a three-dimensional space is defined by three dimensions x, y, and z.

To connect space with time t, Minkowski added one more dimension ict in which $i = \sqrt{-1}$ is the imaginary number. Thus, the modulus of the vector u defining the location of a point on a curved space was defined by the equation:

$$u^2 = x^2 + y^2 + z^2 - (ct)^2$$

Einstein expanded the application of the term spacetime in his general theory of relativity by using a special branch of mathematics called *tensor calculus*. The result was a groundbreaking departure from the Newtonian interpretation of gravity. Gravitation was no longer a force but an intrinsic curvature of spacetime. It was the curved spacetime around the Sun that attracted the planets. In curved spacetime the motions of the planets were represented by the shortest distances, called *geodesics*.

Numerous experiments conducted since 1915 proved that Einstein's theory of general relativity was correct. This theory had also found its application in the contemporary astrophysics. According to his theory, the spacetime can in principle be warped so strongly by a huge mass that any radiation emitted from the mass curves back in again and cannot escape. These huge masses are thought to exist as black holes.

Theodor Kaluza (1873-1916) – In 1918 the German mathematician Hermann Weyl (1885-1955) proposed an extension of Einstein's general theory of relativity. He modified Einstein's spacetime by providing one additional geometric parameter for electromagnetism, making this parameter an equal partner to gravitational curvature. Einstein rejected Weyl's idea, but, in 1919, the German mathematician and physicist Theodor Kaluza took a completely different approach. He managed to leave Einstein's equations unchanged with only one principal modification: these equations were now applied to five dimensions instead of four. This fifth dimension was expected to account for electromagnetism.

Kaluza presented his multidimensional space by wrapping a space of a higher dimension around a space of a lower adjacent dimension. Consider that a few hundred feet of garden hose is stretched across a canyon, and you view it from a far distance away. From this distance the hose will appear as a one-dimensional string. Therefore, an ant located on the hose would have only one dimension in which to walk: the left-right dimension along the hose's length.

From a magnified perspective, we will see that besides the left-right dimension along the hose's length there is the second dimension along which the ant can walk, either clockwise or counterclockwise. We may say that the second dimension of the hose as "wrapped up" around the one-dimensional material. Similarly, it is possible to describe mathematically a three-dimensional hose "wrapped around" the two-dimensional material, and so on and so forth. So, how in the world did Kaluza manage to make Einstein's four-dimensional equations work in a five-dimensional spacetime? The secret was in his technique called *compactification* that allows one to reduce a spacetime of a higher dimension to a spacetime of a lower dimension.

In his wonderful book *Q is for Quantum*, John Gribbin provides us with an excellent example of how compactification works. One can make a three-dimensional space in the form of a spiral pipe from a thin sheet that may be viewed as a two-dimensional space. From a distant vantage point the three-dimensional spiral pipe will look like a one-dimensional line. Similarly, it is possible to describe mathematically a five-dimensional spiral pipe made of a four-dimensional material. There is one condition, though: the "wrapping" must happen on a scale much smaller than that of an atomic nucleus. In 1926, independently of Kaluza, the Swedish theoretical physicist Oskar Klein (1894-1977) refined Kaluza's theory to take into account the requirements of quantum theory. Although the two scientists never worked together, their discoveries became known as the *Kaluza-Klein theory*.

Karl Schwarzschild (1873-1916) – The idea of the existence of black holes was first proposed in 1783 by the British astronomer John Mitchell (c. 1724-1793). If the stars were sufficiently massive and compact, he thought, then light would not be able to escape from the surface of the "dark stars," as the black hole was called at that time. He further suggested that the dark stars might be detectable through their gravitational effect on nearby objects. But his idea was quickly forgotten. It was revived 133 years later by the German astronomer Karl Schwarzschild. In the last year of his life he wrote two papers in which he gave the first solution to the complex partial differential equations of Einstein's general theory of relativity.

His calculations showed that for any object squeezed inside a sphere with a critical radius, now called the *Schwarzschild radius*, space would curve around the object and separate it from the rest of the universe. It would become a self-contained universe, from which nothing (*not even light*) could escape. The Schwarzschild radius R corresponding to any mass m is equal to:

$$R = \frac{2mG}{c^2}$$

According to the above equation, the Earth would become a black hole if compressed to a sphere with the radius $R = 0.88$ cm, while the same would happen to the Sun, when $R = 2.9$ km. Still, in spite of Einstein's support for Schwarzschild's theory, the idea of a black hole remained a mere theoretical curiosity for half a century. It had to wait for conventional ground-based telescopes to be supplemented with the Hubble Space Telescope before astronomers could obtain indisputable observations to confirm the existence of black holes.

Edwin D. Babbitt (1828-1905) - Edwin D. Babbitt was one of the most influential American pioneers and writers in the field of color research and color therapy. In 1878 he published his *The Principles of Light and Color,* in which he outlined spiral structure of an atom that he called *anu.* Babbitt used his structure of this atom to explain bonding, electricity, light, color, friction, psychic power, and nearly everything else.

Charles Leadbeater (1847-1934) – In 1909, a renegade Anglican clergyman Charles Leadbeater and his assistance Mrs. Besant published the book titled *Occult Chemistry.* In their book they described the basic subatomic structure as seven layers of recursive spirals around other spirals.

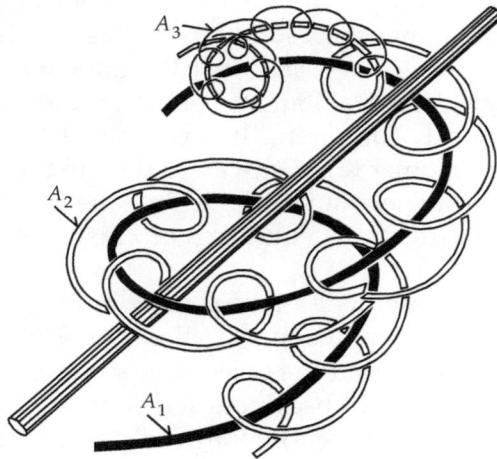

Figure 1.4.2.   Leadbeater's multiple-level spiral.

As shown in Figure 1.4.2, a winding of a spiral A_1 is wound by a spiral A_2 that, in its turn, is wound by a spiral A_3. Similarly, spiral A_3 would be wound by spiral A_4, and so on up to spiral A_7. Although the atomic structures described by Leadbeater and Besant did not make much sense to their contemporaries, these structures became recognizable in some atomic theories developed several decades later. The British physicist Stephen Phillips published, in 1980, a book titled *Extra-Sensory Perception of Quarks* in which he presented the work of Leadbeater and Besant in light of modern theories of atomic structure.

Walter Russell (1871-1963) - Another outstanding amateur self-educated scientist of the 20th century was the American artist and sculptor Walter Russell summarized the results of his revelation in his two masterfully written books: *The Universal One* that was originally published in 1927 and *The Secret of Light* published in 1947. Based on the concept of vortices, he attempted to explain the nature of light, electricity, and magnetism. He went as far as to devise his own Table of the Elements and to predict several new elements unknown during his time. After reading his books, one may only wonder how a man with no formal education in any branch of sciences was able to penetrate intuitively the depths of understanding the secrets of nature. The following quotation from Walter Russell was provided by Glenn Clark in his enlightening book *The Man Who Tapped the Secrets of the Universe:*

> That the man calls matter, or substance, has no existence whatsoever. So-called matter is but waves of the motion of light, electrically divided into opposite pairs, then electrically conditioned and patterned into what we call various substances of matter. Briefly put it, matter is the motion of light, and motion is not substance. It only appears to be. Take motion away and there would not be even appearance of substance.

Ernest Rutherford (1871-1937) – The New-Zealand-British physicist Ernest Rutherford is considered by the scientific community to be the founder of nuclear physics. He is mostly known for his discoveries in radioactivity and atomic structure. His research in radioactivity was triggered by the discovery of X-rays by the German physicist Wilhelm Röntgen (1845-1923) in 1895, the discovery of the radioactivity of uranium by the French physicist Henri Becquerel (1852-1908) in 1896, and the discovery of two new radioactive elements, polonium and radium, made by two French physicists, Pierre Curie (1859-1906) and Marie Skłodowska-Curie (1867-1934) in 1898.

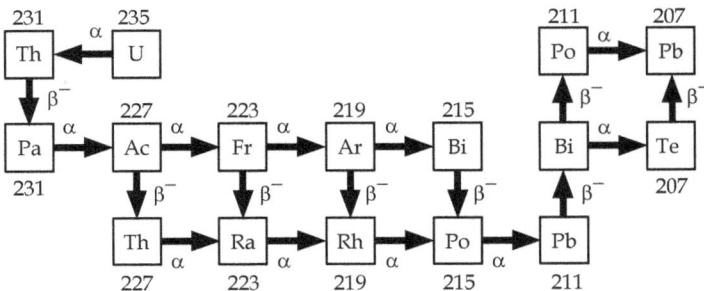

Figure 1.4.3. Sequence of radioactive decay of Uranium 235. Adapted from R.A. Serway (1990).

Based on his extensive experimental research on radioactivity, jointly with Frederick Soddy, Rutherford developed the theory of radioactivity. The theory explained how the decay of radioactive atoms produces the transmutation of a parent into a daughter element. It also identified the role of the three processes: alpha (a) decay, beta (β) decay, and gamma (γ) decay (Fig. 1.4.3). During the alpha decay, a nucleus emits a heavy alpha particle that turned out to be a nucleus of helium. The beta decay involves emission of light charged particles that later were recognized as the electrons and positrons. The gamma decay involves the radiation of high-energy gamma rays. This work brought Rutherford the Nobel Prize in chemistry that he received in 1908.

Rutherford's most sensational discovery came out of the so-called *scattering experiments* that he conducted in 1909. Actually, as a boss and a Nobel Prize winner, he did not have to conduct these tedious experiments by himself, and assigned them to two of his most trusted underlings: a postdoc, Hans Geiger, and an undergraduate student, Ernest Marsden. The main purpose of these experiments was to investigate atomic structures of a metallic samples made of thin foils. A beam of fast moving alpha particles, emitted by a radioactive element, was first passed through a diaphragm and then shot at a thin foil. In passing through the foil, the alpha particles collided with the constituent atoms and some of the particles were scattered in different directions. In other words, the alpha particles, after hitting something inside the metal foil, bounced back. Rutherford rightfully saw the only reason for such a dramatic behavior of the positively charged alpha particles: they were reflected by an even heavier positively charged particle. The nucleus of an atom was now discovered.

In the Rutherford's model of the atom, the positive electric charge was concentrated inside a small nucleus located at the center of the atom, and the negatively charged electrons were moving around it, resembling the solar system. In spite of its attractiveness, the new model soon ran into a serious problem. According to the classical theory of electromagnetic emission, the negatively charged electrons circling the nucleus must emit light waves, carrying with them a part of the electron's energy. Consequently, the electrons would completely lose their energy and spiral into the nucleus. It was a disaster, but nobody could find an immediate solution. The impasse had lasted for several more years before it was resolved, thanks to the contributions made by the three outstanding scientists, the British experimental physicist Henry Moseley, the Danish theoretical physicist Neils Bohr, and the German physicist Arnold Sommerfeld.

Henry Moseley (1887-1915) - During his short life, the British experimental physicist Henry Moseley made the first step toward a better understanding of Rutherford's model by discovering the physical meaning of electric charges of atomic nuclei for different chemical elements. He also provided a completely new interpretation of the periodic table of chemical elements originally developed in 1869 by the Russian chemist Dmitri Mendeleyev (1834-1907). In this table, the elements' relative masses were the principal factors in determining the periodicity of the properties of the chemical elements. Nobody questioned the validity of Mendeleyev's Periodic Table of the Elements for 44 years. This all changed in 1913 after Moseley completed a series of his extraordinary experiments with the X-ray spectra.

In his thorough experiments, Moseley used over thirty different metals (from aluminum to gold) as targets. He found that the X-ray spectra lines changed regularly in position from element to element exactly in accordance with Mendeleyev's Periodic Table. Moseley said of his results: "We have here a proof that there is in the atom a fundamental quantity, which increases by regular steps as we pass from one element to the next. The quantity can only be the charge on the central positive nucleus." After Moseley's discovery, the electric charge of the nucleus replaced the element's relative mass as the principal factor determining the periodicity of the properties of the chemical elements.

Niels Bohr (1885-1962) – In 1913 the Dutch physicist Niels Bohr proposed a model of the hydrogen atom. This was a significant improvement of the model of the hydrogen atom proposed earlier by Ernest Rutherford. Bohr adopted Rutherford's basic premise that in the hydrogen atom an electron orbits around a proton. But, Bohr rightfully wondered, why would the electron remain stable in its path around the proton? Fortunately, he had at his disposal three pieces of knowledge that were critical for finding the answer to this question. Bohr was certainly familiar with the first two: Planck's discovery of quanta and Einstein's theory of photoelectric effect. The third crucial piece of knowledge came to Bohr from the results of investigation of atomic spectra of the hydrogen atom conducted by Balmer and Rydberg.

The phenomenon of atomic spectra is rather simple. When an electric charge passes through a gas and the resulting light is dispersed by a prism, a spectrum of sharp lines of discrete wavelengths, or atomic spectra, is observed. The line spectrum for atomic hydrogen was the simplest one, because its lines had a distinct regularity of spacing.

In 1884 a Swiss school teacher, Johann Jacob Balmer (1825-1898), found a formula that correctly predicted the wavelengths λ_{mn} of four visible lines of hydrogen H_α, H_β, H_γ, and H_δ that became known as *Balmer series* (Fig. 1.4.4). The Swiss spectroscopist

Johannes Robert Rydberg (1854-1919) further extended Balmer's work. Working on atomic emission spectra in 1890, Rydberg found a simple general equation for the frequencies f_{mn} of some of the spectra lines for the hydrogen atom:

$$f_{mn} = R_\infty c\left(\frac{1}{m^2} - \frac{1}{n^2}\right)$$

where
R_∞ = Rydberg constant
m, n = integer numbers.

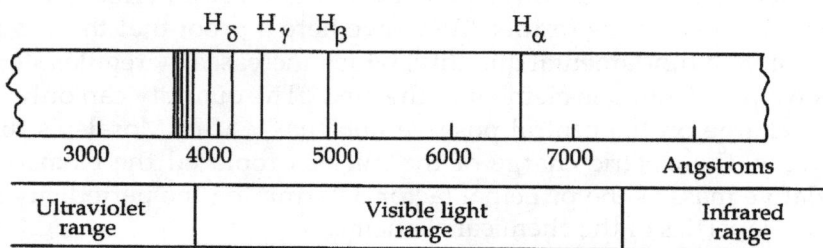

Figure 1.4.4. The line spectrum for atomic hydrogen.
Adapted from G. Gamow (1985).

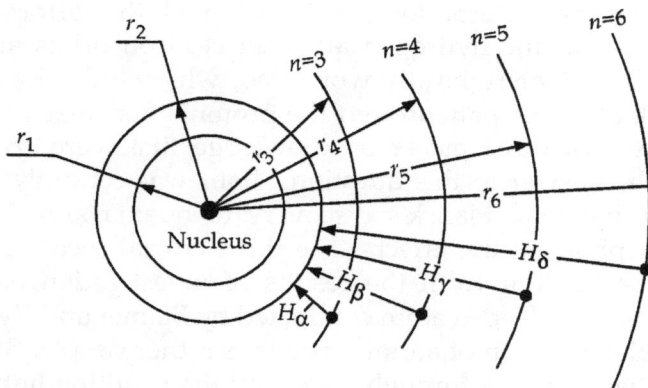

Figure 1.4.5. Electron orbits and spectra lines of a hydrogen atom.
Adapted from H.E. White (1934).

In Bohr's model of the hydrogen atom (Fig. 1.4.5), the electron was stable only on those orbits in which its orbital angular momentum p_a was related to Planck constant h by the equation:

$$p_a = mVr = \frac{nh}{2\pi}$$

where m = electron mass, V = electron orbital velocity, r = radius of electron orbit, and n = quantum energy states 1, 2, 3, . . .

The above formula allowed one to calculate the energies (and frequencies) of the photons that were emitted, when the electron was transferred from higher to lower energy states.

Arnold Sommerfeld (1868-1951) – The German physicist Arnold Sommerfeld further elaborated the Bohr's atomic model. The purpose of his innovation was to account for the so-called *fine structure* of the spectra lines that appeared in close proximity to the main spectra lines H_α, H_β, H_γ, and H_δ.

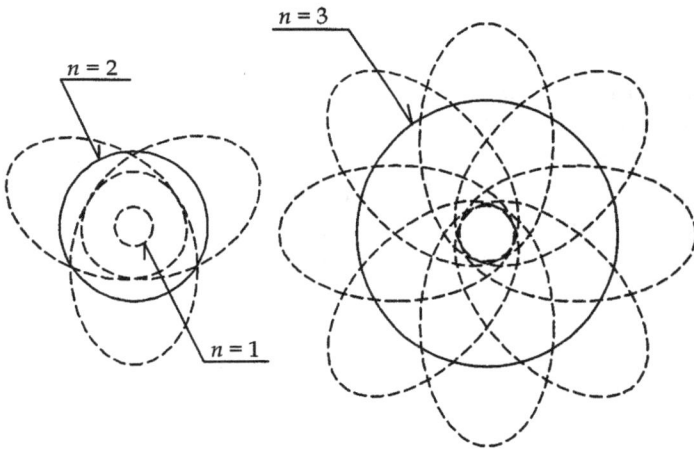

Figure 1.4.6. The quantum electron orbits in the hydrogen atom.

His solution was twofold. Firstly, he proposed that for all orbital quantum states that are greater then the ground state $n = 1$, there must be elliptical electron orbits in addition to the circular orbits described by Bohr's model. As shown in Fig. 1.4.6, in the hydrogen atom there is only one circular electron orbit corresponding to the ground state with the principal quantum number $n = 1$. For $n = 2$, there are four orbits, one circular and three elliptical. For $n = 3$, there are nine orbits, one circular and eight elliptical. To assure stability of electrons on all these orbits, it was necessary to add two additional quantum numbers: l (orbital quantum number) and m_l (orbital magnetic quantum number). Without this additional quantization all electrons would end up at the ground energy state.

Sommerfeld showed that the ellipticity should result in a relativistic effect and discovered that the frequency difference between the fine structure components depends on the so-called *fine structure constant* α given by the equation:

$$\alpha = \frac{e^2}{2\varepsilon_0 hc}$$

Remarkably, the fine structure constant α is dimensionless and approximately equal to the inverse of 137. The fine structure constant showed up later in many models of elementary particles proposed by other physicists. Because the constant does not have dimensions, some scientists believe that it has a much greater significance in nature than the constants with dimensions invented by people.

Leo Szilard (1898-1964) - The Hungarian physicist Leo Szilard proposed the principle of a nuclear reaction based on utilization of the *nuclear fission* during which the atom of a heavy nucleus splits. In such a reaction, the total rest mass of the products is less than the original mass. The decrease in the rest mass is accompanied by a release of energy that is known as the *binding energy*. The binding energy is responsible for holding the *nucleons* (protons and neutrons) together within the atomic nucleus. Discovery of the nuclear fission occurred in 1938 by the German physicists.

A typical bombardment of uranium U by a neutron n yielded, besides the nuclei of barium Ba and krypton Kr, two or three (on the average 2.47) neutrons n as given by the equation:

$$_0^1 n + _{92}^{235}U \rightarrow {}_{56}^{141}Ba + {}_{36}^{92}Kr + 3({}_0^1 n)$$

The fact that the fission reaction yielded either two or three neutrons had a tremendous consequence. The physicists immediately realized that these neutrons could be used to engage other nuclei to undergo fission, producing a so-called *nuclear chain reaction*.

Edward Teller (1908-2003) – The Hungarian physicist Edward Teller played an important role in a practical implementation of another nuclear reaction that became known as the *nuclear fusion*. During nuclear fusion, two light nuclei combine to form a heavier nucleus. In this process, as in the nuclear fission, the mass of the final nucleus is less than the combined rest masses of the original nuclei. Consequently, the loss of mass is accompanied by a release of energy that turned out to be much greater than that in the nuclear fission. Because high temperatures are required to drive the above reactions, they are called *thermonuclear fusion reactions.*

The conventional deuterium-tritium fusion process produces high-energy neutrons which render reactor components radioactive. This problem is eliminated by using the two alternative fusion processes. In one of them, the deuterium is fused with helium 3. In the other one, two helium 3 are fused together. The implementation of the alternative processes is subdued because the helium 3 is very rare on the Earth, but it is expected to be in great amounts on the surface of the Moon.

The physicists are currently investigating a possibility for the development of the so-called *cold-fusion reaction.* Reportedly, one of the processes, known as the *E-cat HT,* uses the high temperature nickel hydride as a fuel. During the cold fusion the nickel and hydrogen atoms are fused into copper. The analysis of the fuel before and after the 32-day burn showed a shift from a natural mix on Ni-58/Ni-60 mix to almost entirely Ni-62 – a reaction that cannot occur without nuclear fusion.

1.5 Models of Elementary Matter Particles

In spite of the great progress made in particle physics during the 20th century, a simple question still remained unanswered: "What is the structure of an electron?" Can it be described by considering it as either a wave or a particle? The presentation as a wave required an explanation of how the wave properties of an electron can be related to its physical properties. If the electron is a particle of a certain size, then how can it withstand the action of the negative electric forces that would try to rip it apart? The quantum theory "solved" the latter problem superficially by presenting the electron as a mathematical point having no dimensions.

Some physicists did not like the presentation of an electron as a charged point mass. Among these scientists was the German physicist Max Born (1882-1970). He was troubled by the fact that electron energy per unit volume becomes infinite when its radius is equal to zero, a paradox he could not accept. According to his calculations based on experimental data and Maxwell's theory of electromagnetism, the radius r_e of an electron with rest mass m_e and rest charge e is equal to:

$$r_e = \frac{0.408e^2}{\pi \varepsilon_0 m_e c^2}$$

In 1903, the German physicist Max Abraham (1875-1922) developed a model of the electron, in which he assumed the electron to be a *rigid sphere*, with a uniform surface charge distribution. He also assumed that the electron's mass originated entirely in its own electromagnetic field. Based on these assumptions, he was able to determine that the radius r_e of the electron is equal to:

$$r_e = \frac{0.125e^2}{\pi \varepsilon_0 m_e c^2}$$

A different formula for the radius of the electron carries the name of the Dutch theoretical physicist Hendrik Lorentz and is known as the Lorentz radius of an electron. It is equal to:

$$r_e = \frac{0.25e^2}{\pi \varepsilon_0 m_e c^2}$$

Neither of the above models, however, accounted for the electron's spin (angular momentum) and magnetic moment, the two extremely important properties of an electron that at that time had not yet been discovered. In any case, these models allowed the electron to have only one degree of freedom, the sphere radius, which limited the models' capability to describe the various properties of the electron.

Alfred Parson (1889-1970) - In 1915, the British chemist and physicist Alfred Parson proposed a more sophisticated model of an electron, in which the electron charge was distributed over the surface of a *toroidal ring* (Fig. 1.5.1).

The ring was very thin, with the radius r of about 5×10^{-11} m, much smaller than the overall ring radius R. The ring produced an overall magnetic field ("spin") due to the current created by the charged elements moving along the ring with the velocity of light c.

In this configuration, the charged elements circled the center of the ring, but the ring as a whole did not radiate, because both electric and magnetic fields remained constant. The model treated electrons or protons as bundles of "fibers" or "plasmoids" with the total charge equal to the elementary charge $\pm e$. The fibers

twisted around the toroidal ring as they wrapped around its surface, forming a slinky-like helix. Each helical fiber was twisted
around the ring an integer number of times to account for quantum values of angular momentum and radiation. The number of
fibers was odd. The helicity of the twist distinguished the electron
from the proton. Reportedly, the model explained how particles
linked together to form atoms.

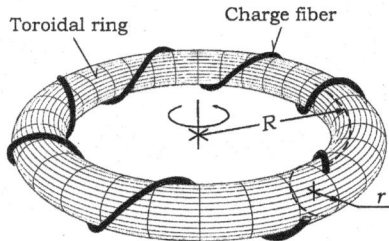

Figure 1.5.1. Parson's toroidal ring model of an electron.

Parson's model was further elaborated by a group of American physicists that included Charles Lucas, Jr., Glen Collins, and
David Bergman. Various models of particles with toroidal topology were also proposed by other scientists, including the British
physicist John Williamson, the Netherlands' physicist Martin van
der Mark, the Russian physicist Phillip Kanarev, the Spanish
physicist Oliver Consa, the American physicist Richard Gauthier,
and the Indian physicist Suraj Kumar.

In 1966, the American physicist Winston Bostick (1916-1991)
proposed a model of an electron composed of toroidal plasmoids
forming vortex loops around a ring. In 2000, two American theoretical physicists, Jean-Luc Cambier and David Micheletti, introduced the toroidal spiral concept of an electron based on the cold-
fluid equations for plasma shell coupled with the self-generated
magnetic field.

Louis de Broglie (1892-1987) – In 1923, while working on his
Ph.D. thesis, the French physicist Louis de Broglie asked himself a
logical question: if waves behave like particles, must not the reverse also be true, that particles behave like waves? De Broglie
generalized this statement by making a profound proposition that
the motion of each particle is accompanied by a wave, which
guides the particle through space. Moreover, it was not an electromagnetic wave but a mechanical wave, propagating faster than
the velocity of light. He then found that the wavelength λ of this

wave is related to the Planck constant h, particle mass m, particle velocity V and particle momentum p by a simple equation:

$$\lambda = \frac{h}{mV} = \frac{h}{p}$$

Now de Broglie was ready to apply his theory to the electron in Bohr's model of the hydrogen atom. His goal was to explain that only certain electron orbits are allowed. Equipped with his idea of the wave accompanying an electron as it moves around a nucleus, de Broglie envisioned this wave as a standing wave similar to a musical sound wave radiating from a violin.

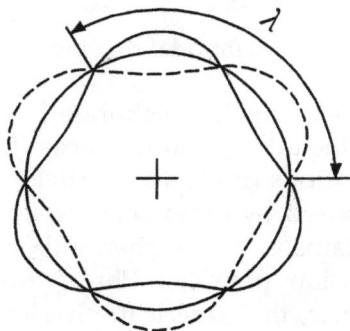

Figure 1.5.2. De Broglie's standing waves.

De Broglie proposed that all of the possible states of the electron in the atom are standing wave states, each with its own wavelength. The electron in the atom no longer appears as a planet in a solar system, but more like a vibrating circular string (Fig. 1.5.2). By using his model, de Broglie provided a "visualizable" explanation of the allowed states of the electron orbits. One simply needs to assume that the allowed states arise because the electron matter waves form standing waves, when an integral number n of wavelengths fits exactly into the circumference of a circular orbit. Thus, for the orbit with radius r we obtain that $nh = 2\pi r$. This relationship leads precisely to the Bohr's equation for quantization of electron orbits that we described earlier.

Erwin Schrödinger (1887-1961) - Soon after de Broglie published his paper in 1925, the Austrian physicist Erwin Schrödinger dealt another devastating blow to the concept of the particle nature of electron. He dared to claim that electrons are not particles at all, but classical waves, like ordinary sound waves, water waves or electromagnetic waves. Moreover, the particle aspect of electrons is merely illusory.

Schrödinger came up with a differential equation in which the electrons were presented as matter waves. The principal parameter in his equation was the so-called *wave function* ψ (Greek: psi) that is dependent on distance and time. Schrödinger solved his equation in application to atomic electrons, producing the wave functions ψ corresponding to all possible orbits of the electrons. In Schrödinger's interpretation, the square of the amplitude of wave function $ψ^2$ described the density distribution of matter waves. Not everyone agreed with the wave theory of electron.

Among the opponents of Schrödinger's wave theory was the German physicist Max Born, who insisted that the electron is still a particle and only the wavy part, still related to the electron, is Schrödinger's wave function. But this function has nothing to do with the electron structure. Moreover, the square of the amplitude of the wave function $ψ^2$ is not the density distribution of matter waves, as Schrödinger believed, but is the probability of finding an electron at various locations in a certain quantum state. So, according to Born, the wave function is a *wave of probability*. The opinions of scientists about the validity of Born's interpretation were not unanimous. Bohr, Heisenberg and Sommerfeld had no issues with Born's idea, but Schrödinger, de Broglie and Einstein found it totally unacceptable. The dramatic split of opinion between these outstanding physicists eventually led to several bitter but productive confrontations over the years following.

Werner Heisenberg (1901-1976) –Amidst the confusion created by the introduction of the idea of wave probability, the German theoretical physicist Werner Heisenberg introduced, in 1927, another even more controversial concept that became known as *Heisenberg's uncertainty principle*. The logical basis for the development of this concept was quite simple. During an experiment we are using sensors to measure certain parameters of an object under investigation. For instance, if we want to measure the water temperature in a swimming pool, we use a temperature sensor that we place into water. We are not concerned if, before we took the measurement, the sensor was in a shady area or under a hot sun, and rightfully so. The amount of heat contained inside the sensor is miniscule compared to the heat contained in the water of the swimming pool. Therefore, the temperature of the sensor has no appreciable effect on the accuracy of our measurement.

But it is a completely different matter if we decide to measure the temperature of water contained in a very small glass tube. In that case, the amount of heat contained in the sensor may become sufficient to affect the temperature of the water in the glass tube, distorting the results of our measurement. A somewhat similar "thought experiment" was probably in Heisenberg's mind when he pointed out that a similar situation occurs when we do our measurements on atomic scale.

Let us suppose that we want to measure a position and the velocity of an atomic electron located at a certain quantum energy state. We will use, of course, the tiniest sensor available to us that utilizes a beam of light. But, as we know from the photoelectric effect, the photons will excite the electron. Therefore, its trajectory will be distorted during the process of measurement. Based on this logic, Heisenberg concluded that it is fundamentally impossible to make simultaneous measurement of a particle's position and velocity (or momentum) with infinite accuracy. He stated in his famous principle:

> If a measurement of position is made with precision Δx and a simultaneous measurement of momentum is made with precision Δp, then the product of the two uncertainties can never be smaller than a number of the order of *h/(2π)*; That is,

$$\Delta x \cdot \Delta p \geq \frac{h}{2\pi}$$

Heisenberg's uncertainty principle became an indispensable addition to quantum theory. But, as with Born's idea of probability waves, not everyone accepted Heisenberg's uncertainty principle. The confrontation between the two different theories of physics stimulated the thinking of many young physicists, and in some cases led to a better understanding of the "spooky" world of quantum physics.

Roger Penrose (b. 1931) - The British theoretical physicist Roger Penrose became well known for his contributions to theories of black holes. His most favorite idea, however, was certainly the *twistor theory* that he conceived in mid-1960s.

At the core of his theory was a 4D mathematical entity called the *twistor* that was capable of defining empty space, all the subatomic particles and the four forces, including gravity. The twistor was supposed to be at nature's most fundamental level. To accomplish this task, Penrose thought, the twistor must combine both translational and rotational motions; that is, it must be both spinning and moving along. In addition, it must be both quantum and relativistic. Penrose envisioned the twistor as a pair of concentric doughnuts with a short shank of rope through its eye (Fig. 1.5.3).

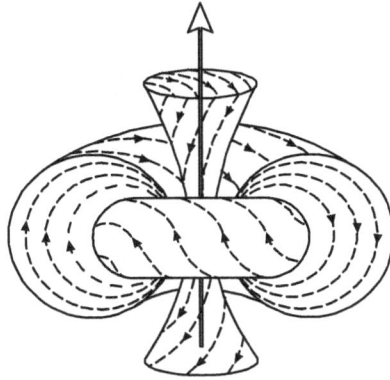

Figure 1.5.3. The Penrose's twistor.
Adapted from J. Boslough (1992).

The size of the twistor was somewhere between an absolute point and the size of an elementary particle. By combining various types of twistors it was possible to produce certain elementary particles. But massless particles, such as photons or neutrinos, could be made from a single twistor.

Similarly to the general theory of relativity, Penrose's twistor theory held that gravity was caused by mass giving rise to a curvature of the universe. To solve the problem mathematically, Penrose employed the theory of complex numbers and other advanced theories of mathematics. Unfortunately, the equations turned out to be very complicated, even for Penrose, one of the world's foremost mathematicians.

Richard Gauthier (b. 1946) - Since 1993, the American theoretical physicist Richard Gauthier has been exploring the hypothesis that matter and energy can be composed of helically moving point-like quantum particles that may move faster than light. According to his hypothesis that he describes below, these particles are called transluminal energy quanta, or TEQs. They may move along open helical paths and form photons - quantum particles of visible light or other electro-magnetic energy such as radio waves, x-rays and gamma rays. Figure 1.5.4 shows transluminal energy quantum (TEQ) models of three photons of different wavelengths. The TEQs are the three black dots moving to the right along open helical trajectories at a velocity 1.414 (the square root of 2) times the velocity of light.

As shown in Figure 1.5.5, the TEQ moves from a minimum velocity of 0.707 times the velocity of light to a maximum velocity of 2.516 times the velocity of light, passing frequently through the velocity of light. The self-intersecting mathematical toroidal surface on which the TEQ moves is partly cut away to show the complete TEQ trajectory.

Figure 1.5.4.
Transluminal energy quantum (TEQ) models of three photons of different wavelengths.

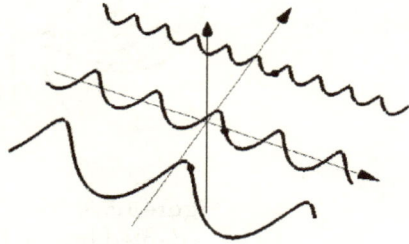

Alternatively, they may move in closed helical paths to produce various particles having mass, such as the electron, the muon and the tau particle (the family of electron-like particles). A TEQ carries energy, momentum and spin and sometimes charge (as with the TEQ electron model.)

When a TEQ moves in a closed helical path, it moves along the mathematical surface of a torus. A TEQ can moves at velocities slower and faster than velocity of light, even passing through the velocity of light many times per second when composing a particle with mass like the electron.

Figure 1.5.5.
Transluminal energy quantum model of the electron.

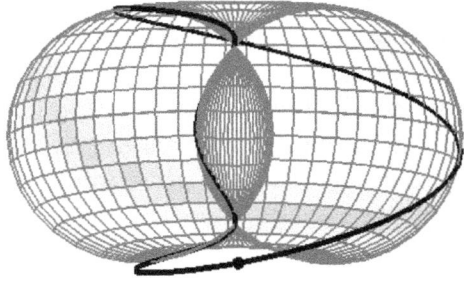

Figure 1.5.6.
Transluminal energy quantum model of the cosmic quantum.

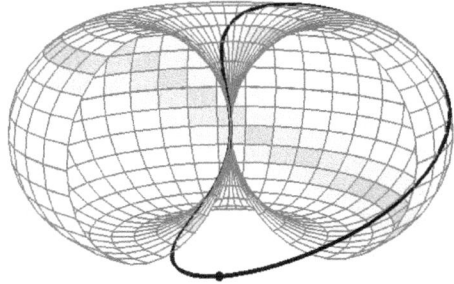

In addition to the photon and the electron, two hypothetical particles have been modeled with the TEQ. The first is the cosmic quantum (Fig. 1.5.6), which is the hypothetical first particle of the universe. The cosmic quantum would have contained all the positive initial energy of the universe in a single tiny quantum particle much smaller than a single atom. It is a spin-1 particle or boson. Here the TEQ's velocity varies from the velocity of light (at the minimum) to 2.236 times the velocity of light (at the maximum).

Figure 1.5.7.
Transluminal energy quantum model of a dark matter particle.

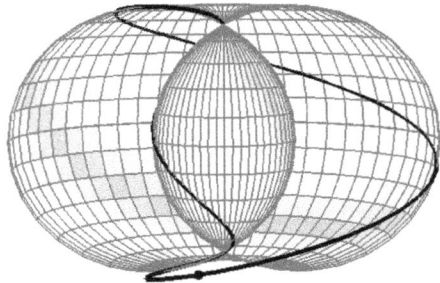

The second hypothetical particle is a particle of dark matter. As shown in Figure 1.5.7, it is a spin-$\frac{1}{2}$ particle or fermion. Here the TEQ's minimum velocity is the velocity of light and its maximum velocity is 3.162 times the velocity of light. Dark matter is considered to compose about 85% of all the matter in the universe. Most of the dark matter in the universe is generally thought to be composed of non-baryonic elementary particles that mainly react with other particles through gravitational attraction. A reader may visit website for further information. http://www.superluminalquantum.org.

1.6 The Standard Model

There was an explosion in the discovery of new elementary particles in the second half of the 20th century, thanks to the construction of powerful particle accelerators. There was a need to find some kind of order inside the ever-expanding "zoo of particles." Many physicists rightfully considered Mendeleyev's Periodic Table of Elements as the model to follow. By the end of the 1970s the knowledge accumulated by particle physicists evolved into the so-called *Standard Model* that provides the description of the world in terms of truly fundamental particles, *leptons* and *quarks*. All other particles, the hadrons, are merely composites of the leptons and the quarks. The word "lepton" came from the Greek word *leptos* meaning "small or light." The name "hadron" came from the Greek word *hadros*, which means "thick and heavy." The hadrons are divided into two classes, called *mesons* and *baryons*. The name "meson" originated from the Greek word *mesos*, which means "in between." The name "baryon" derived from the Greek word *barys* for "heavy." Proton is the lightest baryon.

Based on their principal function, the particles are divided into two groups: the mass-carrying particles called *fermions* and the force-carrying particles responsible for interaction between particles called *bosons*. Standard Model considers five principal forces operating between particles: color, weak, electromagnetic, gravitational and Higgs.

- Color force operates between quarks, and also shows up as the strong force acting between nucleons. The color force is mediated by *gluons*.
- Electromagnetic force operates between electrically charged particles. It is mediated by *photons*.
- Weak force is responsible for particle decay. It is mediated by the *intermediate vector bosons* W^+, W^- and Z^0.

- Gravitational force operates between all particles. It is mediated by *gravitons*.
- Higgs force is responsible for the creation of the masses of all particles. It is mediated by the *Higgs boson*.

Quark model – The quark model is a cornerstone of the Standard Model. The idea of quarks was proposed independently by the American theoretical physicists Murray Gell-Mann (b. 1929) and George Zweig (b. 1937). According to this model, all elementary particles, except for leptons, are made of fundamental particles called *quarks* which have fractional charges. Although Zweig and Gell-Mann were talking about essentially the same physics, they treated quarks in completely different ways. Gell-Mann thought of a quark as a "mathematical" entity helping to find some order among subatomic particles. To Zweig, quarks were real, concrete particles, providing a real challenge for experimenters to find them.

In the original quark model, the quarks came in three *flavors*: u (up), d (down), and s (strange), but by 1977, the model was expanded by several physicists to include three additional flavors: c (charmed), b (bottom) and t (top). The u, d, s, c, b, and t quarks have respective fractional charges Q equal to $+\frac{2}{3}$, $-\frac{1}{3}$, $-\frac{1}{3}$, $+\frac{2}{3}$, $-\frac{1}{3}$, and $+\frac{2}{3}$. The antiquarks have opposite electric charges. All quarks have the same spin $J = \frac{1}{2}$ and the same baryon number $N = \frac{1}{3}$. Additionally, there are four quantum numbers assigned to each quark, strangeness S, charm C, bottomness B, and topness T.

Notably, except for the fractional charges Q and the spin J, the other five quantum numbers did not have any physical meaning.

Formation of hadrons - It is easy to use the quark model for defining the compositions of hadrons. The main feature of hadrons is their ability to react to the strong force. The baryons (including proton and neutron) are identified by the baryon number N that is equal to $+1$ for the particles and -1 for the antiparticles. For all particles other than the baryons, $N = 0$.

One should follow three simple rules when using the quark model (Fig. 1.6.1):

1. Each meson is made of one quark and one antiquark.
2. Each baryon is made of three quarks.
3. Each antibaryon is made of three antiquarks.

Meson π^+ Baryon p Antibaryon p^-
(u$\bar{\text{d}}$)   (uud)   ($\bar{\text{u}}\bar{\text{u}}$d)

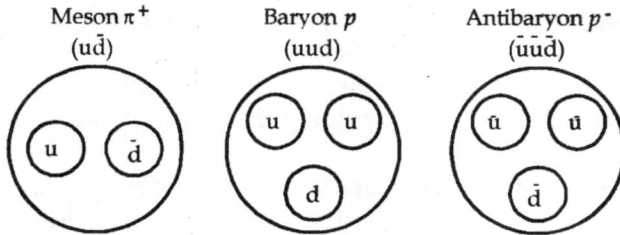

Figure 1.6.1. Quark compositions of several particles.

Leptons include three particles and three antiparticles: electron e, muon μ and tau τ. There is either a neutrino or an antineutrino associated with each lepton. Mesons are made up of two quarks, while baryons are made up of three quarks.

Since the discovery of the first meson in 1937, the number of discovered particles that belongs to the meson class had been increasing at an exponential rate. By the end of the 20th century this class included more than 340 meson particles and antiparticles, and some of them were more than ten times heavier than protons. So, the original meaning of meson as something "in between" became obsolete. To sort out all these particles, it was necessary to split the meson class further into several subclasses.

Notably, among discovered particles only the electron, the positron and the proton are stable, in other words, they never disintegrate into something else. Photons and neutrinos are also believed to be stable. All other particles are not stable, including the free neutron, the average lifetime of which is about 15 minutes.

Higgs mechanism - The quark model has a serious problem with the quark masses. The total mass of the quarks making up an elementary particle is much smaller than the actual mass of that particle. To solve this problem, several particle physicists, including the British cosmologist and particle physicist Peter Higgs (b.1929), proposed in 1960s that there is an undetectable field that is now called the *Higgs field*. The Higgs field is associated with a boson that has a big mass. This particle became known as the *Higgs particle*, or the *Higgs boson*. The Higgs mechanism involves swallowing up the Higgs boson by any photon-like particle, providing the particle with mass. According to different sources, the predicted mass of the Higgs boson varies from 80 to 1000 GeV. The collisions of protons at the CERN Large Hadron Collider in 2012 produced a particle with the mass ranging from 125 GeV to 126 GeV. Some physicists believe that this is the expected Higgs boson. The results of these experiments could be confirmed more conclusively in 2015 when the power of the Large Hadron Collider will be doubled.

Interactions Between Charged Light Particles – The interactions between charged light particles, such as electrons, are described by the *theory of quantum electrodynamics*, or *QED*, that is a part of the Standard Model. Three physicists are mostly credited by the scientific community for the development of QED: the two American theoretical physicists Richard Feynman (1918-1988) and Julian Schwinger (1918-1994), and the Japanese theoretical physicist Sin-Itiro Tomonaga (1906-1979). According to QED, the interactions between charged light particles like electrons are mediated by photons. QED uses very complicated equations of quantum mechanics to describe the interactions between particles.

Feynman introduced his famous *Feynman diagrams* that helped many thousands of particle physicists around the world to describe very sophisticated nuclear reactions by visualizing particle collisions and talking about them in a common language. Consider, for instance, Feynman's spacetime diagram depicting the simplest collision of two electrons, shown in Fig. 1.6.2.

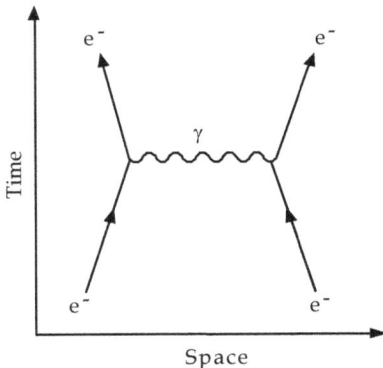

Figure 1.6.2.
Feynmann diagram
depicting the collision
of two electrons.

This is a typical case of an *electromagnetic interaction*. The straight lines represent the paths of the electrons e^- and the wiggly line the path of the virtual photon γ. The virtual photon fires up when the distance between the electrons reaches a certain minimum distance. Consequently, the mediation of the virtual photon produces the electromagnetic interaction that pushes the colliding electrons apart. Thanks to the Feynman diagram, the process of collision of two electrons appears very simple, but a particle physicist knows that behind each line in the diagram hides a specific term in the complex mathematical expression for the probability of this collision. Feynman, Schwinger and Sin-Itiro Tomonaga were involved in attacking the principal problem of QED known as the *War Against Infinities*. According to QED, interactions between charged light particles such as electrons are described as being mediated by photons. The problem arises when one wants to take into account the so-called *self-interaction* between the electron and its own electric field. Based on the laws of interaction between the electric charge and electric field, the mass of the electron must be infinite.

Feynman solved the problem with the aid of a mathematical trick known as *renormalization*. Renormalization was introduced in its complete form in 1947 by the Dutch physicist Hendrik Kramer (1894-1952). It was based on his assumption that the mass of an atomic electron is made of two components: the mass measured at very short range, *bare mass*, and the *infinite mass*, produced by electromagnetic self-interaction. Bare mass can be defined by subtracting the infinite bare mass from the infinite self-interaction mass of the electron. If renormalization of a proposed theory produced correct results, then the physicists considered the theory *renormalizable*, and, thus, acceptable. Eventually, the capability to be renormalized became one of the most important criteria for validating a physical model. Nevertheless, this technique still bothered many physicists.

Weak & Electroweak Interactions Between Particles - The weak force, like the other forces in the subatomic world, operates by the exchange of messenger particles, but with one principal difference. Instead of pushing particles apart or pulling them together, the exchange of the messenger particles modifies the character of the particles that swap them. In the weak interactions, the messenger particles are the charged bosons W^+ and W^-. For instance, during beta decay a neutron n emits an electron e^- and an antineutrino $\bar{v}_e$, converting itself into a proton p according to the equation:

$$n \rightarrow p + e^- + \bar{v}_e$$

The above equation, however, does not provide us with a full description of the weak interaction during the beta decay. Here again, the Feynman diagram becomes very helpful. As shown in Figure 1.6.3, during beta decay, the down quark d inside a neutron interacts with an incoming electron antineutrino $\bar{v}_e$ through a messenger particle W^-.

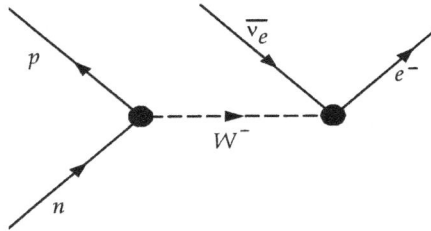

Figure 1.6.3. Feynman diagram depicting neutron decay.

Consequently, the down quark d converts into the up quark u, while the electron antineutrino $\overline{v}_e$ turns itself into the electron e^-.

In 1960, Sheldon Glashow (b. 1932) proposed that the neutral weak current was carried by a massive neutral particle Z^0. This particle would mediate, for instance, the exchange force during the collision between the gauge bosons W^+ and W^-. The old problem with infinities, however, immediately showed up in Glashow's theory. The solution was proposed by another American physicist, Steven Weinberg (b. 1933), who first pointed out on the possibility of getting rid of the infinities in weak interactions by employing the Higgs mechanism. Eventually, jointly with the Pakistani theoretical physicist Abdus Salam (1926-1996) and the Dutch theoretical physicist Gerald't Hooft (b. 1946), Sheldon Glashow and Steven Weinberg proved that by using the Higgs mechanism their theory becomes renormalizable.

Strong Interactions Between Particles - The strong interactions occur between nucleons. It holds nucleons together within the nuclei. The quark model worked very well for the construction of most subatomic particles, but there were serious difficulties with making several particles. The solution of this problem was as always swift: simply propose a new quark to possess one more property. The physicists with artistic minds named this property *color*, and the theory of interaction between quarks became known as *quantum chromodynamics*, or QCD.

In QCD, everything comes in color that occurs in three varieties: *red*, *green* and *blue*. Each quark carries a *color charge*, in analogy to electric charges. For each quark of a certain color there is a quark with the opposite color. In another analogy with the electric force between electric charges, the force between quarks is called the *color force*. The color force is thought to be mediated by massless particles called *gluons*, in analogy with the photons responsible for mediation of electric force. Notably, in QCD we are dealing with the dynamic quarks that continuously mingle with one another by exchanging their colors.

Notably, there is no physical meaning behind the "color." The color is employed merely to allow for more varieties of elementary particles. Instead of different colors one could use, for instance, different kinds of fruits or vegetables.

Symmetry - Intuitively, physicists believed that our world must be symmetrical. In application to the subatomic world they considered several spacetime symmetries that must be conserved during the interactions between particles. The most important among them were related to the conservation of *parity* (P), *charge conjugation* (C), and *time-reversal invariance* (T).

Parity (P) is the operation which reverses the signs of coordinates to describe a system. Thus, a position described in three dimensions by the coordinates x, y, and z is now described as the positions $-x$, $-y$, and $-z$. This is equivalent to studying the mirror image of the original system. For instance, a mirror image of a particle with a certain spin will produce a particle with the opposite spin.

Charge conjugation symmetry (C) states that every subatomic particle has an antiparticle with the same mass and opposite values of all other quantum numbers, such as charge. In respect to charge this symmetry appears to be the same as (P) symmetry, but (C) symmetry is more general because it requires other parameters, besides charge, to be symmetrical. Therefore, it can also be applied to neutral particles.

Time-reversal invariance symmetry (T) states that the properties of particles will not change if the direction of time changes. The (P), (C) and (T) symmetries turned out to be more elusive than the physicists had expected. Although parity is conserved in electromagnetic interactions, it refuses to be maintained for the weak nuclear interactions. This odd behavior of the weak nuclear interactions got the name *broken symmetry*.

1.7 Unified Theories of Physics

As we described before, the first serious attempts to develop a unified theory of physics were made in the beginning of the 20th century by the German mathematician Hermann Weyl and the Swedish theoretical physicist Oskar Klein. Albert Einstein accepted neither Weyl's nor Kaluza-Klein's theory. After numerous unsuccessful attempts to modify them to his satisfaction, he eventually decided to abandon both of them. But, in spite of their demise, these theories certainly stimulated Einstein's efforts to develop a unified field theory, a monumental effort that lasted for more than three decades.

It is well known that Einstein did not succeed. This failure, however, did not discourage many young physicists from following in his footsteps. But in fact, their task was even more difficult. During Einstein's time there were only two kinds of known forces, electromagnetic and gravitational. The new generation of physicists had to deal with three additional forces, color, weak, and Higgs. But the challenge was clear and, in mid-1970s, when the Standard Model was at the peak of its success, it seemed that it was only a matter of time before the quantum mechanics of the Standard Model would be unified with the curved spacetime gravity of the general theory of relativity.

The unified theories of physics developed during the 20th century are divided into two groups, the Grand Unified Theories (GUTs) and the Theories of Everything (TOEs). GUT is a gauge theory that provides the same mathematical tool to describe the electromagnetic, the weak nuclear and the strong nuclear forces. The Standard Model is a typical representative of GUT. The most popular alternative to GUT is the Supersymmetry Theory that carries the cute name *SUSY*. The first realistic version of this theory was proposed in 1981 by the American physicist Howard Georgi (b. 1947) and the Greek physicist Savas Dimopoulos (b. 1952). This theory gets rid of the last major asymmetry in the world of subatomic particles that divides them into fermions (the particles that make matter) and bosons (the particles that carry force).

To accomplish this unification task, SUSY provides a boson partner for every fermion and a fermion partner for every boson. This happy marriage of particles is accompanied by an easy way of naming the partners. For instance, the SUSY partners of electrons and quarks are respectively called *selectrons* and *squarks*, while the SUSY partners of photons and gluons are respectively called *photinos* and *gluinos*. But all this happiness has its price. The downside of SUSY is two-fold. Firstly, it doubles the already excessive number of particles that are present in the Standard Model. Secondly, it works only by adding another four dimensions to the four dimensions of ordinary spacetime. Another GUT, called SU(5), was proposed in 1974 by Howard Georgi and Sheldon Glashow. It includes all forces except gravity. The theory makes a prediction, still unproven, that protons could decay.

The end of the 20th century appeared to be a fair time mark to complete the task. It did not happen. Instead, some scientists have begun abandoning the Standard Model as the basis for a unified theory of physics.

1.8 String, Supergravity & M-Theories

The origin of the string theory goes back to 1968. That year the Italian theoretical physicist Gabriele Veneziano (b. 1942) noticed that a two-hundred-year-old formula, known as the *Euler beta function*, seemed to match the data on the strong force with surprising precision. Was it a pure coincidence? In 1970, the American theoretical physicist Leonard Susskind (b. 1940), the Danish theoretical physicist Holger Nielsen (b. 1941) and the American physicist Yoichiro Nambu (b. 1921) recognized the physical meaning of Veneziano's discovery. They showed that if two particles are connected by a tiny, extremely thin rubber-like string, then the strong forces between these particles are described by Euler's equation.

In 1974, string theory was further extended by one of the earliest string enthusiasts, the American theoretical physicist John Schwarz (b. 1941) and his collaborator, the French theoretical physicist Joël Scherk (1946-1980). The two physicists proposed a string theory of quantum gravity. But it took ten years for this theory to be widely accepted. String theory treats particles in a very special way. According to a conventional theory, such as the Standard Model, the electrons and the quarks are simply the dots with no spatial extent in any direction. This is not so according to string theory: it presents the particles as tiny, vibrating filaments of energy, called *strings*.

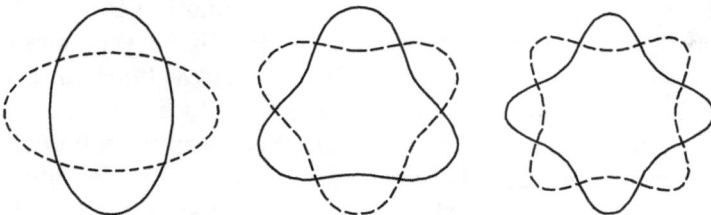

Figure 1.8.1. Some examples of the string vibrational patterns. Adapted from B. Greene (2004).

The strings have no dimensions except for length, making strings one-dimensional entities. At the same time, because the strings are extremely small (about 10^{-35} m), they appear to be points whenever they are detected by the most sensitive devices in the world. The string vibrates similarly to a string of a violin or a cello. But unlike music, where we are mostly concerned with different tones, string theory deals with different vibrational patterns of the strings (Fig. 1.8.1) that make up both the matter particles (fermions) and the force carrier particles (bosons).

Each specific vibrational pattern corresponds to certain values of the particle mass, electric charge, and spin. In addition, John Schwarz and the Joël Scherk showed in 1974 that there is an additional vibrational pattern that has all the properties of a graviton. Thus, string theory included gravity. One of the serious drawbacks of string theory is that it requires many dimensions. The original version, called *bosonic*, required 26 dimensions, while, in the later versions, this number was reduced to ten. The reduction of the dimensions became possible by applying Kaluza's compactification described before. Another difficulty with the original version was that it required a particle called *tachyon* that travels faster than light.

As we know, according to Einstein's special theory of relativity, the velocity of light is a kind of barrier no material body can trespass. However, superluminal velocity is not forbidden by the equations of his theory for the particles already existing on the other side of the light barrier. Einstein's equations predict that the mass of tachyons must be expressed by imaginary numbers, and these particles can never be slowed down to reach the velocity of light. An introduction of the tachyon added another kind of symmetry to the world of particles that now could travel both below and above the velocity of light. The string theorists, however, preferred to abandon this symmetry rather than to deal with a controversial superluminal velocity. The tachyons did not show up any more in the later versions of superstring theories.

The next significant improvement of the string theory came in 1980, when several physicists discovered that patterns of string vibrations come in pairs (super-partner pairs), differing from each other by a half unit of spin. The new version of the string theory that incorporates this novel symmetry was named the *superstring theory*. Still, the new theory was plagued with deadly quantum anomalies that threatened to make it senseless. But in 1984 the American theoretical physicists Michael Green (b. 1946) and John Schwarz successfully demonstrated that the superstring theory would be free from quantum anomalies if two conditions are met. Firstly, the theory must employ ten spacetime dimensions and, secondly, it must use a special type of the quantum gauge symmetry group known as SO(32) or E(8)×E(8).

This breakthrough helped to bring attention of many particle physicists to this theory. The triumphant march of the superstring theory continued and, in 1984, four physicists at Princeton – David Gross, Jeffrey Harvey, Emil Martinec, and Ryan Rohm (dubbed the *Princeton String Quartet*) discovered another class of anomaly-free superstring theory with the quantum gauge symmetry group E(8)×E(8). Their version of the superstring theory had even better properties than the Green-Schwartz version. By the end of the 20th century, three additional versions of the superstring theory showed up along with another promising theory called the *supergravity*. Can be the string theory unified with the theory of supergravity? Some physicists believe that the supergravity and string theory are just parts of the so-called *Membrane*, or *M-theory*, in which the strings are replaced with membranes.

In 2003, the American mathematician and theoretical physicist Edward Witten (b. 1951) proposed a way to unify the string theory with the twistor theory. His model employed the supertwistor space, a supersymmetric extention of twistors introduced by Allen Ferber in 1979. The resulting model has become to be known as the *twistor string theory*.

Only time will tell us if all the above mentioned propositions have some merits.

1.9 Maximum Velocity in the Universe

As the skepticism about the bright prospects of both the Standard Model and superstring theory grew stronger, some of the most daring scientists began questioning the basic assumptions of these theories. Some scientists also started reviewing various aspects of Einstein's theory of relativity. Particularly, they went as far as to reexamine the most fundamental postulate of the theory, that of the maximum velocity in the universe.

The EPR Paradox - Paradoxically, Einstein, the man proclaiming that the velocity of light is the maximum velocity in the universe, initiated a debate that led to the discovery of instantaneous transmission of information between particles.

Figure 1.9.1.
Interaction
between
two electrons.

Einstein was always bothered by Heisenberg's uncertainty principle. Remember his famous saying that God does not play with dice? So, he was periodically creating new "thought experiments" in his attempts to disprove Heisenberg's concept, each time to be defeated by his clever opponent, Niels Bohr. In 1935 he, with the help of his two collaborators, the American physicist Boris Podolsky (1896-1966) and the American-Israeli physicist Nathan Rosen (1909-1995), came up with the idea of the eponymous *EPR experiment*. The essence of their thought experiment is deceptively simple. Imagine two electrons *A* and *B* that, upon colliding, fly a long distance apart (Fig. 1.9.1). Suppose that we were able to measure the total angular momentum of the system that includes both electrons at the time when they interacted. According to the law of the conservation of momentum, the total momentum of two electrons cannot change as long as they do not interact with anything else after their collision. Thus, if we were to measure, much later, the momentum of the electron *A*, then it would be possible to calculate the momentum of the electron *B* by subtracting the measured momentum of the electron *A* from the total.

The same is true if, instead of measuring the momentum of the electron *A*, we measure its position. According to quantum theory, if we observe the precise position of electron *A*, we can, at the same moment, calculate the precise position of electron *B*. Heisenberg's uncertainty principle tells us that we cannot measure precisely both the momentum and the position of the electron. So, if this rule is strictly maintained, then whenever we measure the precise position of the electron *A*, the electron *B* will also have a precise position, but both electrons will have uncertain momentum. Conversely, whenever we measure the precise momentum of the electron *A*, the electron *B* will also have a precise momentum, but the positions of both electrons will be uncertain. Einstein argued that this contradiction can be solved only by accepting Newton's "spooky action at a distance," according to which the information from one electron to another propagates instantaneously. But he could not accept this solution because it contradicted the most fundamental postulate of his special theory of relativity - nothing can move faster than the velocity of light.

In 1952, three years before Einstein's death, the American physicist David Bohm (1917-1992) came up with his version of the EPR experiment. Although it was originally designed by Bohm as a thought experiment, the advances in the technologies of measurement have made it possible to conduct a real experiment. In 1964 the Irish physicist John Bell (1928-1991) demonstrated that Bohm's thought experiment could be practically implemented by measuring the polarization of photons. Crucial to the success of his method was the so-called *Bell's inequality* that provides an objective criterion for the treatment of the results of the experiments.

Several experiments were conducted by following the ideas of Bohm and Bell. The most prominent of them were carried out in Paris in the early 1980s by the French physicist Alain Aspect (b. 1947) and his colleagues. After the publication of their paper in 1982, the idea of action at a distance, no longer spooky, took its legitimate place in the world of quantum mechanics. Lately, a different term, *non-locality*, became more popular in describing the same concept. Non-locality underlines that the behavior of quantum entities is affected not only by what is going on at the point where the entity is located, but also by events that are occurring at other localities, which may in principle be far away, even across the Universe. So, one may rightfully conclude that Einstein turned out to be wrong. The events in the subatomic world can be correlated instantaneously. But at the same time, thanks to his inquisitive mind, by proposing a challenging paradox, he stimulated the thinking of many bright scientists who discovered one of the most puzzling phenomena in the world of quantum physics.

Instantaneous Transmission of Information - The idea of instantaneous transmission of information between subatomic particles leads to another "spooky" concept, that of *negative time*, when time goes backwards. Let us start from Maxwell's theory of electromagnetism. Initially, its differential equations were thought to be applicable only for positive, or conventional, time. But between 1929 and 1932 the Dutch physicist Adriaan Fokker (1887-1972) proved that Maxwell's equations are completely symmetrical in time. Simply saying, whether you will input in these equations the time variable $-t$ or t, the result of your calculations will be exactly the same.

In 1940 Richard Feynman and John Wheeler (1911-2008) applied this concept to their *absorber theory*. The principal assumption of this theory is that an excited electron emits two types of waves, the *retarded waves* that travel forward in time, and the *advanced waves* that travel backwards in time. Consider two electrons separated by a certain distance (Fig. 1.9.2). After its excitation, the first electron emits simultaneously two half waves, a half retarded wave *R1* and a half advanced wave *A1* that propagate in the opposite directions with the velocity of light. After the half retarded wave *R1* reaches the second electron, this electron will emit in response a half retarded wave *R2* and a half advanced wave *A2* that also propagate in the opposite directions with the velocity of light.

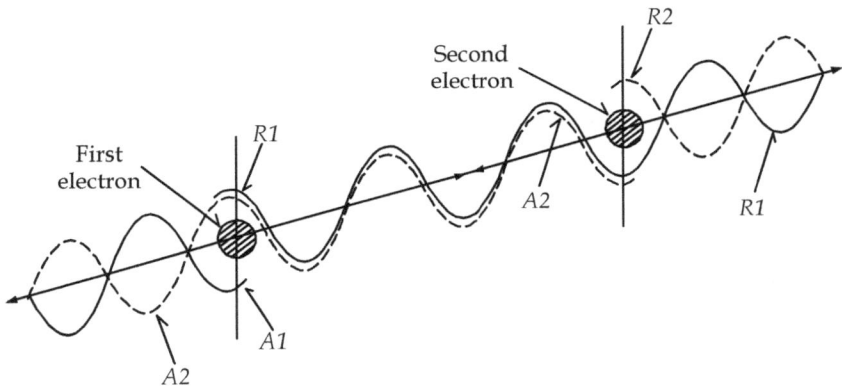

Figure 1.9.2. Instantaneous transmission of light according to Wheeler-Feynman absorber theory. Adapted from J. Gribbin (2000).

The two sets of waves cancel out everywhere except in the region between the two electrons. Since within this region the half retarded wave *R1* travels forward in time and the half advanced wave *A2* travels backwards in time, the connection between the two electrons is made simultaneously.

1.10 Cosmic Messangers

The 20th century was a turning point in astronomy. New technology enabled astronomers to detect cosmic radiation covering a very wide range of wavelengths (Table 1.10).

Radio waves - Radio waves have the longest wavelengths in the electromagnetic spectrum. Radio wavelengths vary widely, ranging from ten centimeters to several hundred meters. Incidentally, AM radio uses the longest wavelengths, while FM radio uses the shortest ones. The main advantage of radio waves over visible light is their capability to penetrate cosmic dust, so one can see not only the galactic center but also the opposite side of the galaxy.

Table 1.10. Regions of the electromagnetic radiation.

Region	Wavelength cm	Frequency Hz
Radio waves	> 10	$< 3 \times 10^9$
Microwave	$10 - 10^{-2}$	$3 \times 10^9 - 3 \times 10^{12}$
Infrared	$10^{-2} - 7 \times 10^{-5}$	$3 \times 10^{12} - 4.3 \times 10^{14}$
Visible	$7 \times 10^{-5} - 4 \times 10^{-5}$	$4.3 \times 10^{14} - 7.5 \times 10^{14}$
Ultraviolet	$4 \times 10^{-5} - 10^{-7}$	$7.5 \times 10^{14} - 3 \times 10^{17}$
X-rays	$10^{-6} - 10^{-9}$	$3 \times 10^{16} - 3 \times 10^{21}$
Gamma rays	$< 10^{-9}$	$3 \times 10^{18} - 3.4 \times 10^{22}$

Microwave radiation – We know that microwaves are used for heating food. They are also good for transmitting information, because they can penetrate haze, light rain and snow, clouds and smoke. One of the most popular cosmological theories of the 20th century is closely related to microwaves. In 1948, the Russian-born American physicist Georgy Gamow (1904-1968) and the American cosmologist Ralph Alpher (1921-2007) published their version of the Big Bang theory. They predicted that, as a result of a cataclysmic explosion, there must be a remnant radiation observable in the microwave region of the spectrum. Seventeen years later, in 1965, two American radio astronomers, Arno Penzias (b. 1933) and Robert Wilson (b. 1936), discovered this radiation, which became known as *cosmic microwave background* (CMB) radiation.

Infrared radiation – Infrared light lies between the visible and microwave portions of the electromagnetic radiation spectrum. They are divided into two sub-ranges, near infrared and far infrared. The wavelength of near infrared is closest to visible light, while the wavelength of far infrared is closer to the microwave region of the electromagnetic spectrum. Therefore, far infrared waves are thermal. In 1989, NASA's COBE satellite detected *cosmic infrared background* (CIB) radiation in the near-infrared range. Some scientists believe that the CIB radiation represents a "core sample" of the universe. It contains the cumulative emissions of stars dating back to the epoch when these objects first began to form.

Ultraviolet waves – Ultraviolet waves are divided into three sub-ranges: the near ultraviolet (NUV) that lies close to visible light, the extreme ultraviolet (EUV) that lies close to X-rays, and the far ultraviolet (FUV) that lies between NUV and EUV. The ozone layer of the Earth's atmosphere absorbs high-energy radiation. Therefore, ultraviolet radiation can be detected only by using either rockets or satellites. Ultraviolet observations have contributed a great deal to the understanding of stellar evolution and formation, as well as the outer atmosphere of stars.

X-rays –X-rays were discovered by the German physicist Wilhelm Roentgen (1845-1923) in 1895, but it took another sixty-seven years before X-ray astronomy was born. The first rocket flight to successfully detect a cosmic source of X-ray emission was launched in 1962 by a group of American scientists headed by Riccardo Giacconi (b. 1931). Later it was found that the X-ray emission of this source is ten thousands times greater than its optical emission. It comes from sources which contain an extremely hot gas at temperatures from a million to one hundred million degrees Kelvin. The X-rays are emitted by compact celestial structures, such as neutron stars and black holes. Like the cosmic microwave background (CMB) radiation and the cosmic infrared background (CIB) radiation, the *cosmic X-ray background* (CXB) radiation has an extremely uniform distribution of intensity across the sky.

The X-rays are nowadays produced by the X-ray machines in which an electron gun inside the X-ray tube shoots high energy electrons at a target made up of heavy atoms. Two methods of making the X-rays are mostly used. In one method called *Bremsstrahlung* (or "braking radiation" in German), the X-rays are emitted after the velocity of the electron shot at the tungsten atom slows down and loses energy. In another method called the *K-emission*, the X-rays are emitted after the electron gun knocks out the K-shell electron corresponding to the lowest energy state of tungsten atom.

Gamma rays – Gamma rays are the most energetic form of electromagnetic waves; they are produced by the hottest regions of the Universe. They are also produced by such violent events as supernova explosions or the destruction of atoms, as well as during the decay of radioactive material. The other sources of gamma rays include neutron stars, pulsars, and black holes. In spite of the many discoveries of gamma-ray bursts, this enigmatic phenomenon continues to surprise astronomers. In December 1997, a gamma-ray burst was detected that turned out to be the largest explosion ever seen since the beginning of astronomic observations. Some astronomers think that this explosion was the result of a neutron star being drawn into a black hole.

Neutrinos - Contemporary theories define neutrinos as electrically neutral particles. Not affected by electromagnetic forces, they are able to pass through great expanses of matter without being affected by it. This makes the detection of neutrinos very difficult. In 1956 two American scientists, Clyde Cowan (1919-1974) and Frederick Reines (1918-1998), detected electron neutrinos near the nuclear plant on the Savannah River, in South Carolina. The first experiments to detect electron neutrinos produced by the Sun began in 1968. The experiments confirmed that the Sun produces electron neutrinos, but only about one-third of the number of neutrinos predicted by theory could be detected. This so-called "solar neutrino puzzle" was solved in 2001-2002 when the scientists working at the Sudbury Neutrino Observatory in Ontario, Canada, found strong evidence that the electron neutrino has the ability *to oscillate*, or transform into either muon or tau neutrinos.

Gravitational radiation – In 1915 Einstein predicted in his general theory of relativity that gravitational waves are emitted as a result of the fluctuation in the curvature of spacetime. He also predicted that gravitational waves propagate with the velocity of light. Because they are very weak, we can expect to directly observe only those gravitational waves on Earth that were generated by very distant and violent cosmological events, such as a collision of neutron stars or a collision of two super-massive black holes. Upon their arrival on Earth, the gravitational waves cause some changes in distances everywhere. But the magnitude of the changes is expected to be very small, on the order of about one thousandth of the diameter of a proton, making direct measurements of gravitational waves extremely difficult.

Although before 2006 gravitational radiation had not been directly detected, there is already significant indirect evidence for its existence. In 1974, two American astrophysicists, Russell Hulse (b. 1950) and Joseph Taylor, Jr. (b. 1941), discovered a binary pulsar made up of a pulsar and a black companion star. Although it was not understood at the time, this was the first of what are now called *recycled pulsars*: neutron stars that have been spun very fast around their centers by the accretion of mass onto their surfaces from a companion star. The orbit of this binary system is slowly shrinking as it loses energy through the emission of gravitational radiation. Hulse and Taylor convincingly demonstrated that the measurements of the shrinking rate of the binary system match the predictions from Einstein's theory with accuracy better than 1%.

1.11 Quantum Vacuum

To explain some bizarre properties of field and matter, quantum mechanics required the existence of the so-called *quantum vacuum*, or *zero-point energy*. The origin of the zero-point energy is closely associated with the name of an outstanding German physical chemist Walther Nernst (1864-1941). In 1916 he proposed that empty space was filled with zero-point energy. Eventually, zero-point energy was defined as the energy which is still associated with a particle or a system (over and above its mass-energy) at the absolute zero of temperature, 0^0 Kelvin. Still, the origin of zero-point energy was not clear. In 1927, after Heisenberg introduced his uncertainty principle, physicists agreed that this principle required the existence of zero-point energy. Belgian astronomer George Lemaître (1894-1966), the originator of the Big Bang theory for the origin of the universe, believed that vacuum energy played an important role in the creation of the universe. In 1930, the British physicist Paul Dirac (1902-1984) presented quantum vacuum as an infinite sea of particles with negative energy that became known as the *Dirac sea*. This presentation helped him to explain the anomalous negative-energy quantum states that his equations predicted for relativistic electrons.

In quantum theory, the vacuum ground state is not completely empty, but contains a seething mass of virtual particles and fields. One can compare these fields with ripples of water in the ocean. This may explain why these fields called *vacuum fluctuations*. In 1948, the Dutch physicist Hendrik Casimir (1909-2000) predicted that the electromagnetic zero-point energy fluctuations of the vacuum can be observable on microscopic objects. He noted that the interactions between two neutral molecules can be interpreted in terms of quantum vacuum fluctuations that produce virtual photons.

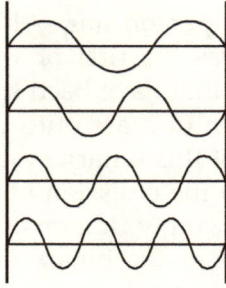

Figure 1.11.1.
Standing waves
between
conducting plates.

Casimir, as a thought experiment, replaced two molecules with two metallic plates with their highly reflective surfaces facing each other in a vacuum. He then envisioned the virtual photons bouncing between the plates. Since the photons behave like electromagnetic waves, the electromagnetic field between the plates would be amplified, if integer multiples of half a wavelength could fit exactly between the plates (Fig. 1.11.1). This phenomenon is now known as the *Casimir effect.*

Thus, the pressure from the photons, as they bounce between the plates, will depend on the distance between the plates. But outside of the plates all of the frequencies are of equal importance. Therefore, the pressure from the photons bouncing off the outer surfaces of the plates will not depend on the distance between plates. Casimir showed that the total pressure P (or force F per unit plate area A) from the inside and the outside of the plates will be attractive, trying to bring the plates closer to each other. This pressure is inversely proportional to the fourth power of the distance a between the plates:

$$P = \frac{F}{A} = \frac{\pi}{480} \frac{hc}{a^4}$$

The notion of the vacuum space filled with radiation and vacuum fluctuations eventually led some physicists and astrophysicists to propose several hypotheses in which the entire universe is merely a particular form of vacuum. The total energy of the Universe may balance out to zero, while locally there may be positive and negative energy regions. According to one of these hypotheses, proposed by the American physicist Edward Tryon in 1973, the universe may be a large-scale quantum mechanical vacuum fluctuation, where positive mass-energy is balanced by negative gravitational potential energy.

Yet another theory of space and its role in the creation of the universe was proposed by the Indian electrical engineer Paramahamsa Tewari (b. 1937). Described in a series of books and articles published since 1977, it was called *the Space Vortex Theory*. The theory considers the medium of space, throughout the universe, as an eternally existing non-material, continuous, isotropic fluid substratum. It has a limiting flow speed equal to the velocity of light relative to the absolute vacuum, and a limiting angular velocity, when in a state of circular motion. It is also eternal and endowed with motion.

The theory reveals the basic relationships between the space and matter pinpointing that space is more fundamental entity than matter. Therefore, the physical significance of mass, inertia, gravitation, charge and light can be revealed only by understanding the substratum of space.

A new kind of instrument called *holometer* was built at the Fermilab Particles Astrophysics Center near Balavia, Ill in 2015 to study the quantum character of spacetime. It measures the quantum coherence of location with unprecedented precision. The basic concept of the instrument was proposed by the American physicist Craig Hogan, the director of the Fermilab and professor of Astronomy & Astrophysics at the University of Chicago.

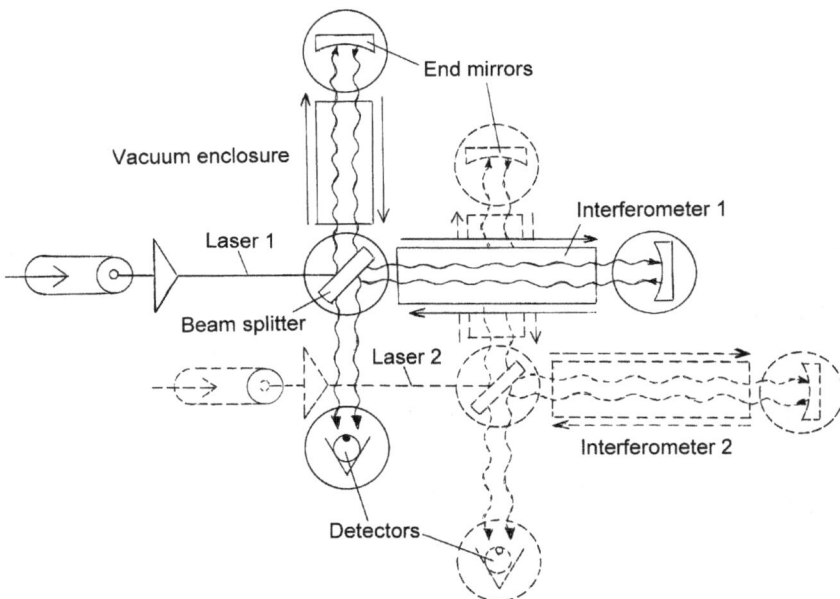

Figure 1.11.2. Basic schematic structure of the holometer. (Adapted from *Scientific American*, Feb. 2012).

The holometer (Fig. 11.1.2) consists of two interferometers. Laser light passing through an arrangement of mirrors will show whether spacetime stands still, or whether it always jitters by a tiny amount, carrying all matter with it, due to quantum fluctuations of spacetime called the *holographic noise*. The experiment may help to understand better the composition of spacetime and how it relates to matter and energy.

1.12 Vortices in Nature

Are there any spiral entities in nature that we can observe without a use of either telescopes or microscopes? Of course, there are plenty of them. Among them are huge rotating storms periodically swirling over our oceans and lands, carrying with them the destruction of property and death. They are called *hurricanes* in the Atlantic and eastern Pacific Oceans, *typhoons* in the western Pacific and *cyclones* in the Indian Ocean.

Much smaller in size but more frequent are the *tornadoes* roaming over the lands. Many people had either direct or indirect encounter with these fearful forces of nature. Incredibly, this unbelievable force is created from weightless air by merely twisting it into a spiral shape, a puzzle still waiting for a complete answer. The Norwegian meteorologist Vilhelm Bjerknes (1863-1951) was one of the first scientists who attacked this problem. In 1904 he became the first scientist to use successfully mathematical models for weather prediction.

During the First World War he established a network of weather stations throughout Norway. Based on the information collected from these stations, a team that included Vilhelm Bjerknes, his son Jacob Bjerknes (1897-1975) and the Swedish meteorologist Tor Bergeron (1891-1977), developed a foundation of the first theory explaining a mechanism of precipitation of rain from clouds. These ideas were soon confirmed by the experimental studies and observations of the German meteorologist Walter Findeisen (1909-1945) and are now known as *Bergeron-Findeisen frontal theory*. This theory explains how hurricanes are generated over Atlantic Ocean, where warm and cold air masses meet (Fig. 1.12). The vortices are also discovered in the atmospheres of other planets of the solar system. The atmosphere of Jupiter is a home to hundreds of vortices with their lifetimes varying from several days to hundreds of years depending on their size.

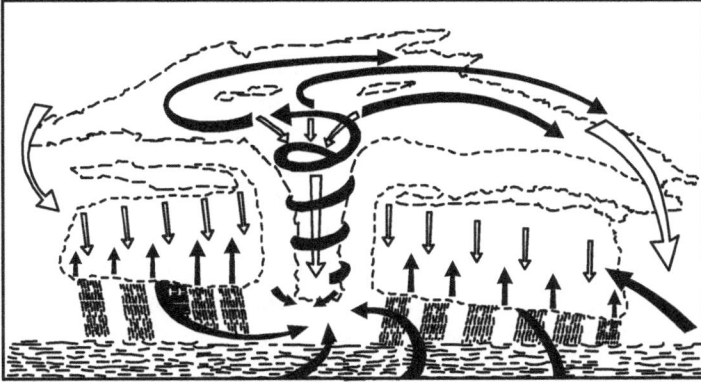

Figure 1.12. Formation of a hurricane by mixing warm air (solid arrows) with cold air (empty arrows). (Adapted from *Scientific American*, Oct. 2004).

A big contribution to the discovery of magnetic vortices in space was made by the Swedish scientist Hannes Olof Gösta Alfvén (1908-1995) who is considered one of the founding father of the field of space plasma physics. He made many contributions and discoveries associated with both terrestrial and cosmic plasmas, (not always acknowledged by his peers), and in 1970 he won the Nobel Prize in Physics for his work developing a theory of plasmas called *magnetohydrodynamics*. Alfvén's contributions to plasma physics includes theories describing the behavior of aurorae, the Van Allen radiation belts, the effect of magnetic storms on the Earth's magnetic field, magnetic and electric fields in cosmic plasmas, the terrestrial magnetosphere, and the dynamics of plasmas in our galaxy.

1.13 The Helical Theory of DNA

In their ground-breaking book *Understanding DNA – The Molecule and How It Works*, the British analytical engineer Christopher Calladine (b.1935) and the American-Australian molecular biologist Horace R. Drew (b.1955) clearly demonstrated that the double-helical structure provides the most favourable conditions for the DNA molecule to perform its important functions, while keeping the two chemically-opposite substances, the acids and the bases, extremely close to one another, but still separate (Fig. 1.13.1). Horace Drew kindly wrote for this book a brief history of the development of the helical theory of DNA between 1953 and 2013.

DNA was soon confirmed to be the genetic material by biochemical methods. Certain triplets of bases on one strand of DNA could code for specific amino acids in a protein, say as "AAA" for lysine, "TTT" for phenylalanine, or "GCX" for glycine where X can be any of A, G, C or T. When cells divide, one strand of the double helix separates from the other, and then provides a template for the enzymatic synthesis of a new complementary strand. Simple mutations in the double helix (say from A to G, C or T), at any of six billion base pairs in a human genome, were envisaged to facilitate Darwinian evolution by random changes followed by natural selection.

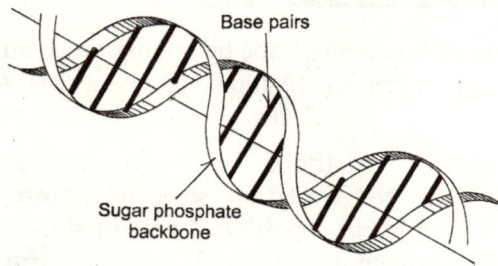

Figure 1.13.1. Schematic presentation of DNA.

Several different helical forms of DNA were subsequently seen by the x-ray diffraction of pulled fibers. Between 1952 and 1980, these helical forms were identified as "A, B, C, D or E" by the British biophysicist Rosalind Franklin (1920-1958), the American biochemist and biophysicist Robert Langridge (b.1933) and the Scottish molecular biologist Struther Arnott (1934-2013). How might such alternative forms of DNA be related to the model made by the American microbiologist James Watson (b. 1928) and the English molecular biologist Francis Crick (1916-2004). In 1980, after making single crystals from short pieces of DNA which had been synthesized chemically, two groups at MIT and Caltech, led by American x-ray crystallographers Alexander Rich (1924-2015) or Richard Dickerson (b.1931) respectively, found a left-handed form of DNA which they called "Z". Other studies of right-handed DNA showed that certain sequences of bases A, G, C, or T could influence the helical structure to some extent, but no one understood why.

Also no one understood how DNA interacts specifically with certain proteins called "transcription factors" or "repressors" to turn genes "on or off", or how DNA wraps tightly about certain proteins called "histones" in the cell nucleus. Still other studies by the American microbiologist Donald Crothers (1937-2014) and others suggested that long pieces of DNA could be naturally curved due to their base sequence in water solution without protein, and again this was poorly understood. Amidst such a proliferation of new data, Christopher Calladine thought in 1983 that it might be useful to describe the DNA double helix in terms of "six degrees of freedom" which had originally been studied by Leonard Euler, namely three of translation and three of rotation (Fig. 1.13.2).

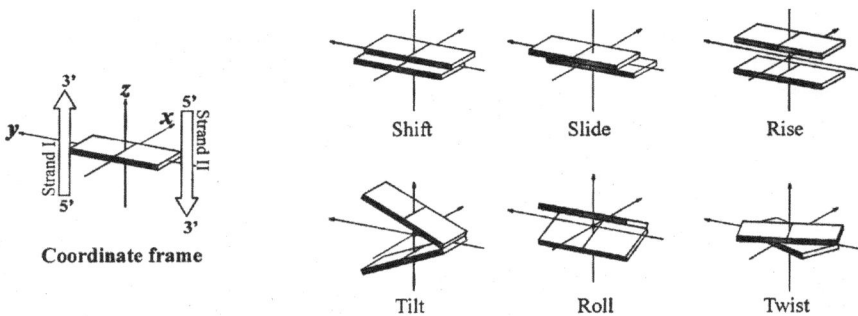

Figure 1.13.2. The DNA double helix in terms of "six degrees of freedom" (three of translation and three of rotation).

As applied to DNA, those six parameters are called "shift, slide, rise" for translation, or "tilt, roll, twist" for rotation. A careful study of atomic coordinates, taken from the x-ray diffraction of single crystals of DNA, next showed that the "B-to-A" transition of DNA seen by Rosalind Franklin (1920-1958) in the 1950's was a combination of just two Euler parameters "roll" and "slide"(Fig. 1.13.3).

Christopher Calladine had previously studied the dynamics of bacterial flagella, which may adopt a variety of helical forms, and where such concerns are equally applicable. When complexes of DNA with various proteins were studied in the 1980's or 1990's, by the x-ray diffraction of single crystals, often we saw that the double helix would curve tightly about a protein. Atomic coordinates became available for many examples. On careful study of those coordinates, many workers found that the DNA double helix would curve smoothly about a protein, using a simple "roll" motion which varied as a Fourier wave (due to the helix twist), as shown in case (Fig. 1.13.4d) below on the right.

The bases A, G, T or C are not very soluble in water, so smooth bending by "roll" (without any "kinks") provides the lowest-energy solution of how to curve DNA without exposing its hydrophobic surfaces to water. In mathematical language, we can say that the local curvature of DNA may be approximated by a Fourier transform of its "roll" angles. This is similar to what Erwin Schroedinger postulated in 1944, when he proposed in his book *"What is Life?"* that the genetic material (not known then) would be an "aperiodic crystal".

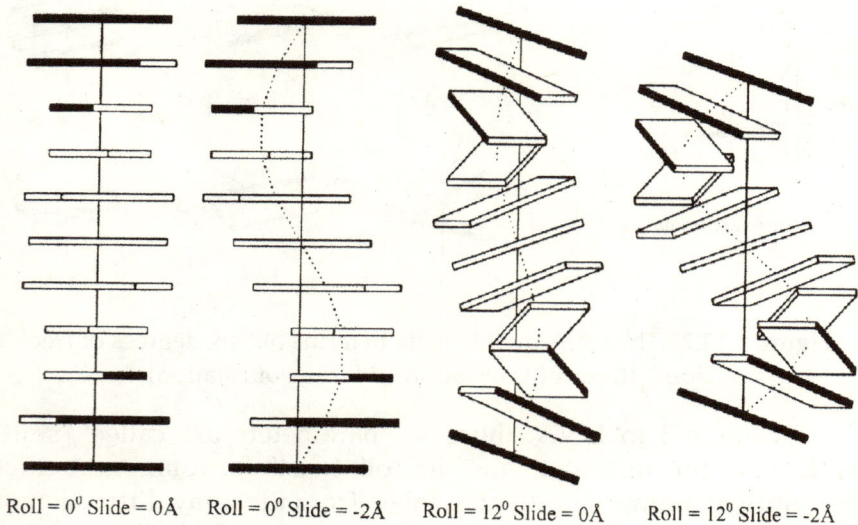

Roll = 0⁰ Slide = 0Å   Roll = 0⁰ Slide = -2Å   Roll = 12⁰ Slide = 0Å   Roll = 12⁰ Slide = -2Å

Figure 1.13.3. The "B-to-A" transition of DNA as a combination of just two Euler parameters "roll" and "slide."

Such "smooth rolling motion" is especially true for a complex between histone proteins in the cell nucleus, and 145 base pairs of DNA wrapped tightly around it into almost two super-helical turns, in what is called the "nucleosome core particle" of chromosomes. The atomic coordinates of such a complex were determined by the British biophysicist Aaron Klug (b.1926), Italian biochemist Daniela Rhodes (b. 1950) and American x-ray crystallographer Tim Richmond (b. 1953) in a series of investigations conducted between 1978 and 1990.

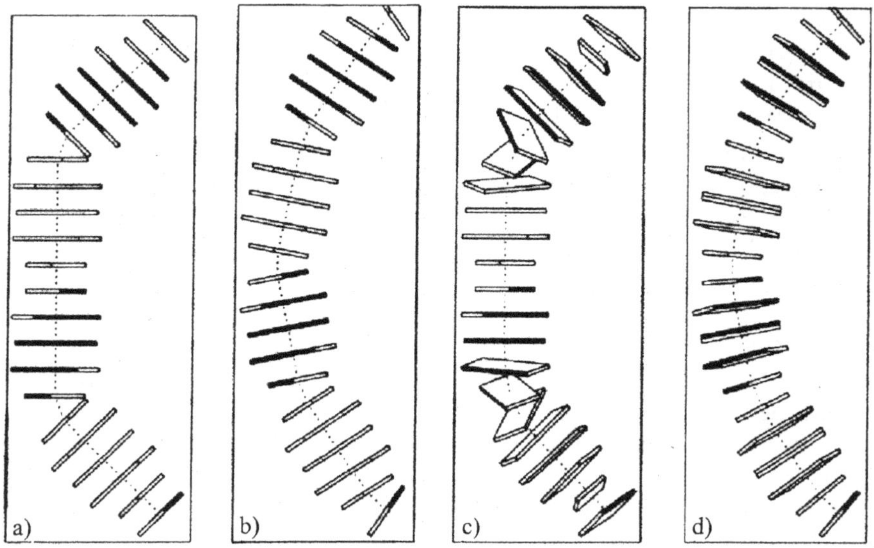

Figure 1.13.4. The DNA double helix curves smoothly about a protein, using a simple "roll" motion (d).

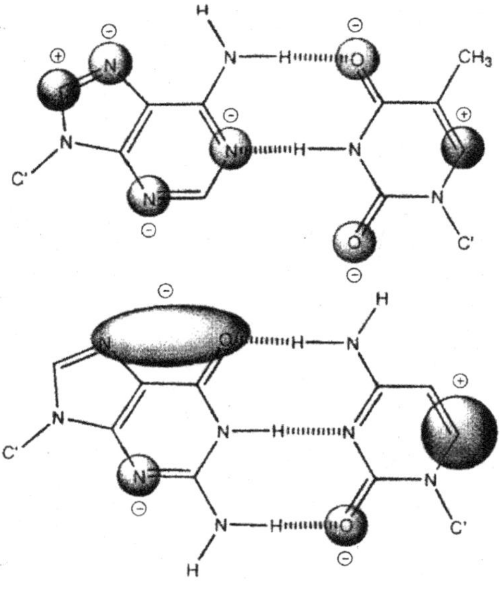

Figure 1.13.5. Partial electrical charges on the base pairs.

This "roll-slide-twist" model for DNA is just a geometrical abstraction, because it does not address any underlying chemical features of the bases A, G, C or T, or of their attached sugars and phosphates, which might give rise to observed behaviours. That deeper level was addressed by a New Zealand-British chemist Chris Hunter (b.1965) in the 1990's, when he made new theoretical calculations which included partial electrical charges on the base pairs (Fig. 1.13.5). We can see that there are negative partial charges (-) in certain places, or positive partial charges (+) in other places. By including detailed chemical information in his models, Hunter was able to successfully model the "roll-slide-twist" behaviour on an atomic or chemical scale.

Outside of his house on Portugal Place in Cambridge, Francis Crick erected a large "wire helix", because he became convinced that helices would be a fundamental aspect of all sciences on Earth. His reasoning was that, if you take any small building block, and then join a series of such blocks end-to-end, you will always form in three dimensions some kind of "helix", by the combination of translation and rotation. In higher dimensions we might see four or five-dimensional "helices" of even more complex shapes, although we do not understand that subject very well at present.

Currently particle physicists are studying "points" or "strings" in an attempt to explain subatomic particles in as many as 11 dimensions. It might not be a bad idea to study "helices" while they are so engaged.

1.14 Vortices in Biology and Medicine

This section is a kind contribution to this book made by the recognized American energy medicine expert Dr. James Oschman (b. 1939) - who is also a physiologist, cellular biologist, and biophysicist, and his naturalist colleague, Nora H. Oschman (b. 1951). It continues our extended and modern inquiry into the geometric structure of space and electromagnetism.

Scholars from ancient times have advanced a Spiral Field Theory (SFT), in which space and light are organized as vortices. The list includes such luminaries as Archimedes, Kepler, Descartes, Leibniz, Swedenborg, Ampère, Fresnel, Stokes, Helmholtz, Beltrami, Schauberger, W. and J.J. Thompson, Tait, Maxwell, Alfvèn, Wheeler, Crick, Watson, Bostick, Penrose, Russell, and

Ginzburg. Vortical models became fashionable and then disappeared and reappeared, in cycles that repeated again and again, down through the ages. Some of the most thoughtful and innovative thinkers have become entranced with spiral and helical forms and dedicated a significant portion of their intellectual lives to exploring them. Oschman has summarized some 35 years of exploration of the biophysics of helices and vortices[1]. The study began when a therapist pointed out how everyone has a swirl of hair at the top of their head, and that this spiral extends down through the body all the way to the feet. This vortex extends to the molecular level in the triple-helical collagen molecules in the connective tissues forming the bulk of the human body. We shall see that anatomists Raymond A. Dart (1893-1922) and Thomas Myers provided a precise description of 'the spiral line' and we will also see how quantum biology explores the issue at the molecular level.

Many medical interventions involve applying therapeutic electric, magnetic or electromagnetic fields. A thorough understanding of the nature of these interventions depends on detailed knowledge of the properties of the space through which physical energies propagate, both within and outside of the body. At the present time, this knowledge is incomplete. Information in this and the other books in this series is crucial for future progress in delineating the properties of space and the propagation of field effects.

One place where this shows up very dramatically is in the apparent chasm separating conventional and alternative medicine. For example, many alternative practitioners rely on carefully honed sensitivity and intuition, rather than sophisticated diagnostic tools such as X-rays, magnetic resonance imaging, or blood tests. These clinicians always seek scientific understandings of what is happening underneath their hands when they work with people, and subtle processes taking place underneath their own skin as they continuously glean relevant information from their inner and outer environments. Most medical doctors have little training in physics and are therefore unable to comprehend the possible basis for therapies that rely on intuition or so-called subtle energies. Both the biology and the clinical applications of natural and artificial electric, magnetic, and electromagnetic fields have had a long and controversial history.

Naturalists have repeatedly documented the great sensitivity of organisms to exceedingly tiny fields in their environments. A host of sensory systems evolved for a variety of survival purposes. Organisms use energetic cues to locate and orient themselves geographically, to set their biological rhythms, to detect prey, predators and mates, and to anticipate meteorological and earth changes, including seasonal variations, weather fronts, hurricanes, tornadoes, and earthquakes. Extreme examples of energy sensitivity have been discovered for virtually all living systems at all levels of organization: viruses, bacteria, algae, higher plants, protozoa, flatworms, insects such as honeybees, snails, fish, birds such as carrier pigeons, green turtles, sharks, whales, and humans.

For several decades, physics seemed to be at odds with these discoveries. The repeated observation of naturalists that geophysical and celestial rhythms influence plant and animal behavior seemed utterly preposterous to all but a few scientists. Healing with natural or artificial energy fields had little scientific credibility. When concern arose about possible health effects of exposure to electromagnetic fields from appliances, power lines, Wi Fi, smart meters, and other technologies, many physicists stated that these energies are far too weak to have any biological effects.

Physicists attempt to study living systems like other forms of matter. Known or measurable physical properties of cells and tissues and the established laws of electricity and magnetism are employed to calculate the currents induced in tissues by environmental fields of various sorts. The calculations are based on the degree that fields of different frequencies penetrate into the body, tissue conductivity, viscosity and dielectric properties, interactions of induced currents with larger currents from other sources, the 'noise' from random thermal agitation at body temperature, and so on. A consistent but entirely incorrect conclusion was that environmental fields can have no biological effects on living matter unless the energy intensity is sufficient to ionize or heat tissues. Radiation safety standards were set on the basis of this inaccurate conclusion. It was recognized that weaker fields may induce microcurrents in living tissues, but these are much smaller than the noise from thermal agitation and normal physiological signaling processes, and should therefore have no biological effects. The biological evidence and experiences of people with electromagnetic sensitivity had to be inaccurate.

In 1996, the scientific consensus shifted from a certainty that weak environmental energies can have no influence on living systems (the behaviorists must be mistaken, and those suffering from environmental illness need psychological help) to agreement that such influences are extremely important and deserving of thorough study to determine the precise mechanisms involved. The turning point was published by the prestigious Neurosciences Research Program and a distinguished biophysicist, W. Ross Adey (1922-2004)[2]:

> There is ... a growing scientific consensus on the cell and molecular biology mediating interactions with environmental electromagnetic fields. Beyond the chemistry of molecules that form the exquisite fabric of living tissues, we now discern a new frontier in biological organization ... based on physical processes at the atomic level, rather than in chemical reactions between biomolecules ... these physical processes may powerfully regulate the products of biochemical reactions.

Hence "physical processes at the atomic level" have taken on great significance in regulatory biology. The conclusions from this approach have only slowly percolated into medical practices, which continue to focus primarily on biochemical, molecular and pharmaceutical medicine. However, all medical issues can be seen as problems in regulation of the organism's inherent systems that maintain healthy balance and vitality. We shall see that the conventional textbook regulatory paradigm involves "signal molecules" diffusing from sources to "targets" where information molecules fit into receptors like keys fitting into locks. This model is being supplemented with one in which signal molecules vibrate and emit fields that propagate to receptors at the velocity of light, setting up virtually instantaneous co-resonance that activates cellular processes. Life is much too rapid and subtle to be explained by slowly diffusing signal molecules and nerve impulses.[3] Evidence accumulates for regulations via biophotonic communications and other processes that are hundreds to thousands of times faster than nerve impulses.[4] To understand the nature of biophotonic communications, information on the structure and properties of space are essential.

Helical structures in living systems - Many organisms have an overall spiral structure and many of the molecules and organelles found in both plants and animals are partly or entirely helical (Figure 1.14.1). To the biologist these architectures are so common that the question arises as to whether these forms evolved because they conferred particular advantages to the organisms in which they are found. If so, what is the exact nature of these advantages? As an example, collagen, the most abundant structural protein in the animal kingdom, is a triple helix. What do these forms tell us about living structure and function, and can these forms teach us something about the physics and quantum physics of light and space?

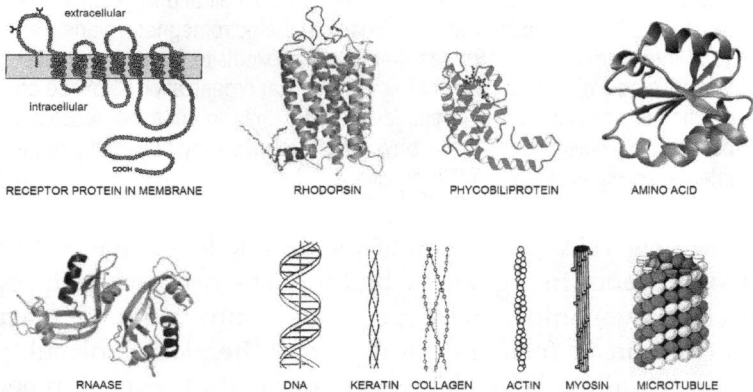

Figure 1.14.1. Many of the molecules and organelles found in both plants and animals are partly or entirely helical.

In 1917, Sir D'Arcy Wentworth Thompson (1869-1948) wrote *On Growth and Form*. Sir Peter Medawar, the 1960 Nobel Laureate in Medicine, called it "the finest work of literature in all the annals of science that have been recorded in the English tongue." The book was revised by Thompson in 1942. He described the mathematics of the spiral and made a few suggestions as to why certain animals had spiral structures, but did not give any unifying principles of the spiral form. Thompson referred to a book by Sir Theodore A. Cook (1867-1928) entitled *The Curves of Life*. The book was first published in 1914, and was reprinted in 1979 by Dover Publications. The book contains 426 illustrations of spirals and helices found in nature, architecture, science and art. The spiral form is fundamental to the structure of shells, leaves, horns, the human body, drawings of Leonardo, the Leaning Tower of Pisa, staircases, and many other structures found throughout nature and architecture. Helical forms are also found in microtubules in cells, actin and myosin molecules in muscles, the ventricles of the heart, hurricanes, whirlpools, bones, single-celled organisms, propellers, screws, springs, ropes and cables, the cochlea of the inner ear, spiral galaxies, growing plants, spiral radio antennas, spiral staircases, traces of subatomic particles in cloud chambers, feathers, bird wings, fingerprints, umbilical cords, the ducts of sweat glands, fly egg cases, elephant tusks, ruminant intestines, myofibers in the walls of arterioles and capillaries, the Milky Way, sunflowers, and so on. Cook made a vague statement that the spiral was the fundamental energy structure in the universe, but he provided no specific basis for his statement.

Eventually it was recognized that the spiral has profound energetic significance because of the way it deals with the intersections of forces.[1] Helical and spiral structures have the ability to allow energy fields or forces or motions to intersect with each other at angles that produce advantages from specific energetic perspectives. Consider, for example, bringing a piano up the stairs and into your home. Yes, you can employ a number of people to lift the piano straight up and then carry it up the stairs. This requires strong people and a lot of effort. But it is much easier to have a ramp – in physics it is called an inclined plane. Once the ramp is in place, one or two people can easily push the piano into the house with modest effort. In the first case, lifting the piano, force must be exerted against the gravity field. In the second case, with the ramp, force is directed at an angle to the gravity field. While you have to push farther, you do not need nearly as much effort to move the piano.

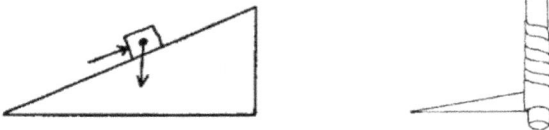

Figure 1.14.2. Inclined plane (left) wrapped around a cylinder.

Figure 1.14.2 is a simple diagram of the inclined plane and what happens if you wrap the inclined plane around a cylinder. If a cone is used instead of a cylinder, the result is a spiral. From this it can be seen how the screw, bolt, or propeller can allow a rotational force to be applied at an angle to overcome a resisting force. A relatively weak force can have a large effect. With a good screw jack, one person can lift a very heavy weight. With a good propeller, a modest wind can be used by a windmill to lift water from a deep well, or to produce substantial electricity. It is much easier to climb a spiral staircase than a vertical ladder. A screw can penetrate straight through hard wood with a little rotational effort applied to a screwdriver. This explains the energetic advantages of elephant tusks, animal horns and seashells, which derive their great strength from their ability to divert or absorb energy applied in one direction at appropriate angles into the strongest part of the structure. The design is such that the prey, rather than the predator, has the mechanical advantage.

A remarkable study by A.A. Schaeffer from the University of Kansas, published in 1928, was entitled *Spiral Movement in Man*. Schaeffer blindfolded people and told them to walk, run, swim, row or drive their automobiles in a straight line. He traced their movements on a large open field in the "flat country" of western Kansas and eastern Colorado. Other studies were done on the ice on the large Wachusett reservoir in Clinton, Massachusetts. Swimming, rowing and paddling studies were done at Cold Spring Harbor, Long Island, New York. The movements were recorded either by drawing them or by using a special camera. When told to follow a straight path, blindfolded persons walk, run, swim, row, or drive automobiles in "clock-spring" paths. They think that they are moving in a straight line, but they are not! Some of these results are shown in Figure 1.14.3.

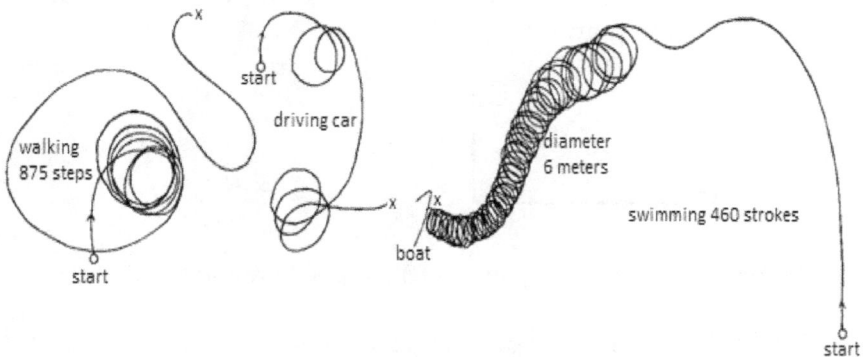

Figure 1.14.3. Spiral movements of people who were blindfolded and told to walk, run, swim, row or drive their automobile in a straight line.

Schaeffer's studies were done in part to try to understand the well-known phenomenon of people who lose their way in the forest, or in a snowstorm, or in fog, and go around in circles. Schaeffer concluded the mechanism that causes the spirality of the path in man and other animals must be located in the central nervous system. He did not explain this in any further detail.

In 1950, a famous anthropologist, Raymond A. Dart MD, Dean of the Medical School at the University of the Witwatersrand in Johannesburg, South Africa wrote an article, *Voluntary Musculature in the Human Body. The Double-Spiral Arrangement.*[5] Dart believed in the universality of spiral movement. He began his article with this statement:

> I recall an elderly otologist, named Miller, 30 years ago in New York City, demonstrating by means of examples ranging from the spiral nebulae to the human cochlea and from the propagation of sound into the propulsion of solid bodies, that all things move spirally and that all growth is helical.

Dart had severe scoliosis (a lateral curvature in the normally vertical line of the spine) and he had a child who had night terrors, and another who was spastic. Searching for help for the health issues in his family, he encountered a practitioner of the Alexander technique, who was able to resolve all of these problems by bringing his patients' awareness to postural twisting and "relationing" of the parts of the body. Through careful observation and anatomical considerations, Dart developed his understanding of the "spiral mechanism of the body." He concluded his paper:

> In the members of my family and in myself, the "undoing" of previously unrecognized leftward twists, whether congenital or acquired, has been a major preoccupation of the past 7 years. That employment has led me to recognize and to correlate many facts about the body and its functioning, both as a whole and in parts, which had previously eluded my notice. Amongst those anatomical facts, the double-spiral arrangement of the voluntary musculature is basic. In ontogeny, as in phylogeny, man grows and moves spirally.

On the basis of Dart's work, anatomist Tom Myers defined the spiral line (Figure 1.14.4):

> The spiral line loops around the body in a double helix, joining each side of the skull across the upper back to the opposite shoulder, and then around the ribs to cross in front at the level of the navel to the same hip. From the hip, the spiral line passes like a 'jump rope' along the anterolateral thigh and shin to the medial longitudinal arch, passing under the foot and running up the back and outside of the leg to the ischium and into the erector myofascia to end very close to where it started on the skull.[6]

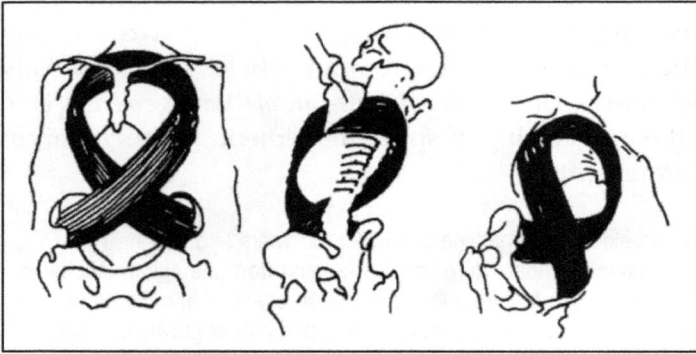

Figure 1.14.4. Diagram of the spiral line as described by Manaka.[12]

Dart explained that the arrangement is such that the pelvis is actually suspended from the occiput and neck vertebrae. This description ties vortex theory to the biotensegrity concept, which is a new way of understanding anatomy, based on the work of Steve Levin.[7] The usual way of viewing the skeletal system is that the bones are more or less a stack of blocks, with each bone supported by the one below it. Biotensegrity recognizes that the soft tissues provide lift, so that each bone is actually suspended and lifted from the one below it. Tom Flemons from Salt Spring Island has made a complete tensegrity model of the human body that demonstrates a remarkable similarity to a human skeleton. It can walk, sit, stretch, and contort; most amazingly, it will stand self-supporting with all of the compression elements floating in the web of tension that is woven around it from top to bottom.[8] Tensegrity was originally conceived by R. Bukckminster Fuller, an American architect, systems theorist, author, designer and inventor, as well as by sculptor, Kenneth Snelson.

Tensegrity represents a common pattern, like the vortex, that is found everywhere in nature, at many different size scales, from subatomic particles to atoms to molecules to cells and tissues to whole organisms and to the forms and motions of celestial bodies. Harvard Professor Donald Ingber (b. 1956) has studied the roles of tensegrity in cells and tissues, and asks, in his famous *Scientific American* article[9] if this recurrent pattern is visual evidence of the existence of common rules for self-assembly. We can ask the same question about the vortex. Perhaps, Ingber suggests, there is a single underlying theme to nature after all. He quotes the Scottish zoologist D'Arcy W. Thompson, who quoted Galileo, who, in turn, cited Plato: *the Book of Nature may indeed be written in the characters of geometry.* Consider the possibility, for example, that the vortex is tensegrity in motion.

Marvin Solit (1931-2006) devoted his life to a penetrating study of both tensegrity and geometry. His work has been summarized on the web site of the organization that continues his investigations, the Foundation for New Directions.[10] A true Pythagorean, Marvin sought to represent universal principles. He modeled structures of the wave/particle, the tensegrity of the cell, the primacy of the "6 great circles," and the basic figures that underlay the Platonic Solids.

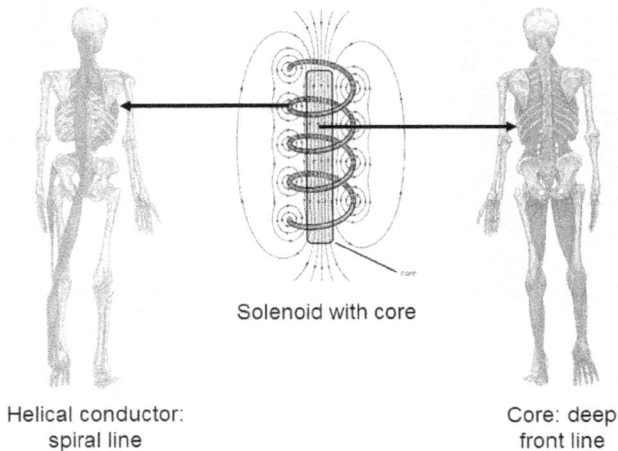

Solenoid with core

Helical conductor:
spiral line

Core: deep
front line

Figure 1.14.5. Yoshio Manaka pointed out the profound energetic significance to the spiral system, because of its analogy to a coil and a core, otherwise known as a solenoid.

He also hosted a workshop entitled Synergetics 3 for the Synergetics Collaborative, an organization continuing the work of Buckminster Fuller. Marvin's work on tensegrity and geometry has been summarized.[11] To quote Vladimir Ginzburg, "He impressed me with his strongly held belief in the 'simple complexity' of nature, and his amazing discovery of a completely unexpected correlation between the 6 great circles and the golden ratio."

A part of the spiral line was also described by Yoshio Manaka

(1911-1989) in his classic acupuncture text, *Chasing the Dragon's Tail*.[13] For Manaka there was profound energetic significance to the spiral system, because of its analogy to a coil and a core, otherwise known as a solenoid (diagram to the right in Figure 1.14.5). Placing a conductive core inside a solenoid greatly increases the strength of the field produced in the surrounding space.

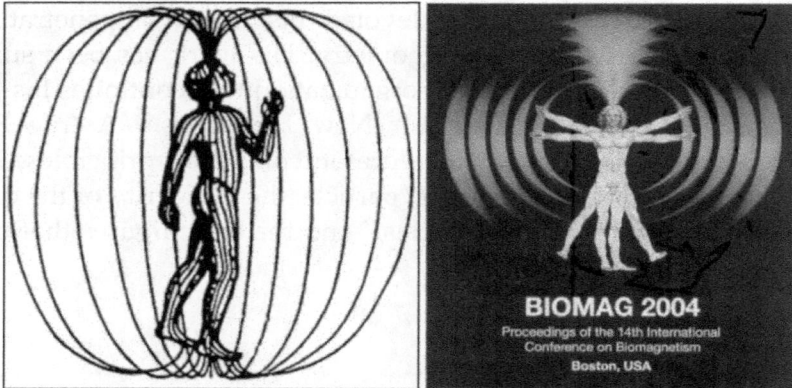

Figure 1.14.6. In the human body, the main source of electrical energy flowing through the spiral musculature is the heart. The descending aorta and ascending vena cava are also part of the energetic core. The resulting biomagnetic field is shown to the left (from Richard Gordon's book entitled *Polarity Therapy*). A similar diagram shown on the right is on the cover of the proceedings of an international conference on biomagnetism held in 2004.

In the human body, the main source of electrical energy flowing through the spiral musculature is the heart. This is the same electricity that gives rise to the electrocardiogram. Since blood is electrically conductive due to its high salt content, the descending aorta and ascending vena cava are also part of the energetic core. The resulting biomagnetic field is shown to the left in Figure 1.14.6, a diagram from Richard Gordon's book on Polarity Therapy. A similar diagram, shown to the right, adorns the cover of the proceeding volumes of a prestigious organization that sponsors conferences every other year on the topic of biomagnetism.

Vortex theory and subtle energies – In 1861 and 1862, James Clerk Maxwell published his famous four-part paper *Physical Lines of Force*, proposing an analogy between lines of force in the electromagnetic field and vortex filaments in a bounded liquid. In Part IV of the series, entitled *The Theory of Molecular Vortices Applied to Statical Electricity*, Maxwell expanded on the significance of the concept of the vortex to the electromagnetic theory of light.[14] In his earlier modeling of the electromagnetic field and the properties of space, Maxwell had relied on his expertise in the field of fluid mechanics.

Maxwell eventually left vortex theory and other physical analogies out of his writings, replacing them with equations and some assumptions regarding the elastic structure of the space medium. In 1865 he published his seminal theory of electromagnetism, *A Dynamical Theory of the Electromagnetic Field*.[15] His 20 equations describing the electromagnetic field consisted of three equations each for magnetic force, electric currents, electromotive force, electric elasticity, electric resistance, total currents; and one equation each for free electricity and continuity. Maxwell expressed electromagnetism with quaternion mathematics and made the electromagnetic potential the centerpiece of his theory. Faraday and Maxwell had agreed that the potentials were the real movers in electricity, magnetism, and electromagnetism, causing things to happen like the puppets are moved by the invisible puppeteers behind a curtain.

Unfortunately, in 1881, two years after Maxwell's death, Oliver Heaviside replaced the electromagnetic potential with force fields (vectors) as the centerpiece of a revised electromagnetic theory. According to Heaviside, the electromagnetic potential was arbitrary and needed to be eliminated. A great debate took place about the relative merits of vector analysis vs. quaternions. In the end, Heinrich Hertz, Josiah Gibbs and Oliver Heaviside prevailed, deleting the potential concepts that both Faraday and Maxwell had considered the most important features of the theory of electromagnetism. By the end of the 1800's the physical insights provided by quaternions had been set aside, and vector algebra was substituted, leading to THE four Maxwell equations as they are taught today. Deleting the potentials deprived physics, biology and medicine of important theoretical tools for nearly a century. As a result, some of the most pressing and unsolved mysteries related to health and disease persist, and some powerful healing technologies were bypassed because there was no adequate electrodynamics theory to explain and advance them.

Regulatory biology – For biology and medicine the long-forgotten potentials may have immediate applications in regulatory physiology. As mentioned above, the classic model of regulations is based on chemical messengers such as hormones or neuropeptides released from an endocrine gland, for example. These messengers are supposed to diffuse randomly through the various fluid compartments within the body until they chance to reach receptors on distant cells, where they interact via a lock-and-key mechanism to trigger a cellular event (Figure 1.14.7A). The interaction of the messenger with the receptor is thought to trigger the release of second messengers such as cyclic adenosine monophosphate (cyclic AMP) or calcium ions inside the cell.

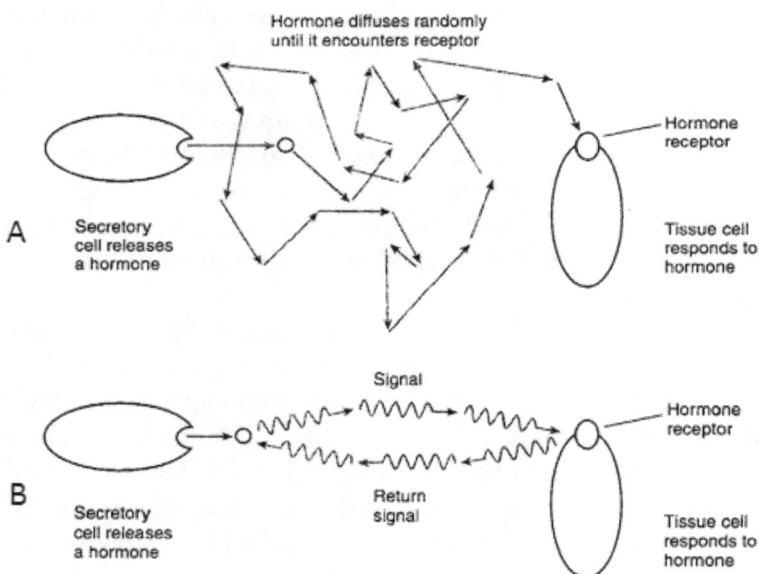

Figure 1.14.7. The widely accepted model of physiological regulations (A) involves random diffusion of a signal molecule such as a hormone from a source to a receptor on a distant cell. A more plausible model (B) involves electromagnetic fields, biophotons instantaneously connecting messenger molecule with target.

This widely accepted model (Figure 1.11.7A) is seriously deficient and outdated, and thwarts progress in many areas, including biomedicine. The model is based on the ancient concepts of Lucretius and Epicurus, who pioneered the idea that matter is composed of hard, indivisible 'billiard-ball' units called atoms. Many biochemists carried this billiard-ball analogy into their descriptions of metabolic processes. And the lock-and-key model dates to Emil Fischer in 1894.[16]

One difficulty with this model is that it does not provide the essential cybernetic feedback from the receptor back to the signal source. The structural complementarity between hormones and receptors, which is traditionally assumed to be the basis for the lock-and-key model, is also a basis for resonant electromagnetic molecular communications, often referred to as "the tuning-fork effect." Structures with similar geometry will tend to resonate with one another. Another difficulty is that diffusion is too random and slow to explain the ability of organisms to quickly adjust their activities when their environment changes. Guenther Albrecht-Buehler wrote a valuable perspective entitled *In Defense of 'Nonmolecular' Cell Biology*, raising major arguments against the view that regulations take place by diffusion of signal molecules to receptor sites on cells. While some cells are in epithelial sheets that are one cell thick, others are packed closely together. If one considers the average volume around a cell, the actual space available is such that hormone molecules at a typical concentration will add up to approximately eight molecules in that space. In the space around the receptor, the hormone concentration, for all practical purposes, is approximately zero. Hence the usual concept of concentration is essentially meaningless.[17]

A final difficulty with regulation by diffusing messenger molecules is the widely accepted miss-conception that the human body and the cells within it are best described as continuous volume conductors. In other words, to simplify the application of the principles of electrical engineering and diffusion theory to living systems, the body and the cells within it are regarded as simple bags containing homogeneous solutions of proteins and electrolytes. The volume conductor assumption disregards significant areas of biomedicine: anatomy, histology and cell biology. The volume conductor assumption, when taken as fact, promotes a distorted perspective on life and health that is incompatible with the actual structure and function of living systems. Over the years, it has been recognized that the cell interior is virtually packed with microtubules, microfilaments, intermediate filaments, vesicles, membranes, RNA particles and many other objects. And every cell has an internal "skeleton" called the cytoskeleton.[18] There is so much cytoskeleton and other structure tightly packed inside of cells that some biologists think there is virtually no space left over for a solution of electrolytes and diffusing signal molecules.[19]

Increasing evidence supports a regulatory model involving electromagnetic resonance between a vibrating signal molecule and its target. Feedback is provided by co-resonance and is instantaneous (Figure 1.14.7B). Biophotonic models of systemic regulation have been reviewed and have a substantial theoretical and research base.[20] The light involved is known as squeezed light, which can carry large amounts of information.[21] The Sandoghdar Lab in Erlangen, Germany has demonstrated that molecules can act as transmitting and receiving antennas for single photons.[22] And Marco Bellini, Constantina Polycarpou and Alessandro Zavatta from the Istituto Nazionale di Ottica in Florence, Italy have measured the shape of individual photons for the first time. Pulses of light can have almost any shape in space and time, and these shapes depend on the amplitudes and phases of the pulse's frequency components. Data can be encoded in light pulses by modulating the amplitude or phase of the light. A single photon shape could represent, for example, any letter in the alphabet, or even a quantum combination (or superposition) of several letters.[23] Taken together, there is strong evidence for a role of biophotons in cellular and organismal regulations, and the model has the additional feature that it makes sense.

A vital issue for biomedicine is whether the so-called "subtle energies," i.e. scalar fields and vector potentials, are also involved in regulations. If electromagnetic fields are indeed used in biological regulations, we can expect scalar fields and vector potentials to be present as well. How much of the signal is carried by the electromagnetic component and how much is carried by the subtle energy fields? This is an important question for investigation.

Light sensing systems – Support for a vortex model of light comes from the design of the light-sensing systems in the eye.[24] The first helically organized structure that light must pass through is the corneal stroma (*substantia propria*) (Figure 1.14.8). The stroma is about 500 μm thick and forms the bulk of the cornea. It combines optical transparency with mechanical resilience. These properties are possible because of an extracellular matrix containing narrow (36 nm diameter) parallel type I collagen fibrils spaced and organized uniformly into 200-250 sequential lamellae or sheets. Each sheet is arranged orthogonal to its neighbor and to the path of light through the cornea.[25-27] Strength arises from the plywood-like architecture. The collagen fibrils are much smaller than the wavelength of light, and their spacing is such that any light they scatter is eliminated by destructive interference in all directions other than forwards into the retina. Nature has provided a superb design for transparency.

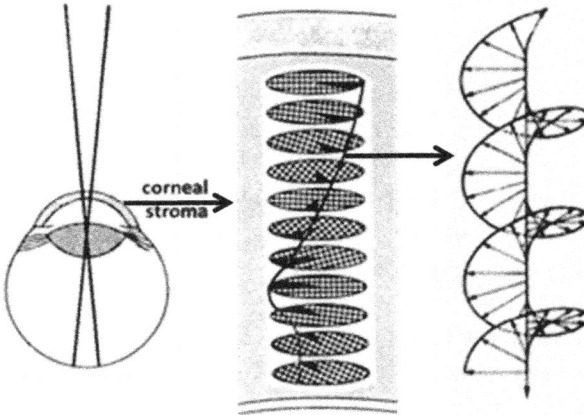

Figure 1.14.8. The structure of the corneal stroma supports a vortical fabric of space and consequent vortical flow of light.

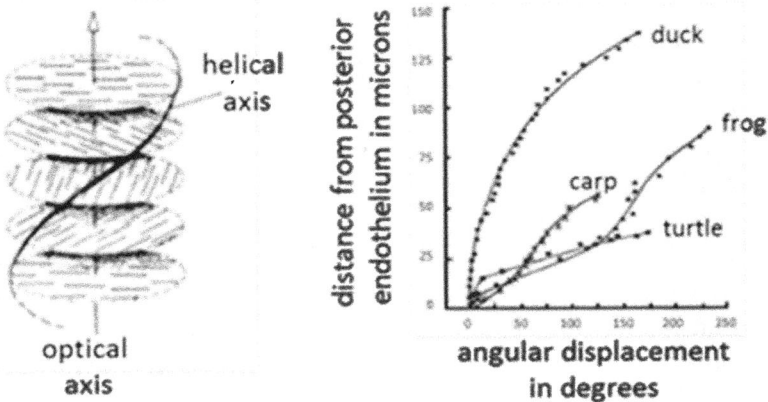

Figure 1.14.9. The collagenous corneal stroma is a cholesteric liquid crystal-like lattice which has the same right handedness in both eyes. Different displacements between adjacent layers in different species) may match the refractive index for life in different environments.

Trelstad[10] analyzed serial sections of corneas of birds, fishes, amphibians, and reptiles, cut perpendicular to the optical axis. The collagenous stroma is a cholesteric liquid crystal-like lattice which has the same right handedness in both eyes and is thus bilaterally asymmetrical (Figure 1.14.9).

Bilateral symmetry is the general rule for the heads of animals, with mirror symmetry in the sagittal plane dividing the body vertically into left and right halves, with one of each of the sense organs and limbs paired on either side. Why would this universal symmetry rule be broken for the corneal stroma? A logical answer is that both eyes must accommodate to light moving vortically in the same right-handed vortical path (Figures 1.14.8 and 1.14.9). Different displacements between adjacent layers in different species (Figures 1.14.9) may relate to matching of the index of refraction for life in different environments.

The alpha-helix — If biophotons from vibrating signal molecules follow vortical paths through the body, an interesting scenario arises at the receptors in cell membranes. Most if not all of the receptors on cell surfaces are composed long proteins that snake back and forth across the membrane from 7 to 24 times. Seven-trans-membrane-helix (7TM) receptors (upper left in Figure 1.14.1) are responsible for transducing information initiated by signals as diverse as photons, odorants, tastants, hormones, and neurotransmitters. Several thousand such receptors are now known, and the list continues to grow. All of this basic molecular research provides substantial validity for the spiral field theory. These molecular sensors are sometimes referred to as serpentine receptors because the single polypeptide chain "snakes" back and forth across the membrane.

The receptor proteins crossing cell surfaces are right-handed alpha helices. What is the nature of the interaction between light, a right handed vortex, and the alpha helices in cell membranes, which are also right handed? The possibility arises that these alpha helical proteins at cell surfaces are actually "light pipes" that facilitate the entry of photonic messages into cells. While this might at first seem to be a preposterous idea, there is abundant supporting evidence from the literature of plant physiology. Specifically, red and blue-green algae have intricate light-absorbing structures called phycobilisomes. These "antenna" complexes contain many alpha helical regions, and are described as "light pipes" funneling excitation energy (photons) into the reaction centers of chlorophyll a of photosystem II. Chlorophyll a, in turn, is another membrane protein with five trans-membrane helices.[28] It is thought that this "light pipe" arrangement enables the algae to survive in weak light environments. The arrangement permits 95% efficiency of energy transfer, as reviewed by Glazer.[29] Moreover, the light sensitive pigment in the human eye, rhodopsin, is also a seven-trans-membrane-helix like the one illustrated in the upper left in Figure 1.14.1.

Much less seems to be known about the functions of alpha helical proteins in eukaryotic cells, including those in mammals. It has been speculated that the alpha helices contribute strength to protein assemblies. Right handed helices are energetically more stable because there is less "steric clash" or "unnatural overlap" of nonbonding atoms between the side chains and the protein backbone.[30] Essentially all helices found in proteins are right-handed.

We suggest that the alpha helical structures in animal cells may serve the same purpose so well-documented in plants, i.e. they provide channels or "light pipes" for biophotons to cross the cell surface so they may interact with enzymes and other molecules within the cell. As with the algae, the alpha helices may permit animal cells to funnel into cells weak electromagnetic signals emitted from vibrating hormones, neurohormones, growth factors, cytokines, etc. It is necessary to qualify the phrase, "weak electromagnetic signals" in the previous sentence. As stated previously, we do not know the proportions of classical electromagnetic signals and scalar or vector components, which could be substantial. We simply do not know completely what "weak" really means in this context.

Some biochemists will assert that a "light pipe" function is impossible because the alpha helices are not hollow and therefore do not provide a pathway for photons. This assertion is based on "space-filling" models of the alpha helix, which show that the amino acids protrude into the channel and therefore occlude it (Figure 1.14.10). This is another example in which the ancient Lucretian concepts create confusion. We now know that atoms consist mainly of empty space. In other words there is virtually nothing in the spaces in "space-filling" models.

Biochemistry texts say that alpha helical proteins are not hollow. For example, "Notice that in this depiction (left in Figure

1.14.10) the center of the helix appears to be hollow. This is not the case (see space filling model to the right). When you illustrate the atoms as space filling objects (atoms represented at their van der Waals radii) then we can see that there is no hollow core to the helix."[31,32] Such descriptions are based on the old Lucretian materialist model and philosophy of the atom as a solid billiard-ball-like structure, an ancient model that continues to pervade many people's thinking about atomic and molecular structure and properties. This is known as the "hard sphere model."

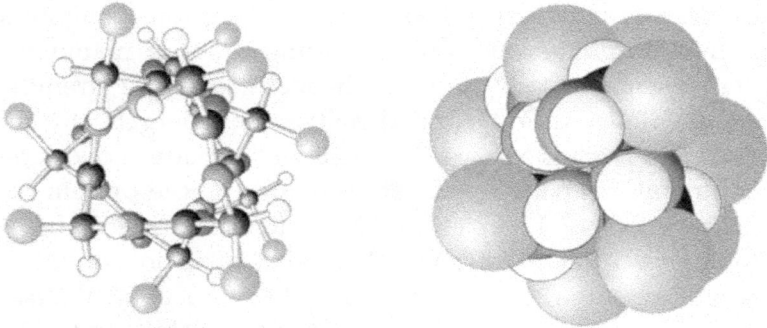

Figure 1.14.10. Some biochemists assert that a "light pipe" function is impossible because the alpha helices are not hollow and therefore do not provide a pathway for photons. However, quantum physics tells us that there is virtually nothing in the spaces in "space-filling" models.

Some recognize that the model is not accurate, but that it has a venerable history and has been useful for some purposes.[33] The term "space filling" implies solidity to the atomic and molecular structure that we now know, some 20 centuries after Lucretius, is inaccurate. The modern concept of atomic and molecular structure, based on quantum physics, is that there is nothing solid about atoms and molecules – they are primarily "composed" of "empty" space. Space-filling models are inaccurate simply because the spaces are not filled. This means that the idea that the alpha helices are not hollow is completely out of sync with quantum physics.

If the alpha-helices are indeed hollow, as suggested here, the model of regulation shown in Figure 1.14.11 is plausible. In the example illustrated here, the "the signal molecule" is a hormone, epinephrine. At body temperature, the molecule is vibrating.

Even in the dark, while one is sleeping, the environment is filled with infrared light, which is ideal for "pumping" energy into molecules and causing them to vibrate and emit electromagnetic fields (biophotons).

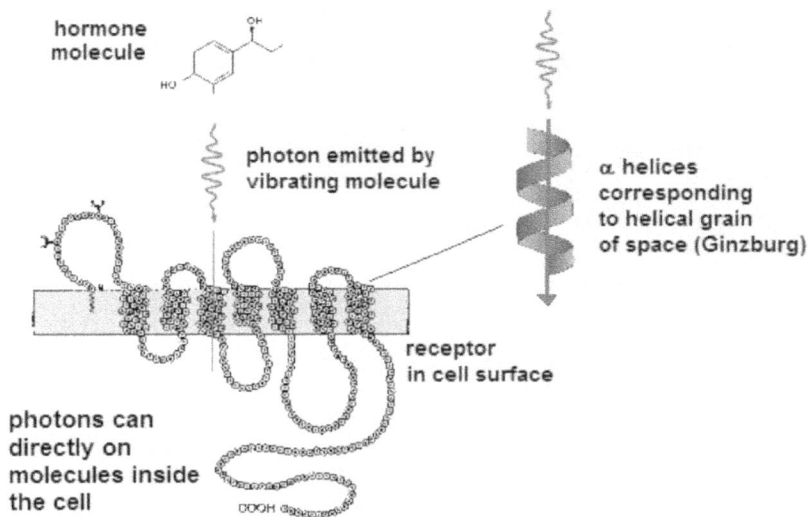

Figure 1.14.11. A model of regulation in which vibrating hormones produce electromagnetic fields that enter the target cells via the alpha helical portions of the protein that extend across the cell surface.

Finally, there is evidence for electromagnetic fields acting as first messengers for activating cellular processes.[34] This idea resolves the issues mentioned above concerning the fact that cells are packed with structures that impede the diffusion of regulatory molecules. Biophotons should have no problem traversing cell membranes. If these speculations are borne out, biophotonic communications provide a new and more realistic model of biological regulations and how they can be optimized for the benefit of biomedicine.

One of the most important receptor proteins is the voltage gated calcium channel because it regulates many cellular activities and responds virtually instantaneously to very weak electromagnetic fields. This receptor protein traverses the cell surface 24 times (Figure 1.14.12A). Note an important error in this illustration. The way the trans-membrane alpha-helical portions of the receptor are drawn gives the impression that helices are more or less parallel to each other and perpendicular to the plane of the membrane.

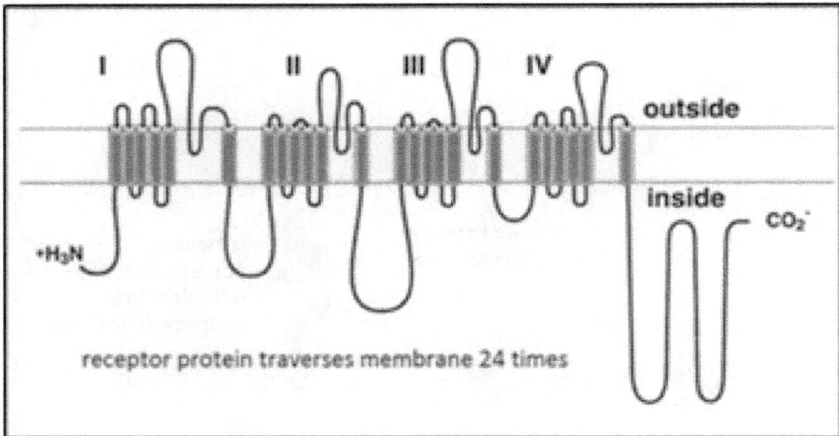

Figure 1.14.12A. The voltage gated calcium channel is a protein that crosses the cell surface 24 times. The diagram gives the incorrect impression that the helices are more or less parallel to each other and perpendicular to the plane of the membrane.

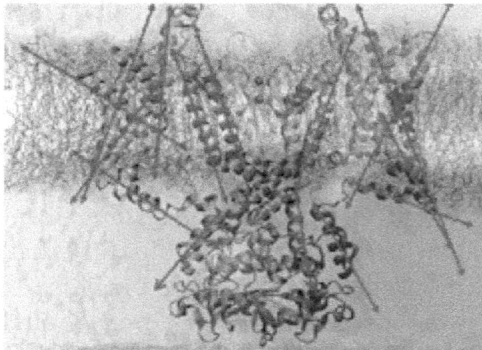

Figure 1.14.12B. The voltage-gated potassium channel and related receptors have alpha-helical segments that traverse the cell surface at a variety of angles, as shown by the arrows.

Modern methods of visualizing membrane structure show that this is not accurate. For example, the Theoretical and Computational Biophysics group at Beckman Institute has used molecular dynamics simulation software they developed called NAMD to create an atomic scale model of the voltage-gated potassium channel, another important cell surface receptor that controls a wide variety of cell functions (Figure 1.14.12B).

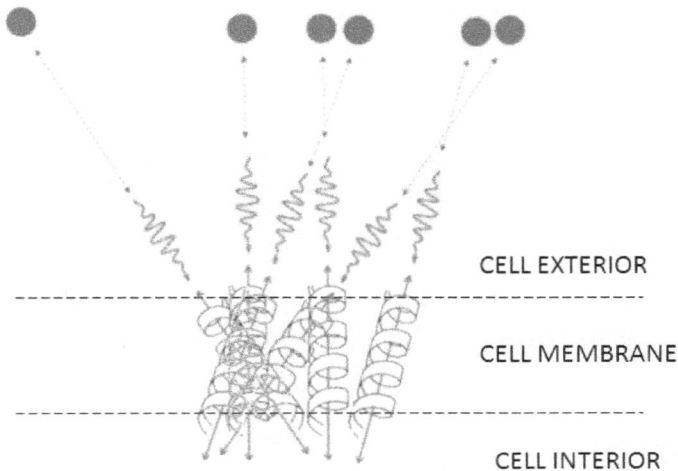

Figure 1.14.13. Hypothetical scheme in which the electromagnetic fields (biophotons) emitted by hormone molecules (black circles) are funneled into the cell interior via alpha-helical "light-pipes" that are portions of the receptor protein.

The helical portions of the receptor are actually at a variety of angles to the cell surface. Similar images have been obtained for other membrane receptors. This arrangement may not be random or accidental. It is an arrangement that makes sense if the alpha-helices are serving as light pipes to bring in signal photons emitted from a variety of sources at a variety of angles to the cell surface of the target cells (Figure 1.14.13). One could ask how such arrangements could be created and maintained. The classic explanation is that the flows of energy through systems organize those systems.[35]

In conclusion, this chapter has documented a number of phenomena relevant to physics, biology and medicine in which vortical structures and vortical flows are involved in fundamental ways. These phenomena occur at a variety of scales, from cosmological to subatomic.

A recent study from MIT and Cambridge University has established a fascinating correspondence between collective vortical movements of bacteria as they flow through a microscopic lattice and vortical flows of electrons in magnetic materials.[36] From an introduction to the study published in *Science Daily*:

> There are certain universal patterns in nature that hold true, regardless of objects' size, species, or surroundings. Take, for instance, the branching fractals seen in both tree limbs and blood vessels, or the surprisingly similar spirals in mollusks and cabbage.[37]

The unexpected shared pattern in the collective or coupled movements of bacteria and electrons were describable with the same lattice field theory that is used to account for the quantum behavior of electrons in magnetic and electronic materials. By adjusting certain dimensions of the microfluidic lattice, the researchers were able to direct billions of microbes to align, spin and swim in the same direction, much the way electrons circulate in the same direction when they create a magnetic field. With slight changes to the lattice, groups of bacteria flowed in opposite directions in adjacent pores, resembling self-cancelling electron flows in a nonmagnetic material.

Living systems can self-organize in highly ordered collective or coherent states strikingly similar to those found in condensed matter at low temperatures (so-called Bose-Einstein condensations). Examples range from the liquid crystal-like arrangements of bacterial colonies, water in cells and tissues, and the remarkable coupled motions of schools of fish and flocks of birds. The vortex is indeed a universal pattern in nature that applies to structures and flows regardless of the size, species, or environment involved. Hence the profound significance of the Universal Spacetime Theory (UST), in which space, light, and many other phenomena are organized as vortices.

References

1. Oschman, J.L., The science of spirals: A thirty-year adventure. In Wordsworth, C.F. *Spiral Up! 127 energizing options to be your best right now.* Resonance Publishing, Scottsdale, AZ, p. 267-282, 2014.
2. Adey, W.R., A growing scientific consensus on the cell and molecular biology mediating interactions with environmental electromagnetic fields. In: Ueno, S. (Ed.), *Biological Effects of Magnetic and Electromagnetic Fields*, Plenum Press, New York, pp. 4S-62 (Chapter 4), 1996.
3. Szent-Gyorgyi, A., personal communication, 1983.

4. Oschman, J.L. and Pressman, M., "An anatomical, biochemical, biophysical, and quantum basis for the unconscious mind," *Energy Psychology* Vol. 5, pp.1-15, 2013.

5. Dart RA., "Voluntary Musculature in the Human Body. The Double-Spiral Arrangement." *The British Journal of Physical Medicine* Vol. 13, No. 12, pp. 265-268, 1950.

6. Myers, T., *Anatomy Trains. Myofascial Meridians for Manual and Movement Therapies,* Churchill Livingston, London, 2001.

7. http://www.biotensegrity.com/

8. http://www.intensiondesigns.com/models.html

9. Ingber, D.E., "The architecture of life," *Scientific American,* Jan 30-39, 1998.

10. http://www.FND.org

11. http://www.fnd.org/pgs/geo/holistic_geometry

12. Manaka, Y., *Chasing the Dragon's Tail: The Theory and Practice of Acupuncture in the Work of Yoshio Manaka,* Paradigm Publishers, 1995.

13. Gordon, R., *Your Healing Hands. The Polarity Experience,* North Atlantic Books, 2004.

14. Maxwell, J.C., "Physical Lines of Force," *Philosophical Magazine,* Vol. 21 and 23, Series 4, Part I and II; Part III and IV, 1861-1862.

15. Maxwell, J.C., "A Dynamical Theory of the Electromagnetic Field," *Philosophical Transactions of the Royal Society of London* Vol. 155, pp. 459-512, 1865.

16. Fischer, E., "Einfluss der configuration auf die wirkung der enzyme," *Ber Ges Dtsch Chem* Vol. 27: 2985-2993, 1894.

17. Albrecht-Buehler, G., "In Defense of 'Nonmolecular' Cell Biology," *Int. Rev. Cytol.* Vol. 120:191-241, 1990.

18. Oschman, J.L., "Perspective: Assume a spherical cow: The role of free or mobile electrons in bodywork, energetic and movement therapies," *Journal of Bodywork and Movement Therapies,* Vol. 12, 40–57, 2008.

19. Luby-Phelps, K., "Cytoarchitecture and physical properties of cytoplasm: volume, viscosity diffusion, intracellular surface area," *International Review of Cytology* Vol. 192, 189–221, 2000.

20. Ives, J.A., van Wijk, E.P.A., Bat, N., Crawford, C., Walter, A., et al. "Ultraweak Photon Emission as a Non-Invasive Health Assessment: A Systematic Review," *PLoS ONE* Vol. 9, No. 2: e87401.

21. Lvovsky, A.I., Squeezed light. Section in book: *Photonics Volume 1: Fundamentals of Photonics and Physics*, Edited by D. Andrews, http://arxiv.org/abs/1401.4118, 2012.

22. Lee, K.G., Chen, X., Eghlidi, H., Kukura, P., Lettow, R., Renn, A., Sandoghdar, V., Götzinger, S., "A planar dielectric antenna for directional single-photon emission and near-unity collection efficiency," *Nature Photonics* Vol. 5, P. 166 (2011).

23. Polycarpou, C., Cassemiro, K.N., Venturi, G., Zavatta, A., & Bellini, M., "Adaptive Detection of Arbitrarily Shaped Ultrashort Quantum Light States," *Phys. Rev. Lett.* Vol. 109, 053602 – Published 3 August, 2012.

24. Oschman, J.L., and Oschman, N.H., "Vortical Structure of Light and Space: Biological Implications." *J. Vortex Sci. Technol.* Vol. 2:1-3, . 2015.

25. Trelstad, R.L., "The bilaterally asymmetrical architecture of the submammalian corneal stroma resembles a cholesteric liquid crystal," *Developmental Biology* Vol. 92:133-134, 1982.
26. Holmes, D.F., Gilpin, C.J., Baldock, C., Ziese, U., Koster, A.J., and Kadler, K.E., "Corneal collagen fibril structure in three dimensions: Structural insights into fibril assembly, mechanical properties, and tissue organization," PNAS Vol. 98, No. 13, pp. 7307–7312, 2001.
27. Standing, S., ed., *Gray's Anatomy, 40th Edition*, Churchill Livingston, Edinburgh., pp. 678-9, 2009.
28. Deisenyhofer, J., Michel, H., Huber, R. "The structural basis of photosynthetic light reactions in bacteria," *Trends. Biochem. Sci* Vol. 10: 243-248, 1985.
29. Glazer, A.N., "Light Harvesting by Phycobilisomes," *Annual Review of Biophysics and Biophysical Chemistry* Vol. 14: 47-77, 1985.
30. Fitzkee, N.C., and Rose, G.D., "Steric restrictions in protein folding: An alpha-helix cannot be followed by a contiguous beta-strand," *Protein Science* Vol. 13:633–639, 2004.
31. Berg, J.M., Tymoczko, J.L., and Stryer. L., *Biochemistry, 7th revised international edition*, Palgrave MacMillan, London, 2011.
32. http://public.csusm.edu/jayasinghe/BiomolTutorials/HelicesAndSheets/HelicesAndSheets.html
33. Richards, F.M., "Areas, volumes, packing, and protein structure," *Ann. Rev. Biophys. Bioeng.* 6: 151–176, 1977.
34. Pilla, A., Fitzsimmons, R., Muehsam, D., Wu, J., Rohde, C., et al. "Electromagnetic fields as first messenger in biological signaling: Application to calmodulin-dependent signaling in tissue repair," *Biochim Biophys Acta* Vol. 1810, pp. 1236-45, 2011.
35. Fuller, R. Buckminster, 1979. *Synergetics 2: Explorations in the Geometry of Thinking*. New York: Macmillan, 1979.
36. Wioland, H., Woodhouse, F.G., Dunke, J., and Raymond E. Goldstein, R.E., "Ferromagnetic and antiferromagnetic order in bacterial vortex lattices," *Nature Physics*, 2016; DOI: 10.1038/nphys3607.
37. Massachusetts Institute of Technology, "Bacteria, electrons spin in similar patterns: Bacteria streaming through a lattice behave like electrons in a magnetic material," ScienceDaily, 5 January 2016. <www.sciencedaily.com/releases/2016/01/160105133132.htm.

2

MULTI-DIMENSIONAL SPACETIME

The dimensions of space are most frequently defined by the minimum number of coordinates needed to specify a location of each point.

2.1 Euclid's Multi-Dimensional Spaces

Here is how to create the multi-dimensional spaces in accordance with Euclid's geometry. The point m (Fig.2.1) has no dimensions at all. Move the point m a distance x along a straight path (a) and you will obtain a line that will have one dimension, length x. Moving this line a distance y in the direction perpendicular to the line will produce a two-dimensional plane (b) with the sides x and y. Now move this plane a distance z in the direction perpendicular to the plane and you will produce a three-dimensional solid (c) with the sides x, y and z.

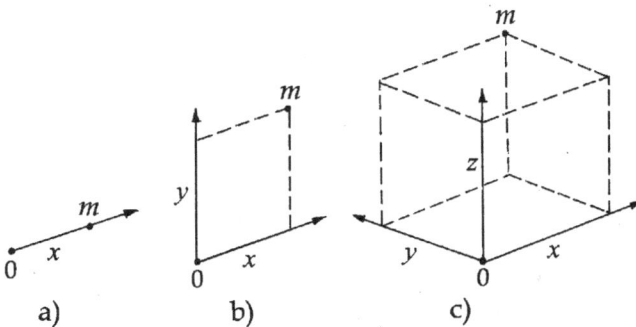

Figure 2.1. Spaces of one (a), two (b) and three (c) dimensions.

Many ancient Greek scientists, including Aristotle and Ptolemy, believed that the fourth dimension is impossible.

2.2 Fitch's Four-Dimensional Hypercube

The four-dimensional *hypercube* was proposed by a member of the Corps of Engineers, U.S.A. Lieutenant-Colonel Graham Denby Fitch. He described his idea in the *Scientific American* July issue of 1909.

Fitch produced the hypercube (Fig. 2.2) by moving each corner of a 3-dimensional cube *ABCDEFGH* in parallel directions *Aa*, *Bb*, *Cc*, *Dd*, *Ee*, *Ff*, *Gg* and *Hh*. These directions are supposed to be perpendicular to our 3-dimensional space. The four-dimensional hypercube was complicated and very difficult to visualize. One may rightfully wonder how complicated must be the spaces of higher dimensions, and could be they "visualizable" at all.

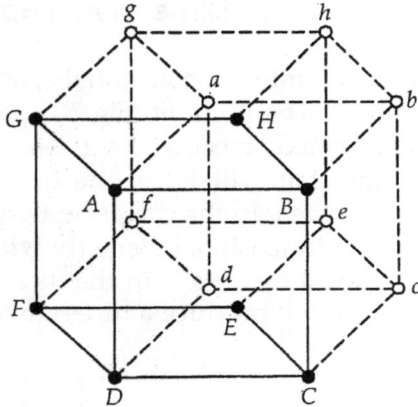

Figure 2.2. The 4-dimensional hypercube.
Adapted from H.P. Manning (1960).

2.3 Riemann's Multi-dimensional Space

The German mathematician Georg Riemann (1826-1866) defined the dimensions of a space by a degree of the space curvature. The higher degree of curvature of space the greater number of coordinates will be required to locate a point in this space.

For instance, the position of a point *A* on a two-dimensional curved line (Fig. 2.3) is determined by two coordinates, *x* and *y*, while three coordinates *x*, *y*, and *z* are needed to define a position of a point *A* on a three-dimensional sphere. In both cases the modulus of the vectors defining the location of the point *A* is given by similar equations:

$$z^2 = x^2 + y^2$$
$$u^2 = x^2 + y^2 + z^2$$

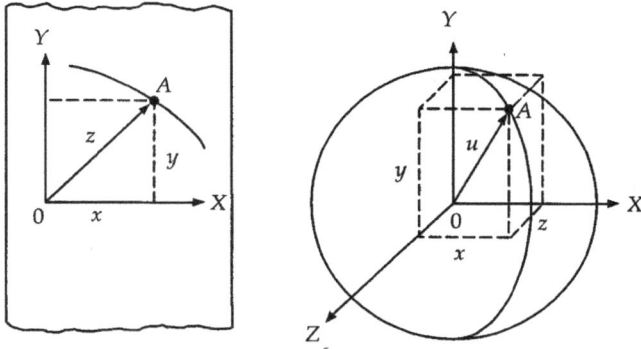

Figure 2.3. Riemann curved spaces of two and three dimensions.

A curved space of any higher order can be described by adding extra terms into the last equation.

2.4 Minkowski's Spacetime

The Russian-German mathematician Hermann Minkowski (1864-1909) made his outstanding contribution to relativity by uniting space and time. When time and space are considered independent of each other, the location of an object in a three-dimensional space is defined by three dimensions x, y, and z. To connect space with time t, Minkowski added one more dimension ict in which $i = \sqrt{-1}$ is the imaginary number. Thus, the modulus of the vector u defining the location of a point on a curved space was defined by the modified Riemann equation:

$$z^2 = x^2 + y^2 - (ct)^2$$
$$u^2 = x^2 + y^2 + z^2 - (ct)^2$$

2.5 Einstein's Spacetime

The German-Swiss-US theoretical physicist Albert Einstein (1879-1955) expanded the application of the term spacetime in his general theory of relativity by using a special branch of mathematics called *tensor calculus*.

The result was a groundbreaking departure from the Newtonian interpretation of gravity. Gravitation was no longer a force but an intrinsic curvature of spacetime. It was the curved spacetime around the Sun that attracted the planets. In curved spacetime the motions of the planets were represented by the shortest distances, called *geodesics*. Figure 2.5.1 illustrates the curvature of spacetime according to Einstein.

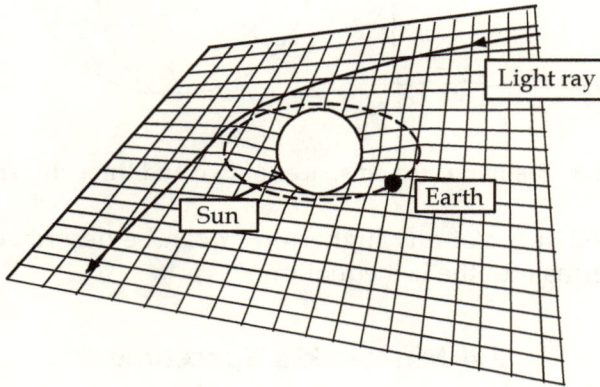

Figure 2.5.1. Einstein's curved spacetime.

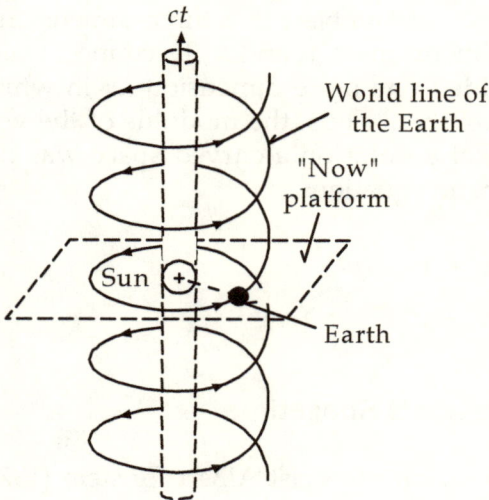

Figure 2.5.2.
The spacetime of
the Sun and the Earth.
Adapted from
B. Hoffmann (1972).

It is difficult to illustrate the curvature of time diagrammatically. In Figure 2.5.2, we omitted the curvature of time to simplify the illustration of Einstein's spacetime. The double-dotted line represents the sun's world line as it moves upwards on the page. The world line of a planet is represented by a helical spiral that is a geodesic in the curved spacetime associated with the Sun. Imagine that the Earth is located at this moment on a platform indicated as the "now" platform. As time passes, the platform will move upwards on this page a distance ct.

2.6 Kaluza's Multi-Dimensional Space

The German physicist Theodor Kaluza (1885-1954) presented the multi-dimensional space by wrapping up a space of a higher dimension around a space of a lower adjacent dimension. Consider that a few hundred feet of garden hose is stretched across a canyon, and you view it from a far distance away. From this distance the hose will appear as one-dimensional string. Therefore, the ant located on the hose would have only one dimension in which to walk: the left-right dimension along the hose's length (Fig. 2.6).

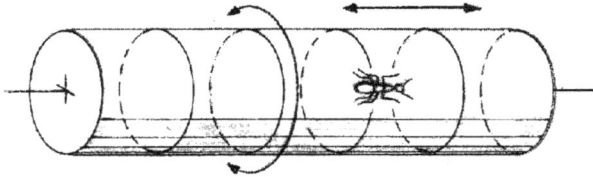

Figure 2.6. Kaluza's multi-dimensional space.

From a magnified perspective, we will see that besides the left-right dimension along the hose's length there is the second dimension along which the ant can walk, either clockwise or counterclockwise. We may say that the second dimension of the hose as "wrapped up" around the one-dimensional material. Similarly, it is possible to describe mathematically a three-dimensional hose "wrapped up" around the two-dimensional material, and so on and so on.

2.7 Spiral Principle

The spiral principle is a purely abstract proposition of UST used to discover a multi-dimensional spacetime called *helicola*. This principle simply states:

Every line is a three-dimensional helical spiral.

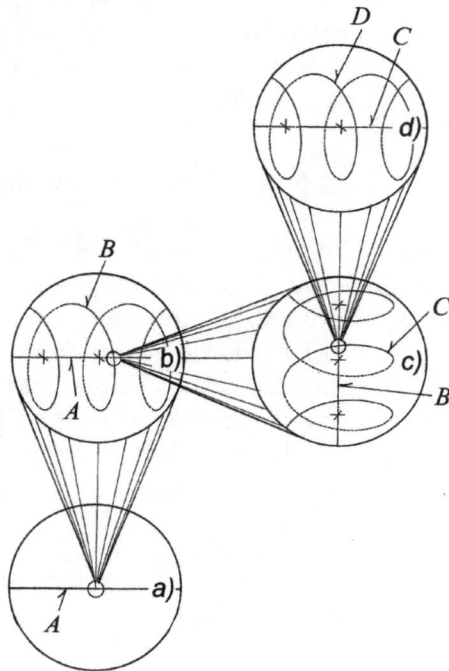

Figure 2.7.1. Multiple-level spiral of diminishing dimensions.

By following this principle, one will discover a multiple-level spiral of diminishing dimensions (Fig. 2.7.1). Start from a straight line A (Fig. 2.7.1a). Then apply the spiral principle to discover that the line A, upon a closer examination, looks like a spiral B wound around the line A (Fig. 2.7.1b). But, the windings of the spiral B are also made of a line. Therefore, according to the spiral principle, this line, under a closer examination, appears as a spiral C wound around the spiral B (Fig. 2.7.1c). A further more detail examinations will reveal a spiral D wound around the spiral C (Fig. 2.7.1d), and so on.

The Spiral Principle allows us also to discover a multiple-level spiral of enlarging dimensions (Fig. 2.7.2). Start from a straight line A (Fig. 2.7.2a). According to the spiral principle, when looking at the line A from a long distance away, it will appear as a part of a large spiral A wound around a line B (Fig. 2.7.2b). Similarly, when looking from a farther distance away from the line B, it will appear as a spiral B wound around a line C (Fig. 2.7.2c). After moving further away from the line C, it will reveal itself as a spiral C wound around a line D (Fig. 2.7.2d), and so on.

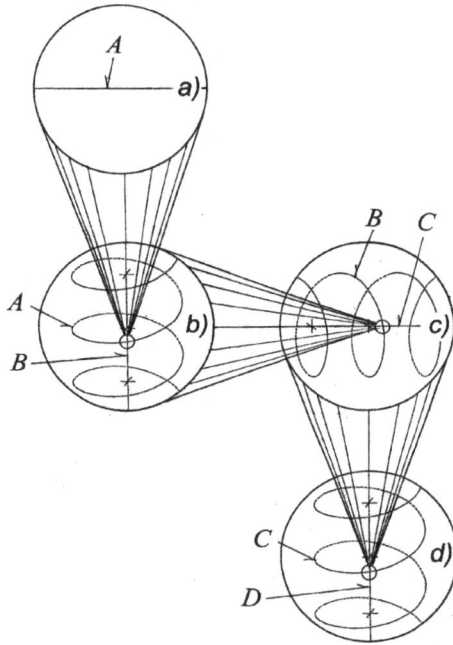

Figure 2.7.2. Multiple-level spiral of enlarging dimensions.

2.8 Helicola

The helicola is the universal multi-dimensional spacetime which geometry is based on the spiral principle. At each level of helicola there are two strings, a *leading string* and a *trailing string*. In Figure 2.7.1b, the leading string A is accompanied by a trailing string B. At the next level of helicola shown in Figure 2.7.1c, the trailing string B of the previous level of helicola becomes a leading string, and it is accompanied by the trailing string C. Similarly, at the next level of helicola shown in Figure 2.7.1d, the trailing string C of the previous level of helicola becomes a leading string, and it is accompanied by the trailing string D, and so on. The dimensions of helicola are defined by the number of parameters required to locate a point on its outer string.

2.9 Helicola of Odd Dimensions

Figure 2.9 shows the examples of helicola of one, three and five dimensions.

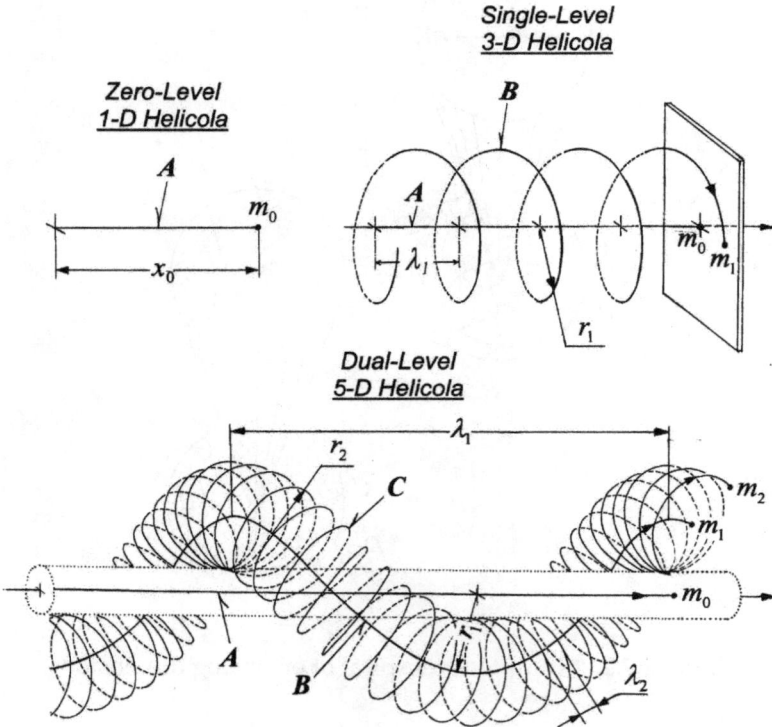

Figure 2.9. The odd-dimensional helicola.

- The 1-D zero-level helicola contains only a straight-line leading string A, and may be viewed as an incomplete helicola. Only one parameter x_0 is required to define the location of the point m_0 on this string.

- In the 3-D single-level helicola, the point m_1 belongs to the trailing string B, and it propagates synchronously with the point m_0. To define the location of the point m_1 it is necessary to know two additional parameters of the trailing string B, the wavelength λ_1 and the radius r_1.

- In the 5-D dual-level helicola, the point m_2 belongs to the second trailing string C, and it propagates synchronously with two points, m_1 and m_0. To define the location of the point m_2 it is necessary to know two more parameters of the second trailing string C, the wavelength λ_2 and the radius r_2.

2.10 Helicola of Even Dimensions

Figure 2.10 shows the examples of helicola of two, four and six dimensions.

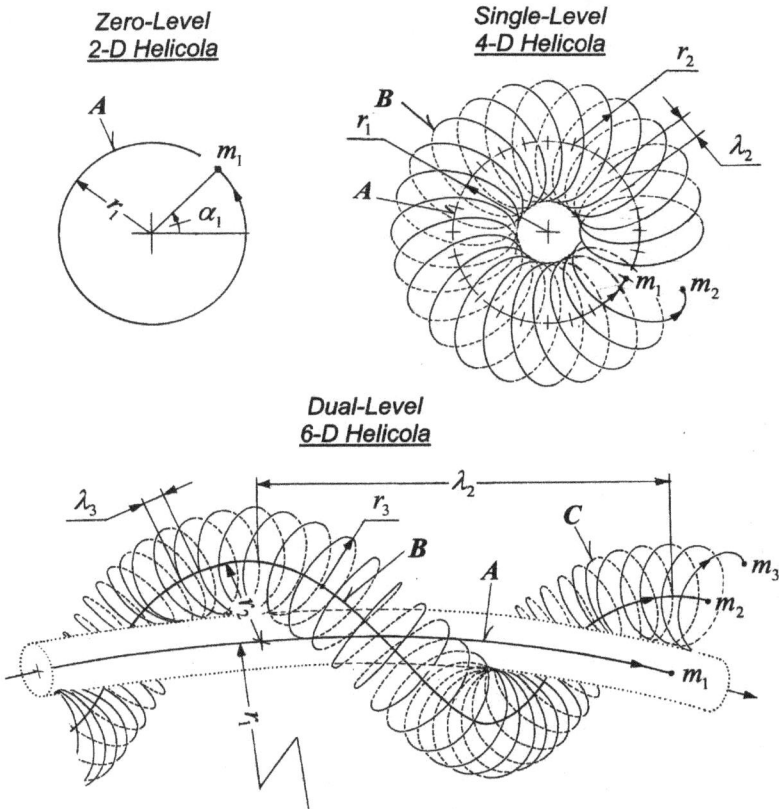

Figure 2.10. The even-dimensional helicola.

- In the 2-D zero-level helicola, the point m_1 belongs to the circular leading string A with the radius r_1. To define the location of the point m_1 it is necessary to know two parameters of leading string A, the radius r_1 and the angle α_1.
- In the 4-D single-level helicola, the point m_2 belongs to the toroidal trailing string B, and it propagates synchronously with the point m_1. To define the location of the point m_2 it is necessary to know two additional parameters of the trailing string B, the wavelength λ_2 and the radius r_2.
- In the 6-D dual-level helicola, the point m_3 belongs to the second trailing string C, and it propagates synchronously with two points, m_1 and m_2. To define the location of the point m_3 it is necessary to know two more parameters of the second trailing string C, the wavelength λ_3 and the radius r_3.

3

STATIC & DYNAMIC HELICAL SPIRALS

We do not have to go too far to see how helical spiral looks like. Many of us saw it already in a popular toy called "slinky." We will construct two kinds of helical spirals, static and dynamic. Nothing moves in the static helical spiral, and we can easily make it. The dynamic spiral is formed by the motion of its strings. Therefore, we have to use our imagination to envision it.

3.1 Static Single-Helical Spiral

Here is one of the ways of making a static single-helical spiral. It involves using three easy steps.

Step 1 – Take a round object like a beverage can shown in Figure 3.1a. In geometry, this round object is known as a *cylinder*. They describe the cylinder geometry by using its two dimensions: the radius r_1 and the height λ_1.

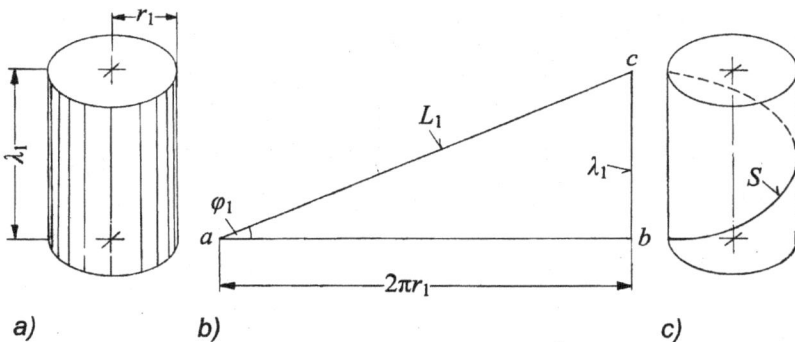

Figure 3.1. Construction of a static single-helical spiral.

Step 2 - Cut a right triangle *abc* out of a piece of paper as shown in Figure 3.1b. Make the length of one side *ab* equal to the perimeter of the can $2\pi r_1$ and the length of the other side *bc* equal to the height of the can λ_1

Step 3 - Finally, wrap the prepared triangular piece of paper around the can as shown in Figure 3.1c. As the side *ab* will wrap around the bottom of the can, the diagonal *ac* will appear on the can surface as one winding of a helical spiral.

We will call the cylinder radius r_1 the *radius of helical spiral* and the can height λ_1 the *wavelength of helical spiral*. These are the basic spiral parameters. There are three derivative parameters of the helical spiral, the *length of one spiral winding* L_1, the *spiral steepness angle* φ_1 and the *spiral volume* U_1. These three parameters are called *derivative*, because they can be derived from the basic parameters. Main space parameters of the static single-helical spiral are:

L_1 = spiral length
r_1 = spiral radius
U_1 = volume
λ_1 = wavelength
φ_1 = steepness angle.

Based on Figure 3.1b, we can determine the spiral length L_1 by using the Pythagorean Theorem:

$$L_1 = \sqrt{(2\pi r_1)^2 + \lambda_1^2} \qquad (3.1\text{-}1)$$

We can determine the cosine of the spiral steepness angle φ_1 by using conventional trigonometry,

$$\cos \varphi_1 = \frac{2\pi r_1}{L_1} \qquad (3.1\text{-}2)$$

The volume U_1 is equal to the volume occupied by the static single-helical spiral within one wavelength λ_1, it is equal to:

$$U_1 = \pi r_1^2 \lambda_1 \qquad (3.1\text{-}3)$$

Importantly, because the leading string of the helical spiral is in the form of a straight line, the values of spacetime parameters of the helical spiral are independent on the position of a point of measurement located along the trailing string.

3.2 Dynamic Single-Helical Spiral

To construct the dynamic single-helical spiral depicted in Figure 3.2, you may simply use tips of pointing fingers of your hands. Call the pointing finger tip of one of your hands the point m_0 and the pointing finger tip of the other hand the point m_1. Then use your imagination and follow the three easy steps shown below.

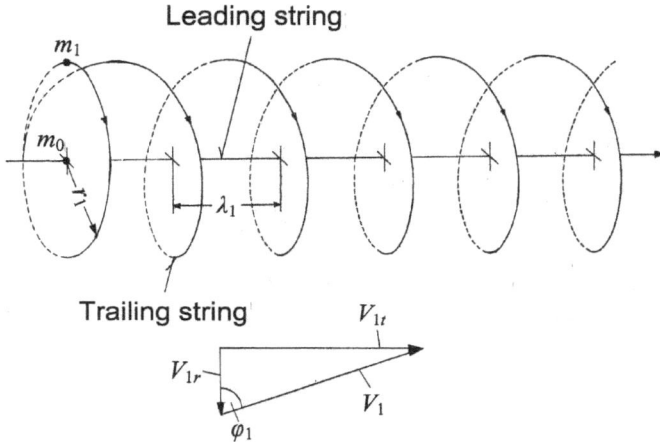

Figure 3.2. Construction of a dynamic single-helical spiral.

Step 1 – Stand up and stay firmly on the floor. Use a pointing finger of one of your hands to locate the point m_0 in front of you. Use a pointing finger of your other hand to locate the point m_1 at the distance r_1 from the point m_0.

Step 2 – Move the point m_1 around the stationary point m_0 along a circle with the radius r_1. Imagine a circular trace left in the air by the moving point m_1. Make this circular trace with any speed you feel comfortable with, keeping in mind that this speed is called the *rotational velocity* V_{1r}.

Step 3 – As you continue making the circular trace left by moving point m_1, begin walking forward by maintaining the same distance from the point m_0 in front of you. Move forward with any speed you feel comfortable with, keeping in mind that this speed is called the *translational velocity* V_{1t}.

An imaginary straight-line trace left by the point m_0 is called the *leading string* of the single-helical spiral. As the point m_1 rotates around the point m_0 and simultaneously moves forward together with the point m_0, it leaves an imaginary trace called the *trailing string* of single-helical spiral. The trailing string of this spiral will propagate along its spiral path with the *spiral velocity* V_1. We can describe the relationship between the spiral velocity V_1 and the velocities of its two components, the rotational velocity V_{1r} and the translational velocity V_{1t}, by using the Pythagorean Theorem.

Notably, the spiral velocities are the *spacetime parameters* expressed in the units of space and time, such as meters per second (*m/s*). The other two parameters of the dynamic single-helical spiral, the spiral period T_1 and the spiral frequency f_1, are the *time parameters* expressed in the units of time, such as second (*s*) or 1/second (*s^{-1}*).

The space parameters of the dynamic single-helical spiral are the same as these of the space parameters of the static single-helical spiral and its space equations are defined by Eqs. (3.1-1) – (3.1-3). The spacetime parameters of this spiral are:

f_1 = frequency
T_1 = period
V_1 = spiral velocity
V_{1r} = rotational velocity
V_{1t} = translational velocity.

From Figure 3.2 and based on Pythagorean Theorem, the spiral velocity V_1 is related tor the rotational velocity V_{1r} and the translational velocity V_{1t} by the equation:

$$V_1 = \sqrt{V_{1r}^2 + V_{1t}^2}$$
(3.2-1)

The period T_1 is the time required for the dynamic single-level helical spiral to propagate through one its full cycle. Based on this definition, the period T_1 is equal to:

$$T_1 = \frac{L_1}{V_1} = \frac{2\pi r_1}{V_{1r}} = \frac{\lambda_1}{V_{1t}}$$
(3.2-2)

The frequency f_1 is the inverse of the period T_1. Therefore, we obtain from Eq. (3.2-2) :

$$f_1 = \frac{V_1}{L_1} = \frac{V_{1r}}{2\pi r_1} = \frac{V_{1t}}{\lambda_1} \qquad (3.2\text{-}3)$$

Importantly, because the leading string of the static single-helical spiral is in the form of a straight line, the values of space-time parameters of this spiral are independent on the position of a point of measurement located along the trailing string.

3.3 Dynamic Double-Helical Spiral

The dynamic double-helical spiral is made up of two dynamic single-helical spirals. It is wide-spread in the universe. Consider, for instance, a binary star that is a star system consisting of two stars orbiting around their common center of gravity. The paths of these two stars correspond to the dynamic helical spiral. To construct the dynamic double-helical spiral depicted in Figure 3.3, you may use a pencil. Call one end of the pencil the point *m* and the other end the point *n*. Then use your imagination and follow the three easy steps shown below.

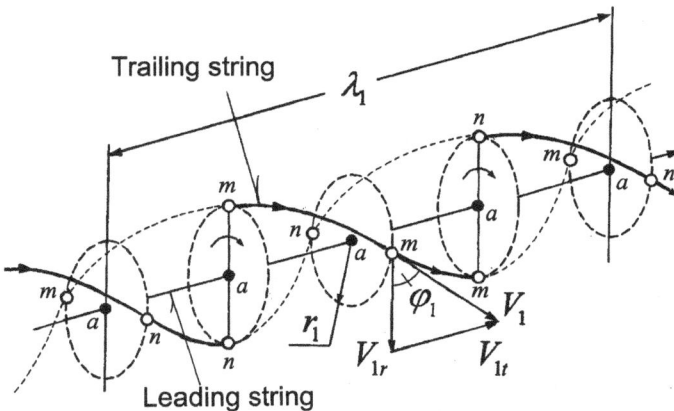

Figure 3.3. Construction of a dynamic double-helical spiral.

Step 1 - Stand up and stay firmly on the floor. Use fingers of one of your hand to hold the pencil at its middle point *a*. Keep the pencil in front of you.

Step 2 - Move the pencil around its middle point a. Imagine a double-circular trace left in the air by the pencil's end points m and n. Make this double-circular trace with any speed you feel comfortable with, just keep in mind that this speed is called the *rotational velocity* V_{1r}.

Step 3 - As you continue making the double-circular trace left by the pencil's end points m and n, begin walking forward by maintaining the same distance from the pencil in front of you. Move forward with any speed you feel comfortable with, keeping in mind that this speed is called the *translational velocity* V_{1t}.

An imaginary straight-line trace left in the air by the pencil's end point a is called the *leading string* of the double-helical spiral. The double-helical trace left in the air by the pencil's end points m and n are called the *trailing string* of this spiral. The trailing string will propagate along its spiral path with the *spiral velocity* V_1.

Notably, the spacetime properties of each branch of the dynamic double-helical spiral are described by the same equations that are applied to the dynamic single-helical spiral. Space-wise the branches are separated by a distance of one half of the wavelength λ_1. Time-wise the branches are separated by one half of the period T_1.

4

Universal Helicola

As many other theories, the UST had a clear starting point. It all started in December 1992 during a dinner conversation with my son Gene, who was then finishing his last year of high school. While looking at the crumb of bread, Gene made a casual observation that there is probably another world inside that piece. Even though I heard this notion more than once before, somehow it registered in my mind differently this time. I started to wonder if there is any geometrical shape that would perfectly represent the infinitude of the world and its repetitious character in both enlarging and diminishing directions. This shape turned out to be the endless multidimensional helical spiral which I called the *helicola.*

At first I thought that the helicola was a purely abstract spacetime construction having nothing to do with our universe. But, a few days later I realized that the helicola outlines the paths of real celestial bodies. The first example that came to my mind was a helicola created by the paths of the center of our galaxy the Milky Way, the Sun, the Earth and the Moon.

4.1 Spiral Path of the Sun

As shown in Figure 4.1, the helicola of the first level describes paths of the Sun and the center of the Milky Way. The leading string *1* of helicola represents a path of the center of the Milky Way toward the constellation Hydra at the translational velocity V_{1t}. The Sun is located at the distance r_1 from the galaxy center. As the Milky Way moves toward the constellation Hydra, the Sun orbits the galaxy center at the rotational velocity V_{1r}.

A combination of the rotational motion of the Sun with the translational motion of the galaxy center makes the Sun moving along a spiral path of the trailing string *1* of helicola. The Sun spiral velocity V_1 is equal to the geometrical sum of the rotational velocity V_{1r} and translational velocity V_{1t}. The trailing string *1* of the first level of helicola also serves as a leading string *2* of the second level of helicola described in the next section.

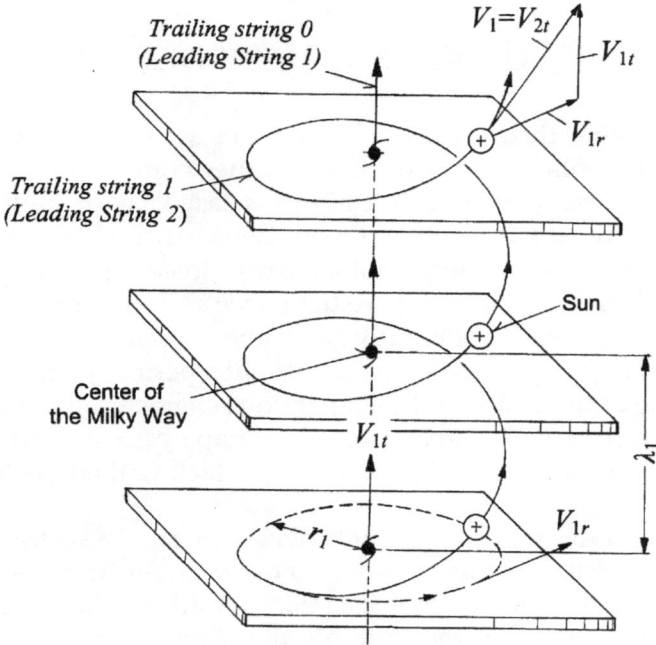

Figure 4.1. Spiral path of the Sun around the center of Milky Way.

4.2 Spiral Path of the Earth

The helicola of the second level (Fig. 4.2) describes paths of the Earth and the Sun. At this level, the leading string *2* is a former trailing string *1* of the first level of helicola; it outlines the path of the Sun along its spiral path at the translational velocity V_{2t} that is equal to the spiral velocity V_1 of the trailing string *3*.

A combination of the rotational motion of the Earth and the translational motion of the Sun makes the Earth moving along a spiral path of the trailing string *2* of helicola. The Earth spiral velocity V_2 is equal to the geometrical sum of the rotational velocity V_{2r} and translational velocity V_{2t}. The trailing string *2* of the second level of helicola also serves as a leading string *3* of the third level of helicola described in the next section, so $V_{2t} = V_1$.

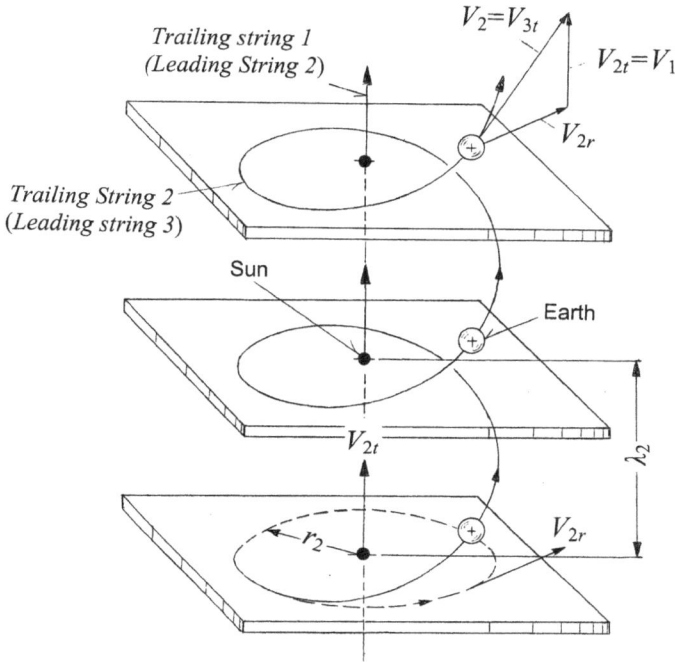

Figure 4.2. Spiral path of the Earth around the Sun.

4.3 Spiral Path of the Moon

The helicola of the third level (Fig. 4.3) describes the paths of the Moon and the Earth. At this level, the leading string *3* is a former trailing string *2* of the second level of helicola; it outlines a path of the Earth along its spiral path at the translational velocity V_{3t} that is equal to the spiral velocity V_2 of the trailing string *3*.

A combination of the rotational motion of the Moon and the translational motion of the Earth makes the Moon moving along a spiral path of the trailing string *3* of helicola. The Moon spiral velocity V_3 is equal to the geometrical sum of the rotational velocity V_{3r} and translational velocity V_{3t}. The trailing string *3* of the third level of helicola also serves as a leading string *4* of the fourth level of helicola that could be a path of a satellite orbiting the Moon.

The traces left by the center of the Milky Way, the Sun, the Earth and the Moon form a three-level helicola. I thought that the levels of the helicola will increase as we will consider the motion of the centre of the Milky Way around the center of even more gi-

Universal Helicola

gantic entity, and then the motion of the center of that gigantic entity around the center of the next super-gigantic entity, and so on and so on.

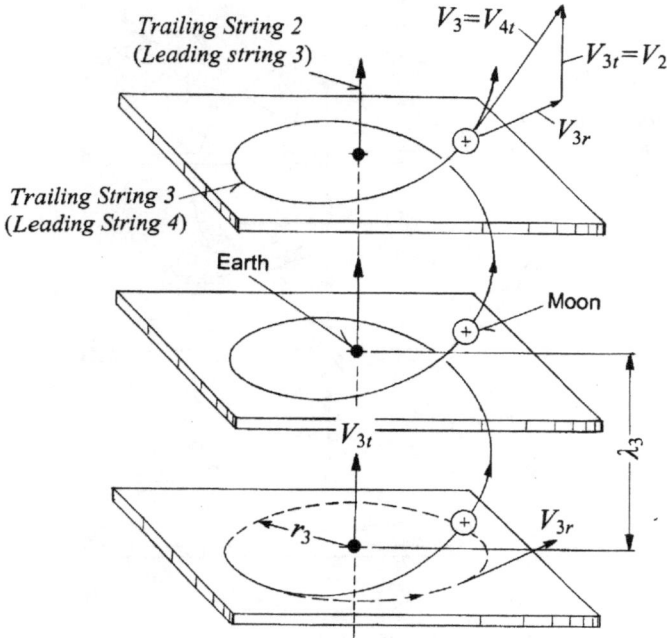

Figure 4.3. Spiral path of the Moon around the Earth.

4.4 Solar Planetary System

After the discovery of the celestial helicola it became obvious to me that the commonly-known illustration of the solar system in which the stationary Sun is surrounded by the planets moving along elliptical paths (see Figure 4.4a) is outdated.

Proposed by the German astronomer and mathematician Johannes Kepler (1571-1630) four centuries ago, it was based on the assumption that the stationary Sun was located at the center of the universe. Figure 4.4b provides us with a contemporary view of our planetary system in which the planets spiral around the Sun, while the Sun orbits the center of the Milky Way.

Ironically, this spiral view of our planetary system was known to the astronomers and astrophysicists since they discovered many years ago that the Sun moves around the center of the Milky Way. However, the image of elliptical planetary orbits still remains in the minds of many ordinary people, and many publications still show outdated illustrations of our solar system.

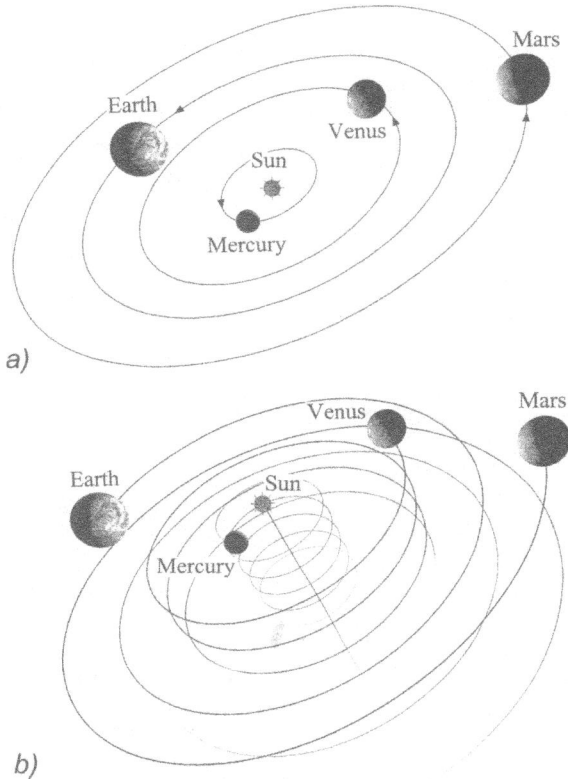

Figure 4.4. Presentations of a solar planetary systems:
a) outdated and b) updated.

4.5 Helicola of Macro- & Micro-Worlds

The discovery of the celestial helicola made me think that the spiral motion may be at the core of creation and existence of the entire universe. I began to view hurricanes, tornadoes, whirlpools and all other spiral structures, including DNA, as merely transitional entities that may exist in nature all the way down to the prime elements of nature forming elementary particles. I wondered if the spacetimes of the prime elements of nature are merely particular cases of helicola. To illustrate this proposition I made a

picture of a well-known structure of a hydrogen atom in which
the atomic electron B orbited the proton A. In this picture shown
in Figure 4.5, the propagation of the atomic electron was accompanied by a toroidal spiral. This spacetime arrangement was a
particular case of helicola, the single-level 4-D helicola shown in
Figure 2.10. In this helicola, the path of the electron was leading
string and the path of the toroidal spiral the trailing string.

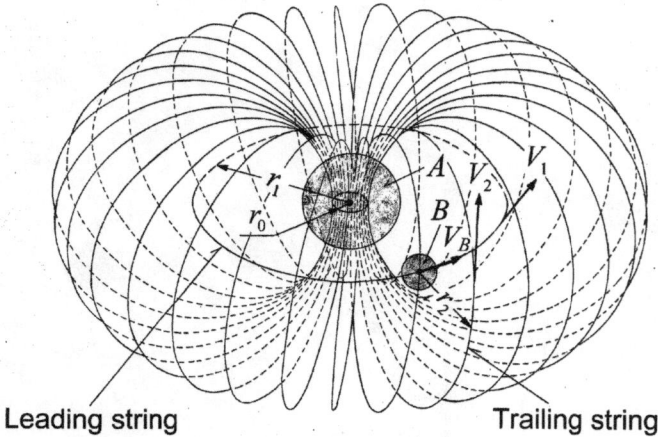

Leading string   Trailing string

Figure 4.5. Toroidal spiral accompanying atomic electron B in
the hydrogen atom.

The helicola turned out to be very suitable for modeling the
spacetime properties of elementary particles. The single-level 4-D
helicola provided required spacetime properties for the constituents of *elementary matter particles* called the *toryces*. Another particular case of helicola, the dual-level 5-D helicola shown in Figure
2.9 provided required spacetime properties of constituents of *elementary radiation particles* called the *helyces*.

PART 2

Elementary
Matter Particles

5

BASIC CONCEPT OF TORYX

The spiral part of the toryx is the single-level 4-D helicola described in Chapter 2 and shown in Fig. 2.10.

5.1 Toryx Basic Structure

The toryx consists of two main parts: a double-circular leading string and a double-toroidal trailing string as shown in Figs. 5.1.1 – 5.1.3. Notice that for the sake of simplicity these figures show only one branch of leading and trailing strings.

Double-circular leading string – The double-circular leading string of a toryx is a particular case of a dynamic double-helical spiral shown in Figure 3.3 in which the translational velocity V_{1t} is equal to zero and the rotational velocity V_{1r} is equal to the spiral velocity V_1. Consequently, the double-circular leading string appears like two circular traces with the radius r_1 left by the moving points m and n shown in Figure 5.1.1.

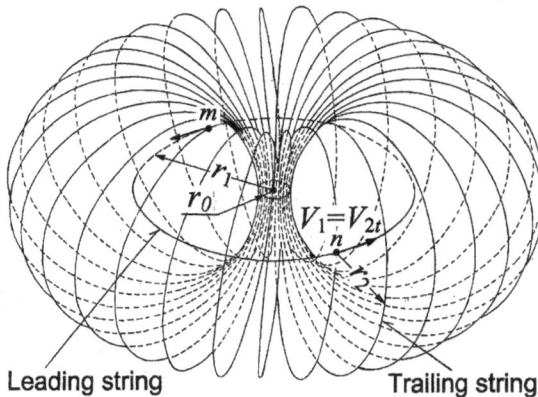

Figure 5.1.1. Isometric view of a toryx.

Double-toroidal trailing string – The double-toroidal trailing string with the radius r_2 propagates along its toroidal spiral path with the spiral velocity V_2 synchronously with the leading string. A circular opening in the toryx center with the radius r_0 is called the *toryx eye*.

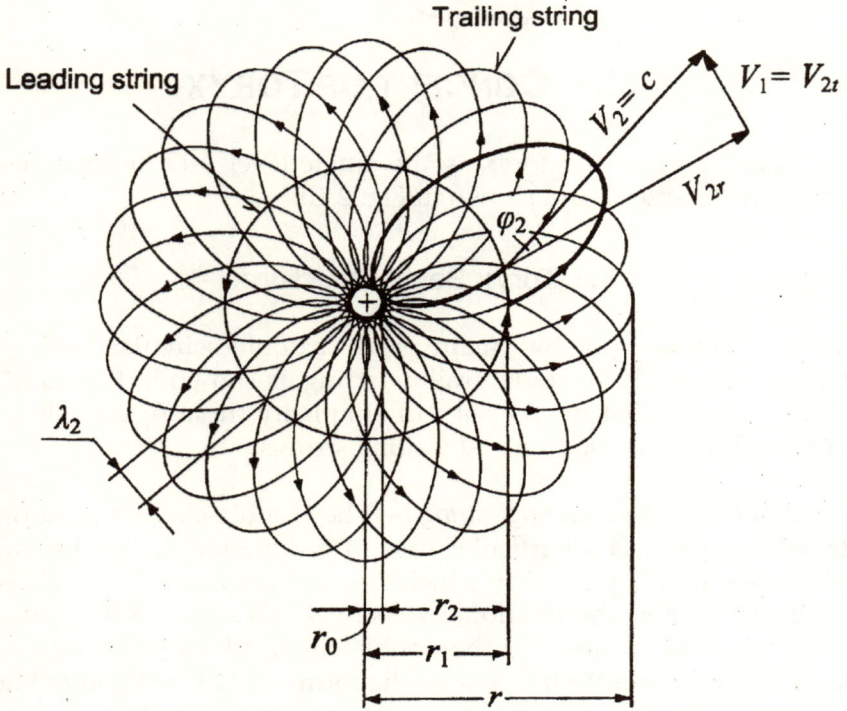

Figure 5.1.2. Top view of a toryx.

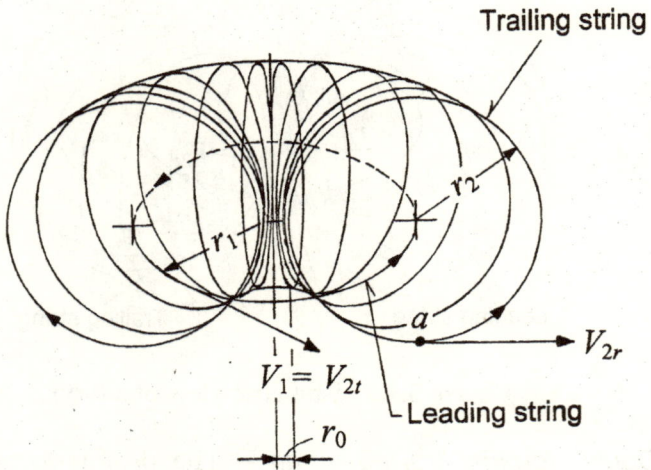

Figure 5.1.3. Cross-section of a toryx.

There are two kinds of *toryx spins*: *orbital* and *satellite*. Both toryx spins are defined by the right-hand rule. The *up* spins correspond to the clockwise rotation of the toryx strings, while the *down* spins correspond to the counterclockwise rotation of the toryx strings.

- The orbital toryx spin is associated with the direction of a circular motion of the toryx leading string propagating with the rotational velocity V_{1r}.
- The satellite toryx spin is associated with the direction of a circular motion of the toryx trailing string propagating with the rotational velocity V_{2r}.

5.2 Toryx Spacetime Parameters in Absolute Units

Similarly to the helical spiral, the relationships between toryx spacetime parameters are also described by the Pythagorean Theorem – see Figure 5.2.

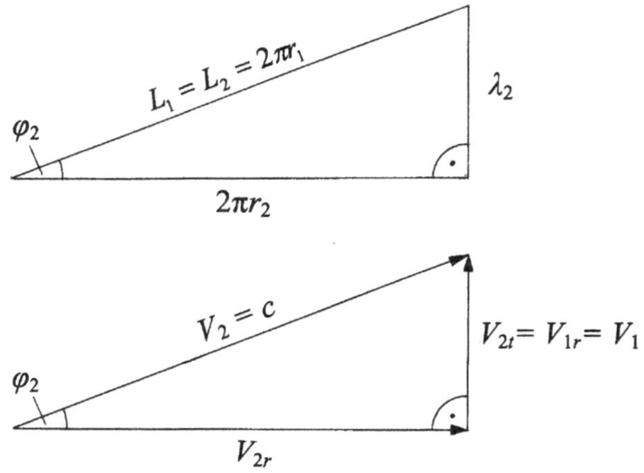

Figure 5.2. Hodographs of the toryx absolute spacetime parameters corresponding to the middle point *a* of the trailing string (Fig. 5.1.3).

Both branches of the toryx trailing string propagate along their toroidal spiral paths at the same spiral velocity V_2. In each branch, the spiral velocity V_2 has two components: rotational velocity V_{2r} and translational velocity V_{2t}. Because the propagations of trailing and leading strings are synchronized, the translational velocity of the trailing string V_{2t} is equal to the spiral velocity of the leading string V_1. Application of the Pythagorean Theorem to a toryx is more complicated than to a helical spiral. In the helyx, the leading string is in the form of a straight line, making its spacetime parameters the same along its trailing string. In the toryx, the leading string is circular, making some of its spacetime parameters to vary along its trailing string.

Consequently, the values of some toryx spacetime parameters are dependent on the position of a point of measurement of these parameters along the trailing string. For instance, as one can see in Figure 5.1.2, distances between windings of the toryx trailing string are shortest near the toryx eye and longest at the periphery of the trailing string.

Position-independent toryx spacetime parameters include:

f_0 = toryx base frequency
f_1 = frequency of the toryx leading string
f_2 = frequency of the toryx trailing string
L_1 = spiral length of one winding of the toryx leading string
L_2 = spiral length of one winding of the toryx trailing string
r = toryx peripheral radius
r_0 = toryx eye radius
r_1 = radius of the toryx leading string
r_2 = radius of the toryx trailing string
T_0 = toryx base period
T_1 = period of the toryx leading string
T_2 = period of the toryx trailing string
U_1 = volume of the toryx leading string
U_2 = volume of the toryx trailing string
V_1 = spiral velocity of the toryx leading string
V_{1r} = rotational velocity of the toryx leading string
V_{1t} = translational velocity of the toryx leading string
V_2 = spiral velocity of the toryx trailing string
w_1 = the number of windings of the toryx leading string
w_2 = the number of windings of the toryx trailing string
λ_1 = wavelength of the toryx leading string
φ_1 = steepness angle of the toryx leading string.

Position-dependent toryx spacetime parameters include:

V_{2r} = rotational velocity of the toryx trailing string
V_{2t} = translational velocity of the toryx trailing string
λ_2 = wavelength of the toryx trailing string
φ_2 = steepness angle of the toryx trailing string.

5.3 Toryx Spacetime Postulates in Absolute Units

The toryx spacetime postulates is a set of three fundamental equations limiting the degrees of freedom of several toryx parameters. This makes possible to establish the relationships between all spacetime parameters of the toryx.

Toryx Spacetime Postulates
(in absolute units of the toryx spacetime parameters)

- The length of one winding of the trailing string L_2 is equal to the length of one winding of the leading string L_1:

$$L_2 = L_1 = 2\pi r_1 \quad (+\infty > r_1 > -\infty) \tag{5.3-1}$$

- The toryx eye radius r_0 is equal to a real positive constant:

$$r_0 = r_1 - r_2 = const. \quad (-\infty < r_1 < +\infty) \tag{5.3-2}$$

- The spiral velocity of the trailing string V_2 is constant and equals to the velocity of light in vacuum c at each point of its spiral path. Its components, the translational velocity V_{2t} and the rotational velocity V_{2r}, relate to the spiral velocity V_2 by the Pythagorean Theorem:

$$V_2 = \sqrt{V_{2t}^2 + V_{2r}^2} = c = const. \quad (-\infty < r_1 < +\infty) \tag{5.3-3}$$

The toryx base frequency f_0 is equal to:

$$f_0 = \frac{c}{2\pi r_0} \tag{5.3-4}$$

The toryx base period T_0 is equal to:

$$T_0 = \frac{2\pi r_0}{c} \qquad\qquad (5.3\text{-}5)$$

In spite of their outmost simplicity, the toryx spacetime postulates provide toryces with amazing capabilities, including:

- A capability to exist in topologically-polarized states as described in Chapter 8.
- A capability to be unified with oppositely polarized toryces as described in Chapter 9.

5.4 Hidden Spacetime Properties of Toryx

Several hidden spacetime properties of toryx arise from Eqs. (5.3-1) and (5.3-2). Let us start from Eq. (5.3-1). This equation is valid within a range of radii of the toryx leading string r_1 extending from positive to negative infinity. Therefore, when r_1 is negative the toryx lengths L_1 and L_2 are also negative. We can easily envision a toryx in which all three parameters r_1, L_1 and L_2 are positive. In that case toryces will appear like the ones shown in Figs. 5.1.1 – 5.1.3. Similarly, according to Eq. (5.3-2), the radius of the toryx trailing string r_2 is positive when the radius of the toryx leading string r_1 is greater than the radius of the toryx eye r_0. But, when the radius r_1 is less than r_0 the radius of the trailing string r_2 becomes negative.

Consequently, as the radius of the toryx leading string r_1 changes from positive to negative infinity, the toryx transforms into four completely different topological shapes. These transformations involve inversions (or turning inside out) of the toryx leading string, trailing string and wavelength of trailing string. Changes of signs of r, r_1, r_2, L_1 and L_2 coincide with changes of the toryx inversion states.

Additional hidden spacetime properties of toryx arise from Eq. (5.3-3) limiting the toryx velocities. You may see it for yourself. This equation tells us that the spiral velocity of the trailing string V_2 is constant and equals to the velocity of light c. At the same time, it says absolutely nothing about the values of its components, the translational velocity V_{2t} and the rotational velocity V_{2r}. Therefore, Eq. (5.3-3) allows these velocities to be subluminal, luminal and superluminal.

Thus, the concept that the velocity of light is the maximum velocity in the universe is applied only to the spiral velocity of the toryx trailing string V_2. But, the components of the spiral velocity of the toryx trailing string, the translational velocity V_{2t} and the rotational velocity V_{2r}, can be either subluminal or superluminal, and can also be expressed with either real or imaginary numbers.

5.5 Toryx Spacetime Postulates in Relative Units

The toryx spacetime postulates are simplified by expressing the toryx spacetime parameters in relative values in respect to the toryx eye radius r_0 and the velocity of light c.

$b = r/r_0$ = toryx relative peripheral radius
b_0 = toryx relative eye radius
$b_1 = r_1/r_0$ = relative radius of the toryx leading string
$b_2 = r_2/r_0$ = relative radius of the toryx trailing string
$l_1 = L_1/r_0$ = relative length of the toryx leading string
$l_2 = L_2/r_0$ = relative length of the toryx trailing string
$t_1 = T_1/T_0$ = relative period of the toryx leading string
$t_2 = T_2/T_0$ = relative period of the toryx trailing string
$u_2 = U_2 / 2\pi^2 r_0^3$ = relative volume of the toryx trailing string
$\beta_1 = V_1/c$ = relative spiral velocity of the toryx leading string
$\beta_{1r} = V_{1r}/c$ = relative rotational velocity of the toryx leading string
$\beta_{1t} = V_{1t}/c$ = relative translational velocity of the toryx leading string
$\beta_2 = V_2/c$ = relative spiral velocity of the toryx trailing string
$\beta_{2r} = V_{2r}/c$ = relative rotational velocity of the toryx trailing string
$\beta_{2t} = V_{2t}/c$ = relative translational velocity of the toryx trailing string
$\delta_1 = f_1/f_0$ = relative frequency of the toryx leading string
$\delta_2 = f_2/f_0$ = relative frequency of the toryx trailing string
$\eta_2 = \lambda_2/r_0$ = relative wavelength of the toryx trailing string.

Toryx Spacetime Postulates in relative values of the toryx spacetime parameters are shown below.

Toryx Spacetime Postulates
(in relative units of the toryx spacetime parameters)

- The relative length of one winding of the trailing string l_2 is equal to the relative length of one winding of the leading string l_1:

$$l_2 = l_1 \quad (+\infty > b_1 > -\infty) \tag{5.5-1}$$

- The toryx relative eye radius b_0 is equal to 1:

$$b_0 = b_1 - b_2 = 1 \quad (+\infty > b_1 > -\infty) \tag{5.5-2}$$

- The relative spiral velocity of the trailing string β_2 is equal to 1 at each point of its spiral path; it relates to the translational velocity β_{2t} and the rotational velocity β_{2r} by the Pythagorean Theorem:

$$\beta_2 = \sqrt{\beta_{2t}^2 + \beta_{2r}^2} = 1 \quad (+\infty > b_1 > -\infty) \tag{5.5-3}$$

Figure 5.5 shows hodographs of toryx relative spacetime parameters expressed in relation to the middle point a of the trailing string (Fig. 5.1.3).

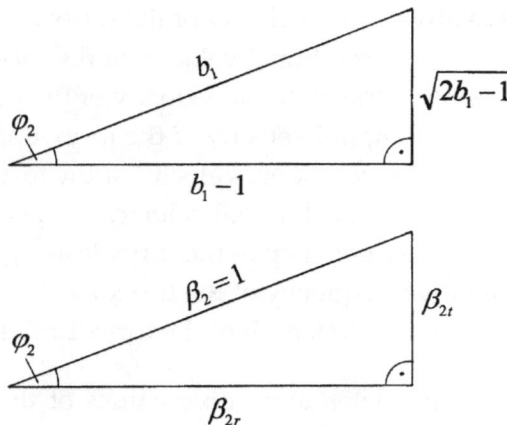

Figure 5.5. Hodographs of the toryx relative spacetime parameters expressed in relation to the middle point a of the trailing string (Fig. 5.1.3).

5.6 Toryx Derivative Spacetime Equations

Table 5.6.1 provides a summary of derivative equations for relative spacetime parameters of the toryx leading and trailing strings as a function of the relative radius of leading string b_1.

Table 5.6.1. Derivative equations for relative spacetime parameters of the toryx leading and trailing strings as a function of the relative radius of leading string b_1.

Relative parameter	Leading string Eq.(a)	Trailing string Eq.(b)
Radius Eq. (5.6-1)	$b_1 = \dfrac{r_1}{r_0}$	$b_2 = \dfrac{r_2}{r_0} = b_1 - 1$
Wavelength Eq. (5.6-2)	$\eta_1 = \dfrac{\lambda_1}{r_0} = 2\pi b_1$	$\eta_2 = \dfrac{\lambda_2}{r_0} = 2\pi\sqrt{2b_1 - 1}$
Length of one winding Eq. (5.6-3)	$l_1 = \dfrac{L_1}{r_0} = 2\pi b_1$	$l_2 = \dfrac{L_2}{r_0} = 2\pi b_1$
Steepness angle Eq. (5.6-4)	$\varphi_2 = 0$	$\cos s\varphi_2 = \dfrac{b_1 - 1}{b_1}$
The number of windings Eq.(5.6-5)	$w_1 = 1$	$w_2 = \dfrac{b_1}{\sqrt{2b_1 - 1}}$
Translational velocity Eq.(5.6-6)	$\beta_{1t} = \dfrac{V_{1t}}{c} = 0$	$\beta_{2t} = \dfrac{V_{2t}}{c} = \dfrac{\sqrt{2b_1 - 1}}{b_1}$
Rotational velocity Eq.(5.6-7)	$\beta_{1r} = \dfrac{V_{1r}}{c} = \dfrac{\sqrt{2b_1 - 1}}{b_1}$	$\beta_{2r} = \dfrac{V_{2r}}{c} = \dfrac{b_1 - 1}{b_1}$
Spiral velocity Eq.(5.6-8)	$\beta_1 = \dfrac{V_1}{c} = \dfrac{\sqrt{2b_1 - 1}}{b_1}$	$\beta_2 = \dfrac{V_2}{c} = 1$
Frequency Eq. (5.6-9)	$\delta_1 = \dfrac{f_1}{f_0} = \dfrac{\sqrt{2b_1 - 1}}{b_1^2}$	$\delta_2 = \dfrac{f_2}{f_0} = \dfrac{1}{b_1}$
Period Eq.(5.6-10)	$t_1 = \dfrac{T_1}{T_0} = \dfrac{b_1^2}{\sqrt{2b_1 - 1}}$	$t_2 = \dfrac{T_2}{T_0} = b_1$

Note: $\cos s\varphi_2$ is the cosine of the spacetime trigonometric function described in Chapter 6.

The toryx relative peripheral radius b and the relative radius of the toryx leading string b_1 are related to one another by the equation:

$$b = 2b_1 - 1 \qquad (5.6\text{-}11)$$

Table 5.6.2 provides a summary of derivative equations for relative spacetime parameters of the toryx leading and trailing strings as a function of the steepness angle of trailing string φ_2.

Table 5.6.2. Derivative equations for relative spacetime parameters of the toryx leading and trailing strings as a function of the steepness angle of trailing string φ_2.

Relative Parameter	Leading string Eq.(a)	Trailing string Eq.(b)
Radius Eq. (5.6-12)	$b_1 = \dfrac{1}{1 - \cos s\varphi_2}$	$b_2 = \dfrac{\cos s\varphi_2}{1 - \cos s\varphi_2}$
Wavelength Eq. (5.6-13)	$\eta_1 = 0$	$\eta_2 = 2\pi\sqrt{\dfrac{1 + \cos s\varphi_2}{1 - \cos s\varphi_2}}$
The number of windings Eq.(5.6-14)	$w_1 = 1$	$w_2 = \dfrac{1}{\sin s\varphi_2}$
Translational velocity Eq.(5.6-15)	$\beta_{1t} = \dfrac{V_{1t}}{c} = 0$	$\beta_{2t} = \sin s\varphi_2$
Rotational velocity Eq.(5.6-16)	$\beta_{1r} = \sin s\varphi_2$	$\beta_{2r} = \cos s\varphi_2$
Spiral velocity Eq.(5.6-17)	$\beta_1 = \sin s\varphi_2$	$\beta_2 = 1$
Frequency Eq. (5.6-18)	$\delta_1 = \sin s\varphi_2(1 - \cos s\varphi_2)$	$\delta_2 = 1 - \cos s\varphi_2$

5.7 The Spacetime Law of Planetary Motion

The first strong indication that the derived toryx spacetime equations were relevant to a real world came from Eq. (5.6-8a). This equation establishes a relationship between the relative velocity of the toryx leading string β_1 and the relative radius of this string b_1 that is rewritten below:

$$\beta_1 = \frac{V_1}{c} = \frac{\sqrt{2b_1 - 1}}{b_1} \qquad (5.7\text{-}1)$$

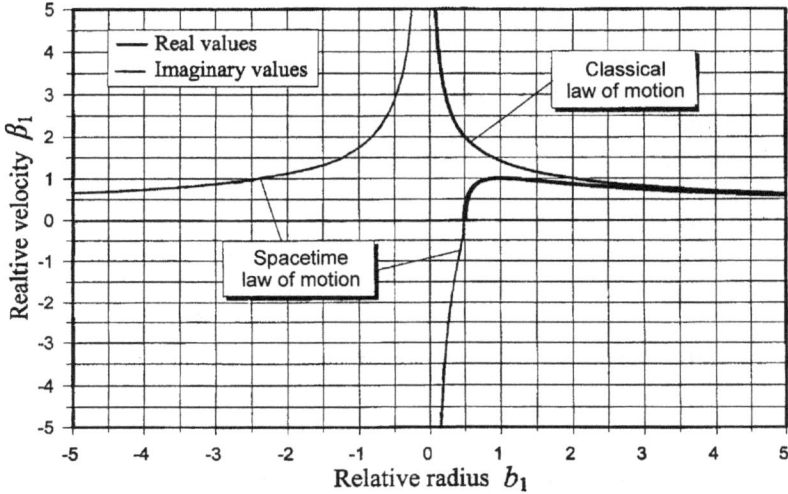

Figure 5.7. The Spacetime Law of Planetary Motion versus the classical law of planetary motion.

For the case when $b_1 \gg 1$, the above equation reduces to the form:

$$\beta_1 = \frac{V_1}{c} = \sqrt{\frac{2}{b_1}} \qquad (5.7\text{-}2)$$

As shown in Chapters 16 and 17, Eq. (5.7-2) yields Kepler's third law of planetary motion. This equation could also be derived by considering a state of equilibrium between electric and centripetal forces defined by classical mechanics. Therefore, the law of planetary motion based on classical mechanics is a particular case of the more general Spacetime Law of Planetary Motion.

Figure 5.7 shows plots of Eqs. (5.7-1) and (5.7-2) expressing respectively the Spacetime and classical laws of planetary motion. The highlights of these plots are:

- The Spacetime Law of Planetary Motion is described by Eq. (5.7-1). It is applied to a range of b_1 extending from negative to positive infinity; within a range of b_1 extending from 0.5 to positive infinity the values of β_1 are expressed with real numbers, while within the remaining range of b_1 with imaginary numbers.

- The classical law of planetary motion is described by Eq. (5.6-2). It is applied to the range of b_1 extending from zero to positive infinity with all values of β_1 expressed with real numbers.

- When $b_1 > 5$, the difference between the values of β_1 calculated based on the Spacetime and classical laws of planetary motion becomes small and it decrease as b_1 increases. As b_1 decreases from 5 to 2, this difference progressively increases.

- According to the classical law of planetary motion, as b_1 decreases from 2 to 0, β_1 sharply increases and approaches positive infinity $(+\infty)$. According to the Spacetime Law of Planetary Motion, as b_1 decreases from 2 to 0, β_1 initially slightly increases and then, after reaching its maximum value of 1 at $b_1 = 1$, it sharply decreases.

5.8 Toryx Fine Structure

As shown in Figs. 5.8.1, fine structure of a negative real toryx toryx is formed by a *standing wave* oscillating inside *toryx fine-structure spherical boundary* with the radius r_3. The standing wave is made up of a leading string with the radius $r_4 \rightarrow \infty$ and a double-helical trailing string with the radius r_5.

The radius of toryx fine-structure spherical boundary r_3 is equal to:

$$r_3 = \sqrt{r_1^2 - r_2^2}$$

$$(5.8\text{-}1)$$

The relative radius of toryx fine-structure spherical boundary b_3 is equal to:

$$b_3 = \eta_2 = \sqrt{2b_1 - 1}$$

$$(5.8\text{-}2)$$

The perimeter $2\pi r_3$ of toryx fine-structure spherical boundary is equal to the wavelength λ_2 of toryx trailing string:

$$2\pi r_3 = \lambda_2 \qquad (5.8\text{-}3)$$

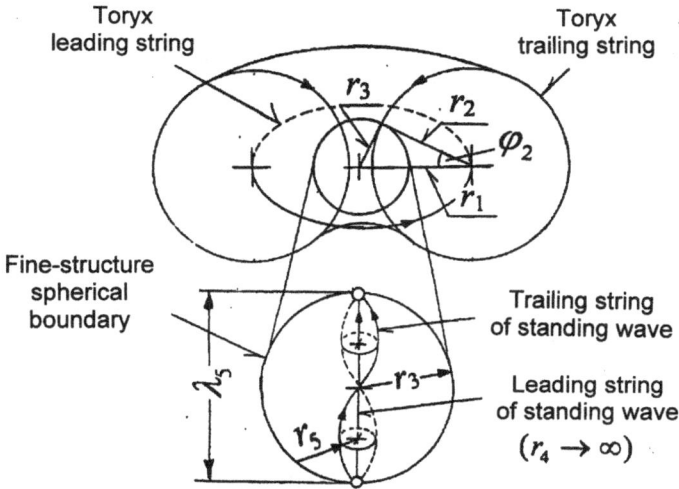

Figure 5.8.1. Fine structure of a real negative toryx.

The radius of leading string of toryx fine-structure standing wave r_4 approaches infinity:

$$r_4 \to \infty \qquad (5.8\text{-}4)$$

The wavelengths of leading and trailing strings of toryx fine-structure standing wave, λ_4 and λ_5, are the same and equal to:

$$\lambda_4 = \lambda_5 = 2r_3 \qquad (5.8\text{-}5)$$

The spiral length of trailing string of toryx fine-structure standing wave L_5 is related to its radius r_5 and wavelength λ_5 by the equation:

$$L_5 = \sqrt{\lambda_5^2 + (2\pi r_5)^2} \qquad (5.8\text{-}6)$$

To define the radius r_5 of trailing string of toryx fine-structure standing wave, it is necessary to introduce the limitation of its degree of freedom shown in Exhibit 5.8.

Exhibit 5.8. Limitation of a degree of freedom of toryx fine-structure standing wave.

The aspect ratio of fine-structure standing wave $\lambda_5/2r_5$ is equal to the toryx quantization constant Λ:

$$\frac{\lambda_5}{2r_5} = \Lambda = const. \tag{5.8-7}$$

From Eqs. (5.8-6) and (5.8-7), the spiral length of trailing string of toryx fine-structure standing wave L_5 is equal to:

$$L_5 = \lambda_5 \frac{\alpha_s^{-1}}{\Lambda} \tag{5.8-8}$$

where α_s^{-1} is the *spacetime inverse fine-structure constant* related to the spacetime quantization constant Λ by the equation:

$$\alpha_s^{-1} = \sqrt{\Lambda^2 + \pi^2} \tag{5.8-9}$$

Notably, Eq. (5.8-9) is similar to the equation proposed by T.J. Burger as an approximation of the inverse fine structure constant. For the case when $\Lambda = 137$, this equation yields the value of $\alpha_s^{-1} = 137.036015720$. The relative difference between this values and the experimental value of the inverse fine structure constant $\alpha^{-1} = 137.035999074$ provided by 2011 CODATA is about one part per ten million.

6

SPIRAL SPACETIME MATH

Equations describing the toryx spacetime parameters are mostly based on the elementary math commonly taught in high schools. However, to reveal all potential spacetime properties of the toryx, it is necessary to modify several aspects of elementary math, including the definitions of zero, number line and elementary trigonometric functions. The modified elementary math is called the *spiral spacetime math.*

6.1 Elementary Number Line

Elementary math expresses polarization by employing positive and negative numbers. Most of us are familiar with an elementary number line. As shown in Figure 6.1, the elementary number line is a set of real numbers n extending along a straight line from zero to the right side towards positive infinity $(+\infty)$ and to the left side towards negative infinity $(-\infty)$.

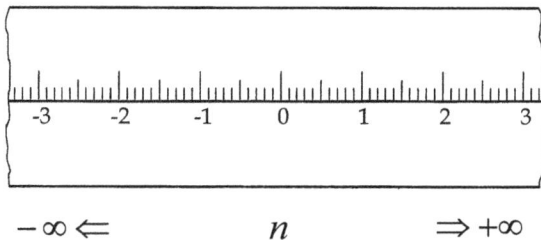

Figure 6.1. Elementary number line.

We use elementary zero as a number in two ways. Firstly, we employ zero for counting of non-divisible entities in the form of an integer immediately preceding 1 in an elementary number line. Secondly, we use zero to represent an absolute absence of any quantity and quality. But the last definition of zero creates a big problem when we try to divide, for instance 1 by ever increasing numbers, such as 10, 100, etc.:

$$1 : 10 = 0.1$$
$$1 : 100 = 0.01$$

$$1 : 1000 = 0.001$$

$$\cdots\cdots\cdots\cdots\cdots$$

$$1 : \infty \quad = 0$$

Thus, in the elementary math zero (0) is mathematically defined as an inverse of infinity (∞). This, however, defies any reasonable logic. Judge yourself. After we divide the ***exact number*** one (1) by an ***uncertain number*** (∞) we will obtain the ***exact number*** zero (0). This also contradicts with our experience. We know that the values of natural entities, however infinitesimally small, never reach absolute zero. For instance, after removing all apples from a basket there will presumably be no, or zero, apples left in there. But this would not be true if we consider a fact that the apples are made up of zillions of molecules. Many of these molecules will still be left in the basket, giving away their characteristic smell.

This is not all. The molecules are made up of atoms, so after the removal of molecules some of their atoms will still be left in the basket. The same will happen with nucleons containing in the atoms and with quarks containing inside nucleons, and so on and so on. Thus, considering a sophisticated structure of apples, it would be impossible to make the basket "absolutely" empty. The above example with apples explains why physicists are not able to reach either the "absolute zero temperature" or the "absolute vacuum." In both cases, it would require to spend an infinite amount of energy.

6.2 Infinility Versus Infinity

The spiral spacetime math clearly separates the two applications of zero described above. The zero is still considered as an integer for counting of non-divisible entities and still retains its old symbol (0). But, in application to the real entities the zero is replaced with a quantity that is infinitely approaching to it. This quantity is called *infinility*, from the "infinite nil." Notably, the term infinility is similar to the known math term *infinitesimal*, but, besides to be much easier to pronounce, the "infinility" also more appropriately defines this term as a counterpart of "infinity." The role of infinility in nature is as important as the role of infinity.

Figure 6.2.1 shows symbolically positive and negative infinities ($\pm\infty$) and also positive and negative infinility (±0) as equal counterparts in respect to the positive and negative unities (±1).

Infinility $\Leftarrow$ $\Rightarrow$ *Infinity*
(±0) $(\pm\ddot{1})$ (±∞)
                Unity

Figure 6.2.1. Infinity $(\pm\infty)$, infinility (± 0) and unity (± 1).

The spiral spacetime math determines infinility in a similar same way that was used by the mathematicians to determine infinity. Most of us learned about infinity in school. Unlike the finite quantities such as 1, 2, 100, 2500, etc., an infinite quantity cannot be completely counted or measured. So, infinity is *larger* than any finite quantity. We can arrive to infinity, for instance, by multiplying a finite quantity 1 by 10, 100, 1000, etc.:

$$1 \times 10 = 10^1 \qquad\qquad 1 \times (-10) = -10^1$$
$$1 \times 100 = 10^2 \qquad\qquad 1 \times (-100) = -10^2$$
$$1 \times 1000 = 10^3 \qquad\qquad 1 \times (-1000) = -10^3$$
$$\cdots\cdots\cdots\cdots \qquad\qquad \cdots\cdots\cdots\cdots\cdots$$
$$\rightarrow +\infty \qquad\qquad\qquad \rightarrow -\infty$$

Infinity can be either real positive $(+\infty)$ or real negative $(-\infty)$. It can also be either imaginary positive $(+\infty i)$ or imaginary negative $(-\infty i)$ where $i = \sqrt{-1}$.

Similarly to infinity, infinility cannot be completely counted or measured. But, unlike infinity, infinility is *smaller* than any finite quantity. We can logically arrive to infinility, for instance, by dividing 1 by 10, 100, 1000, etc.:

$$1 : 10 = 10^{-1} \qquad\qquad 1 : (-10) = -10^{-1}$$
$$1 : 100 = 10^{-2} \qquad\qquad 1 : (-100) = -10^{-2}$$
$$1 : 1000 = 10^{-3} \qquad\qquad 1 : (-1000) = -10^{-3}$$
$$\cdots\cdots\cdots\cdots \qquad\qquad \cdots\cdots\cdots\cdots\cdots$$
$$\rightarrow +0 \qquad\qquad\qquad \rightarrow -0$$

Infinility can be either real positive $(+0)$ or real negative (-0). It can also be either imaginary positive $(+0i)$ or imaginary negative $(-0i)$. Let us summarize the above discussion about infinity and infinility:

$$\text{Real infinility: } \pm 0 = \frac{1}{\pm\infty}; \quad \text{Imaginary infinility: } \pm 0i = \frac{1}{\pm\infty i}$$

$$\text{Real infinity: } \pm\infty = \frac{1}{\pm 0}; \quad \text{Imaginary infinity: } \pm\infty i = \frac{1}{\pm 0i}$$

The replacement of zero with infinility requires us to introduce new definitions of three other conventional geometrical terms, *point, line* and *surface*. According to the conventional geometry (Fig. 6.2.2), the points are zero-dimensional, so they do not have volume, area, length, or any other higher-dimensional analogue. The lines are one-dimensional, so they have no width but can be extended on and on forever in either direction. The surfaces are two-dimensional, so they have width and length, but do not have thickness.

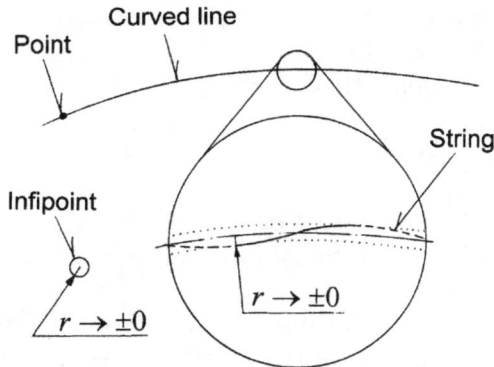

Figure 6.2.2. Infipoint and string versus point and curved line.

The UST replaces conventional definition of the point, line and surface with the respective terms called the *infipoint, string* and *membrane* as summarized below:

- The infipoint is either a circle or a sphere which radius approaches infinility (± 0).

- String is a helical spiral which radius approaches infinility (± 0).

- Membrane is a three-dimensional surface which thickness approaches infinility (± 0).

6.3 Spacetime Trigonometry

Similarly to the elementary trigonometry, the spacetime trigonometry is based on the relationships between the sides of a right triangle. The definitions of elementary trigonometric functions are based on the transformations of the right triangle as a function of the non-right angle φ_2 (Fig. 6.3.1):

$$\cos\varphi_2 = x; \quad \sin\varphi_2 = y; \quad \tan\varphi_2 = y/x$$

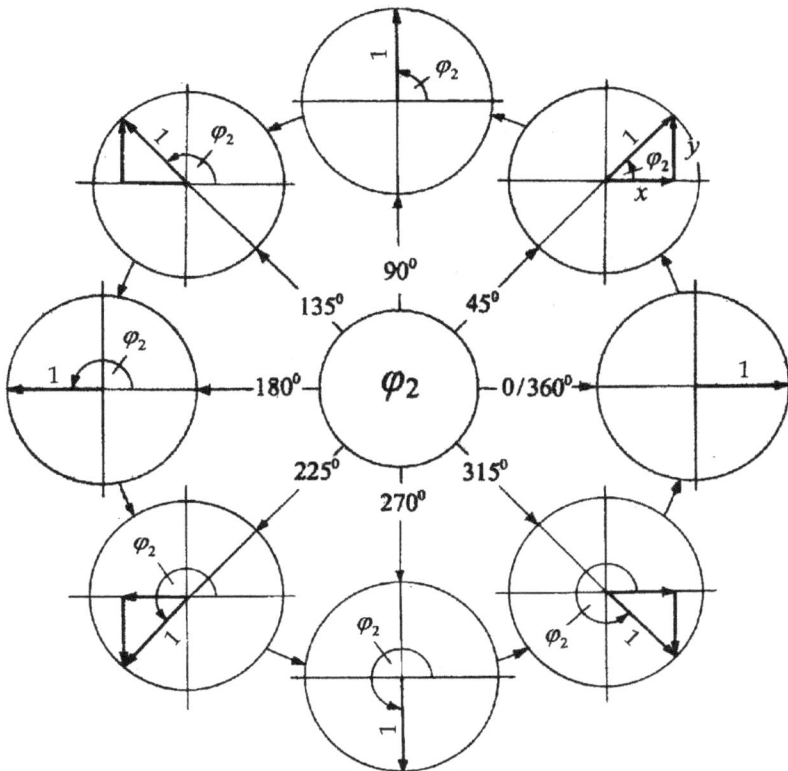

Figure 6.3.1. Transformations of a right triangle corresponding to the elementary trigonometry.

Figure 6.3.1 shows the transformations of the right triangle corresponding to the elementary trigonometry. The main features of these transformations are:

- When the length of the hypotenuse of the triangles is equal to 1, the ranges of the lengths of its sides x and y are between 1 and -1.
- The triangles located at the two left quadrants are the mirror images of the triangles located at the two right quadrants.
- The triangles located at the two bottom quadrants are the mirror images of the triangles located at the two top quadrants.

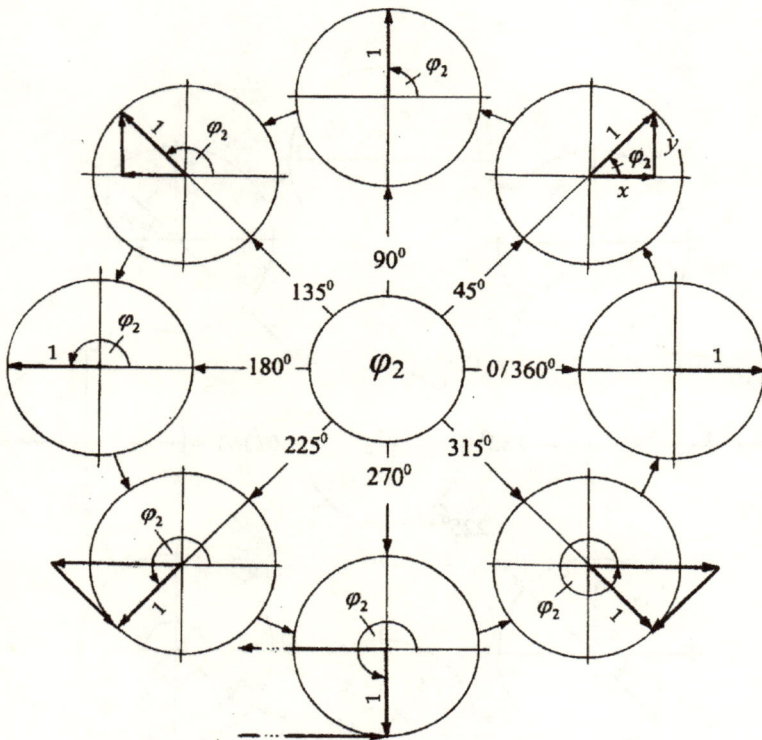

Figure 6.3.2. Transformations of a right triangle corresponding to the spacetime trigonometry.

The spacetime trigonometry employs the following relationship between the spacetime trigonometric function $\cos s\varphi_2$ and the number b_1 that extends from positive infinity $(+\infty)$ to negative infinity $(-\infty)$ as the angle φ_2 changes from 0 to 360^0:

$$\cos s\varphi_2 = x = \frac{b_1 - 1}{b_1} \qquad (0^0 < \varphi_2 < 360^0) \qquad (6.3\text{-}1)$$

The letter "s" at the end of the symbols of the spacetime trigonometric functions differentiates them from the elementary ones. Figure 6.3.2 shows the transformations of the right triangle as a function of the angle φ_2 according to Eq. (6.3-1).

- When the angle φ_2 is between 0 and 180^0, the right triangles in spacetime trigonometry look the same as in elementary trigonometry. Thus, within this range of the angle φ_2 the elementary and spacetime trigonometry are based on the same principle.
- When the angle φ_2 is between 180 and 360^0, the right triangle of the spacetime trigonometry becomes *outverted*. Consequently, the length of its horizontal side x becomes greater than 1, while the length of the other side y is expressed with imaginary numbers.
- When the angle φ_2 approaches 270^0 from the angle smaller than 270^0, the length of its horizontal side x approaches the real positive infinity $(+\infty)$, while the length of the other side y approaches imaginary positive infinity $(+\infty i)$.
- When the angle φ_2 approaches 270^0 from the angle greater than 270^0, the length of its horizontal side x approaches the real negative infinity $(-\infty)$, while the length of the other side y approaches the imaginary negative infinity $(-\infty i)$.
- When the angle φ_2 approaches 360^0 from the angle smaller than 360^0, the length of its horizontal side x approaches 1, while length of the other imaginary side y approaches the imaginary negative infinility $(-0i)$.

Table 6.3 Relationship between the spacetime and elementary trigonometric functions.

Spacetime trigonometry	Elementary trigonometry	
$0^0 < \varphi_2 < 360^0$	$0^0 < \varphi_2 < 180^0$	$180^0 < \varphi_2 < 360^0$
$\cos s\varphi_2$	$\cos \varphi_2$	$\sec \varphi_2 = 1/\cos \varphi_2$
$\sin s\varphi_2$	$\sin \varphi_2$	$-i \tan \varphi_2$

From the Pythagorean Theorem, Eq. (6.3-1), Figs. 6.3.1 and 6.3.2, we obtain the relationships between the spacetime and elementary trigonometric functions shown in Table 6.3.

The spacetime trigonometry yields two kinds of number lines, the *spacetime real number line* and the *spacetime complex number line*. Both of them are circular and present numbers as the functions of the polar angle φ_2.

6.4 Spacetime Real Number Line

As shown in Figure 6.4, in the spacetime real number line, the real numbers V are extended clockwise along a circle from the real positive infinity $(+\infty)$ to the real negative infinity $(-\infty)$. The relationship between the numbers V and the polar angle φ_2 are expressed by the equation:

$$V = -\cos s\varphi_2 \qquad (0^0 < \varphi_2 < 360^0) \qquad\qquad (6.4\text{-}1)$$

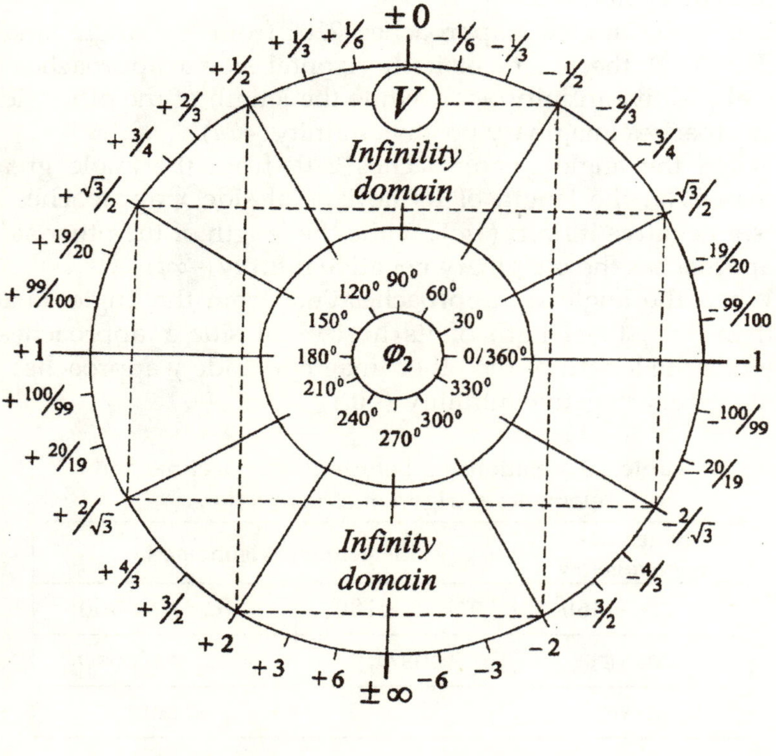

Figure 6.4. Spacetime real number line.

The spacetime real number line is divided into two domains, the *infinility domain* and the *infinity domain*, which occupy equal sectors of the circular number line.

- Infinility domain occupies two top quadrants; it contains the values of V extending clockwise from the real positive unity $(+1)$ and passing through infinility (± 0) to the real negative unity (-1).

- Infinity domain resides in two bottom quadrants; it contains the values of V extending counterclockwise from the real positive unity $(+1)$ and passing through infinity $(\pm\infty)$ to the real negative unity (-1).

Notably, the real positive infinility $(+0)$ merges with the real negative infinility (-0) at $\varphi_2 = 90^0$, while the real negative infinity $(-\infty)$ merges with the real positive infinity $(+\infty)$ at $\varphi_2 = 270^0$.

There are two kinds of symmetries between the numbers V that belong to the four quadrants of circular diagram, the *inverse V-symmetry* and the *reverse V-symmetry*.

- In the inverse V-symmetry, the magnitudes of the numbers V located in the top quadrants are inversed (reciprocated) in respect to the magnitudes of the numbers V located in the bottom quadrants.

- In the reverse V-symmetry, the numbers V located in the right quadrants and the left quadrants have the same magnitudes but reversed signs.

6.5 Spacetime Complex Number Line

The spacetime complex number line can be either positive or negative. Shown in Fig. 6.5 is the positive complex number line in which the real and imaginary numbers R are extended counterclockwise along a circle from the real positive infinity $(+\infty)$ to the imaginary positive infinity $(+\infty i)$.

The relationship between the numbers R and the polar angle φ_2 of the spacetime number line is described by the equation:

$$R = \pm\sqrt{\frac{1 + \cos s\varphi_2}{1 - \cos s\varphi_2}} \tag{6.5-1}$$

The spacetime complex number line is divided into two domains, the *infinity domain* and the *infinility domain*, which occupy equal sectors of the circular number line.

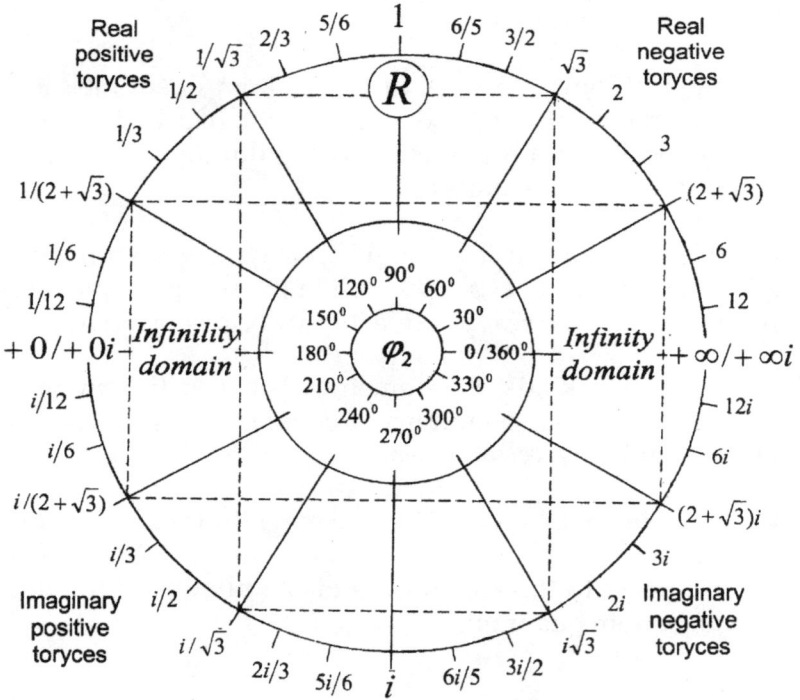

Figure 6.5. Spacetime positive complex number line.

- <u>Infinility domain</u> occupies two left quadrants; it contains the values of R extending counterclockwise from the real positive unity (1) and passing through infinility $(+0/+0i)$ to the imaginary positive unity (i).

- <u>Infinity domain</u> resides in two right quadrants; it contains the values of R extending clockwise from the real positive unity (1) and passing through the real positive and imaginary positive infinities $(+\infty/+\infty i)$ to the imaginary positive unity (i).

Notably, the real positive infinility (+0) merges with the imaginary positive infinility (+0i) at $\varphi_2 = 180^\circ$, while the real positive infinity (+∞) merges with the imaginary positive infinity (+∞i) at $\varphi_2 = 360^\circ$.

There are two kinds of symmetries between the numbers R that belong to the four quadrants of circular diagram, the *inverse R-symmetry* and the *reverse R-symmetry*.

- In the inverse R-symmetry, the magnitudes of the numbers R located in the left quadrants are inversed (reciprocated) in respect to the magnitudes of the numbers R located in the right quadrants.

- In the reverse R-symmetry, the numbers R located in the top and bottom quadrants have the same magnitudes but reversed realities and signs.

6.6 Application of Spiral Spacetime Math in the UST

As we will describe in the subsequent chapters, the spiral spacetime math is directly applied to the description of the toryx spacetime parameters:

- The numbers b_1 in Eq. (6.3-1) of the spacetime trigonometry are equal to the relative radii of the toryx leading string.
- The numbers V in Eq. (6.4-1) of the real number line are equal to the *toryx vorticity V*.
- The numbers R in Eq. (6.5-1) of the complex number line are equal to the *toryx reality R*.
- The polar angle φ_2 of the spacetime trigonometry used in the spacetime real and complex number lines is equal to the steepness angle of the toryx trailing string φ_2.

Figure 6.6 shows the application of the spiral spacetime math for the calculation of the relative velocities of the toryx trailing string corresponding to the middle point of the trailing string as its steepness angle φ_2 increases from 0 to 360°. In each right triangle of velocities of trailing string one side represents the relative translational velocity β_{2t} and the other side the relative rotational velocity β_{2r}, while its hypotenuse represents the relative spiral velocity $\beta_2 = 1$.

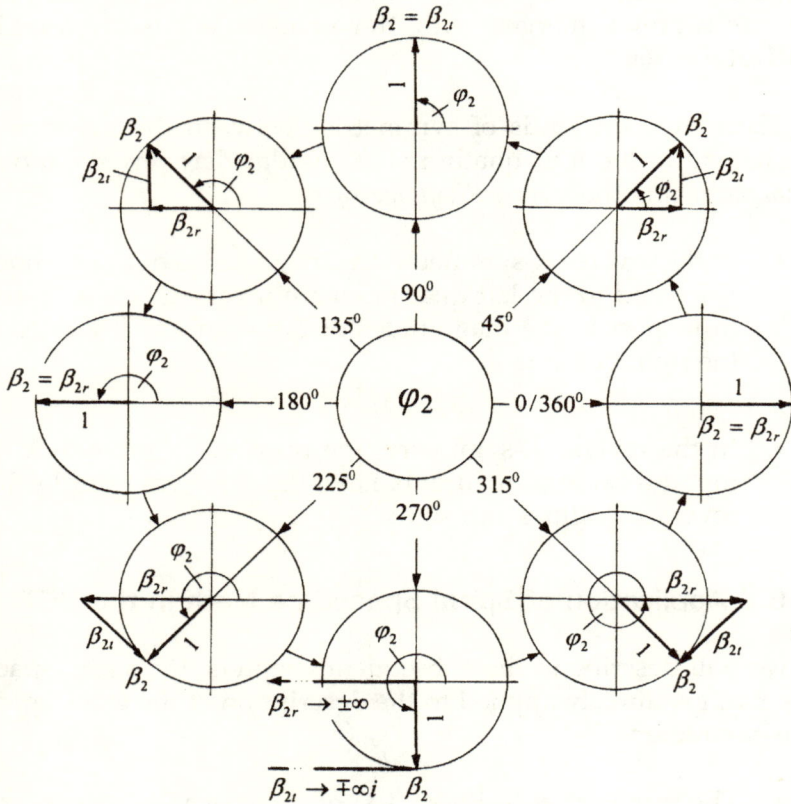

Figure 6.6. Transformations of the relative velocities of the toryx trailing string at the middle point of trailing string.

There is a clear similarity between the transformations of the velocities of the toryx trailing string shown in Figure 6.6 and the transformations of a right triangle corresponding to the spacetime trigonometry shown in Figure 6.3.2.

7

TRENDS OF TORYX SPACETIME PARAMETERS

Two integrated space-time parameters of toryces define their main classification: the *toryx reality R* and the *toryx vorticity V*.

7.1 Toryx Reality

The toryx reality R is equal to the ratio of the radius of spherical membrane r_3 to the toryx eye radius r_0:

$$R = \frac{r_3}{r_0} = b_3 = \pm\sqrt{b} = \pm\sqrt{2b_1 - 1} = \pm\sqrt{\frac{1 + \cos s\varphi_2}{1 - \cos s\varphi_2}} \qquad (7.1\text{-}1)$$

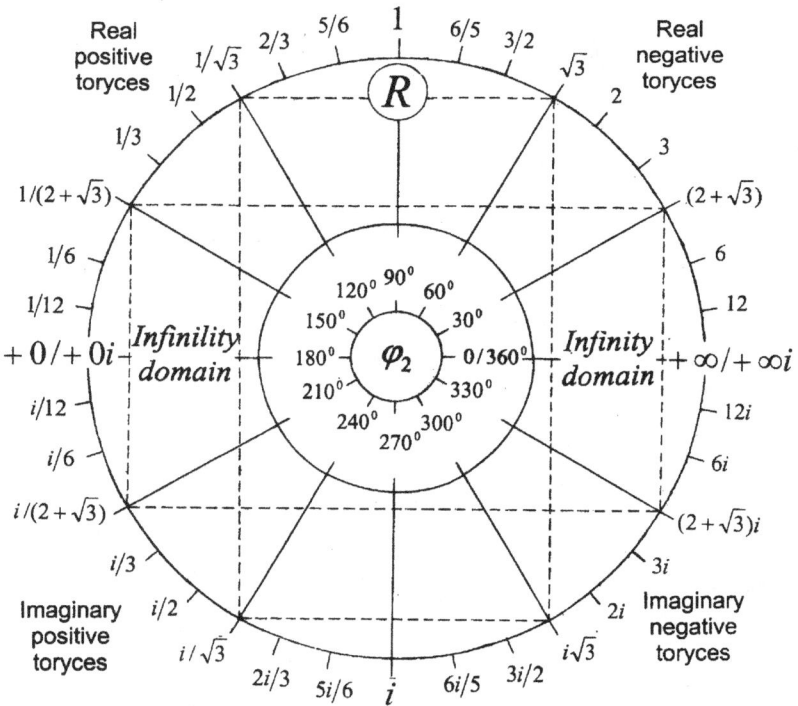

Figure 7.1. Toryx positive reality R as a function of the steepness angle of the toryx trailing string φ_2.

Figure 7.1 shows a circular diagram of the toryx reality R as a function of the steepness angle of the trailing string φ_2. This diagram can be treated as the spacetime complex number line shown in Figure 6.5.

7.2 Toryx Vorticity

The toryx vorticity V is equal to the ratio of the radius of the toryx trailing string r_2 to the radius of the toryx leading string r_1 with the opposite sign:

$$V = -\frac{r_2}{r_1} = -\frac{b_2}{b_1} = -\frac{b_1 - 1}{b_1} = -\frac{b - 1}{b + 1} = -\cos s\varphi_2 \qquad (7.2\text{-}1)$$

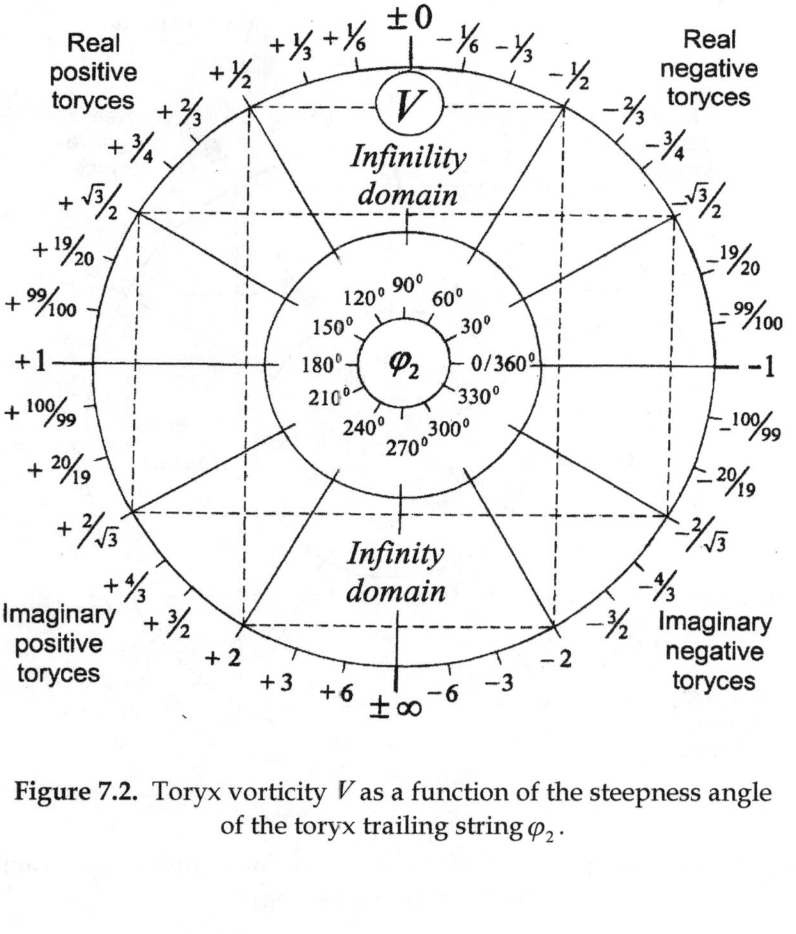

Figure 7.2. Toryx vorticity V as a function of the steepness angle of the toryx trailing string φ_2.

Figure 7.2 shows a circular diagram of the toryx vorticity V as a function of the steepness angle of the toryx trailing string φ_2. This diagram can be treated as the spacetime real number line shown in Figure 6.4.

7.3 Main Classification of Toryces

Toryces are divided into four main groups according to their reality R and vorticity V shown in circular diagrams of Figs. 7.1 and 7.2 and summarized in Table 7.3.

Real negative toryces – The real negative toryces are located in the top right quadrants of the circular diagrams. In these toryces, the toryx realities R are expressed with real numbers, and the toryx vorticities are expressed with negative numbers.

Real positive toryces – The real positive toryces are located in the top left quadrants of the circular diagrams. In these toryces, the toryx realities R are expressed with real numbers, and the toryx vorticities are expressed with positive numbers.

Imaginary positive toryces – The imaginary positive toryces are located in the bottom left quadrants of the circular diagrams. In these toryces, the toryx realities R are expressed with imaginary numbers, and the toryx vorticities V are expressed with positive numbers.

Imaginary negative toryces – The imaginary negative toryces are located in the bottom right quadrants of the circular diagrams. In these toryces, the toryx realities R are expressed with imaginary numbers, and the toryx vorticities V are expressed with negative numbers.

Table 7.3. Main classification of toryces.

Toryx name	φ_2	R	V
Real negative	0^0 - 90^0	real	$(-)$
Real positive	90^0 - 180^0	real	$(+)$
Imaginary positive	180^0 - 270^0	imaginary	$(+)$
Imaginary negative	270^0 - 360^0	imaginary	$(-)$

We show below the trends of the toryx spacetime parameters as a function of the steepness angle of its trailing string φ_2. In the graphs, the real and imaginary parameters are depicted with thick and thin lines respectively. Also shown in each graph are the values of the toryx spacetime parameters corresponding to the steepness angles $\varphi_2 = 0^0$, 90^0, 180^0, 270^0 and 360^0.

7.4 Toryx Peripheral Radius

The toryx relative peripheral radius b is given by the equation:

$$b = \frac{1 + \cos s\varphi_2}{1 - \cos s\varphi_2} \qquad (0^0 < \varphi_2 < 360^0) \qquad (7.4\text{-}1)$$

Figure 7.4 shows a plot of the above equation.

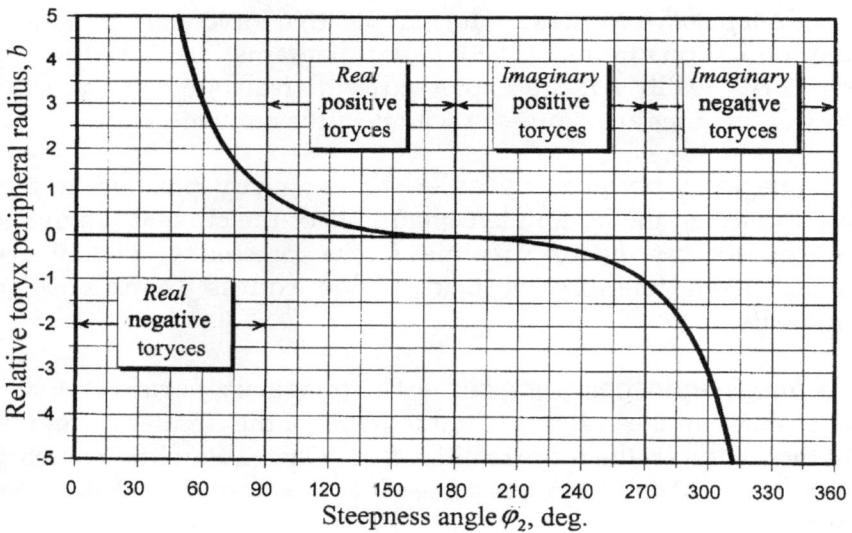

φ_2	$360/0^0$	90^0	180^0	270^0
b	$-\infty/+\infty$	$+1$	$+0/-0$	-1

Figure 7.4. Toryx relative peripheral radius b as a function of the steepness angle of trailing string φ_2.

7.5 Radii of Leading & Trailing Strings

The relative radii of the leading string b_1 and the trailing string b_2 are given by the equations:

$$b_1 = \frac{1}{1 - \cos s\varphi_2} \qquad (0^0 < \varphi_2 < 360^0) \qquad (7.5\text{-}1)$$

$$b_2 = \frac{\cos s\varphi_2}{1 - \cos s\varphi_2} \qquad (0^0 < \varphi_2 < 360^0) \qquad (7.5\text{-}2)$$

Figure 7.5 shows a plot of the above equations.

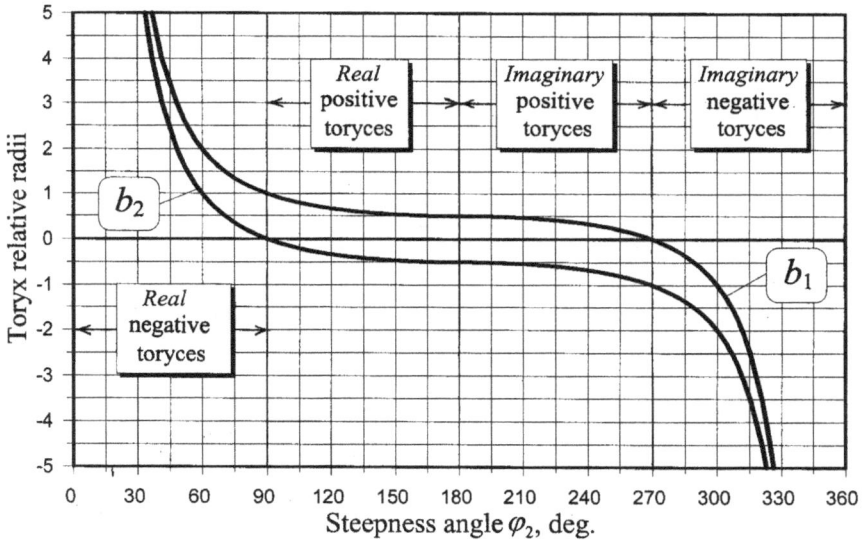

φ_2	$360/0^0$	90^0	180^0	270^0
b_1	$-\infty/+\infty$	$+1$	$+\frac{1}{2}$	$+0/-0$
b_2	$-\infty/+\infty$	$+0/-0$	$-\frac{1}{2}$	-1

Figure 7.5. Relative radii of the leading and trailing strings, b_1 and b_2, as a function of the steepness angle of trailing string φ_2.

7.6 Wavelength of Trailing String

The relative wavelength of the trailing string η_2 is given by the equation:

$$\eta_2 = \pm 2\pi \sqrt{\frac{1 + \cos s\varphi_2}{1 - \cos s\varphi_2}} \quad (0^\circ < \varphi_2 < 360^\circ) \qquad (7.6\text{-}1)$$

Figure 7.6 shows a plot of the above equation.

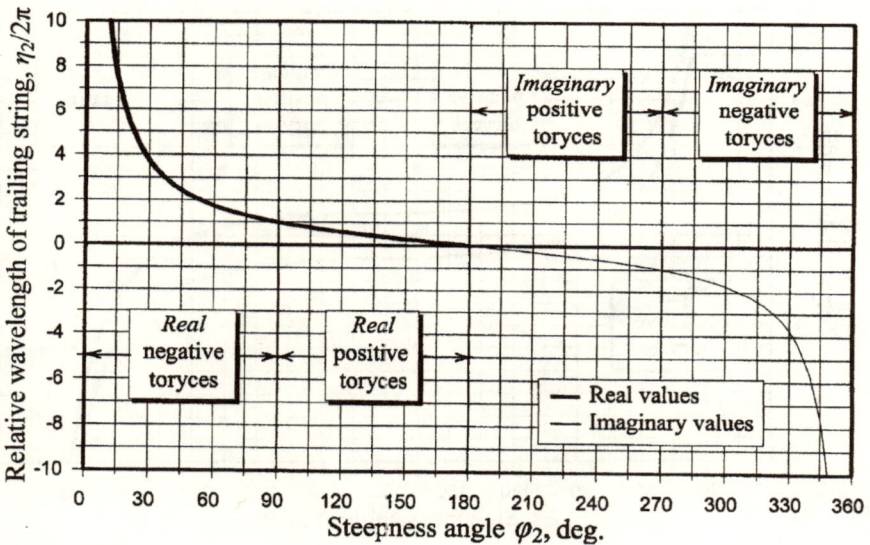

φ_2	$360/0^0$	90^0	180^0	270^0
$\eta_2/2\pi$	$-\infty i\,/+\infty$	$+1$	$+0/-0i$	$-i$

Figure 7.6. Relative wavelength of the trailing string η_2 as a function of the steepness angle of the trailing string φ_2.

7.7 The Number of Windings of Trailing String

The number of windings of the trailing string w_2 is given by the equation:

$$w_2 = \frac{1}{\sin s\varphi_2} \qquad (0^0 < \varphi_2 < 360^0) \qquad (7.7\text{-}1)$$

Figure 7.7 shows a plot of the above equation.

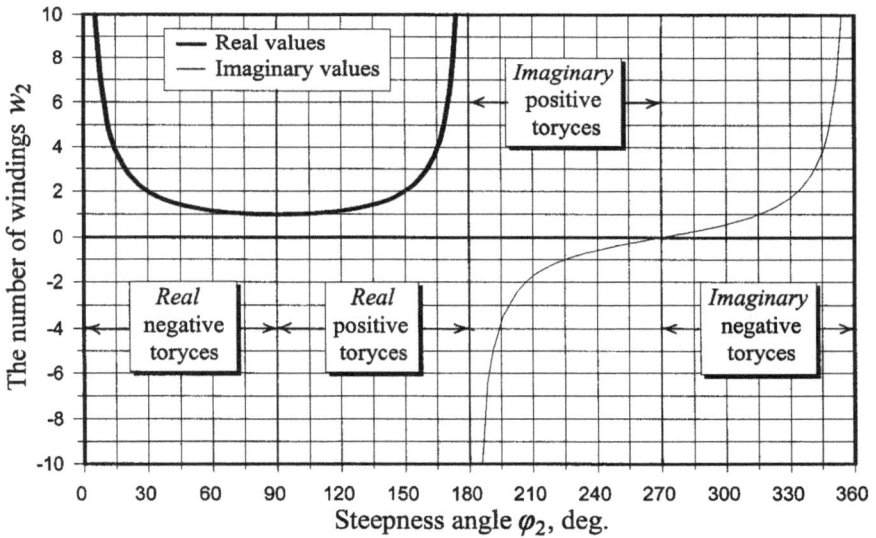

φ_2	$360/0^0$	90^0	180^0	270^0
w_2	$+\infty i/+\infty$	$+1$	$+\infty/-\infty i$	$-0i/+0i$

Figure 7.7. The number of windings of the trailing string w_2 as a function of the steepness angle of the trailing string φ_2.

7.8 Translational Velocity of Trailing String

The relative translational velocity of the trailing string β_{2t} is given by the equation:

$$\beta_{2t} = \sin s\varphi_2 \quad (0^0 < \varphi_2 < 360^0) \qquad (7.8\text{-}1)$$

Figure 7.8 shows a plot of the above equation.

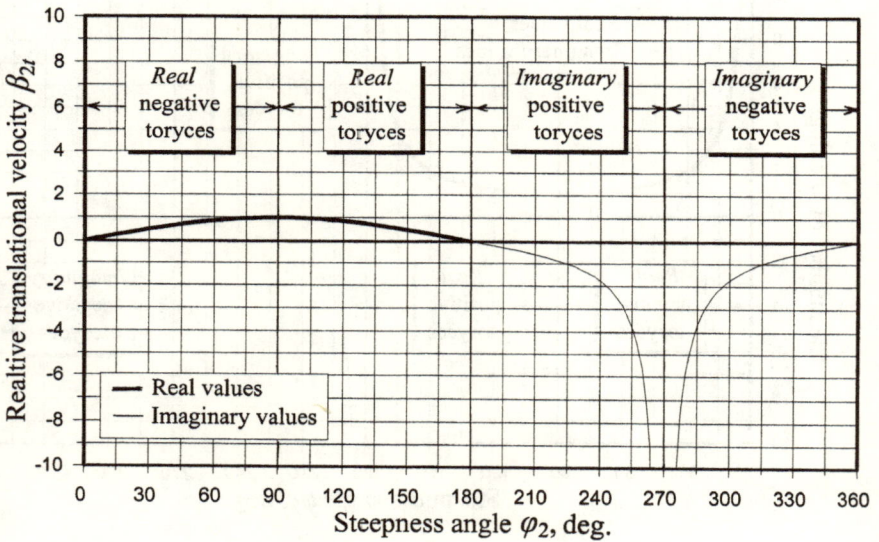

φ_2	$360/0^0$	90^0	180^0	270^0
β_{2t}	$-0i/+0$	$+1$	$+0/-0i$	$-\infty i$

Figure 7.8. Relative translational velocity of the trailing string β_{2t} as a function of the steepness angle of the trailing string φ_2.

7.9 Rotational Velocity of Trailing String

The relative rotational velocity of the trailing string β_{2r} is given by the equation:

$$\beta_{2r} = \cos s\varphi_2 \qquad (0^0 < \varphi_2 < 360^0) \qquad (7.9\text{-}1)$$

Figure 7.9 shows a plot of the above equation.

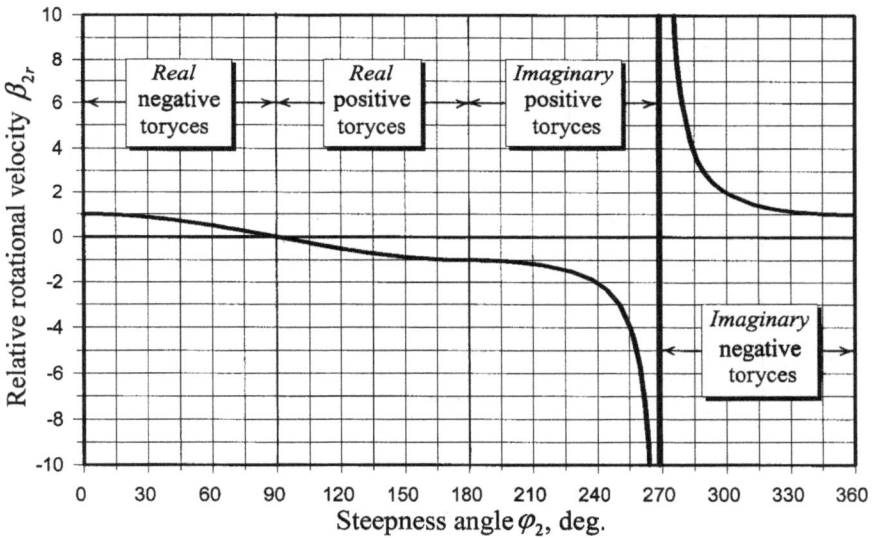

φ_2	$360/0^0$	90^0	180^0	270^0
β_{2r}	$+1$	$+0/-0$	-1	$-\infty/+\infty$

Figure 7.9. Relative rotational velocity of the trailing string β_{2r} as a function of the steepness angle of the trailing string φ_2.

7.10 Frequency of Leading String

The relative frequency of the leading string δ_1 is given by the equation:

$$\delta_1 = \sin s\varphi_2 (1 - \cos s\varphi_2) \quad (0^0 < \varphi_2 < 360^0) \qquad (7.10\text{-}1)$$

Figure 7.10 shows a plot the above equation.

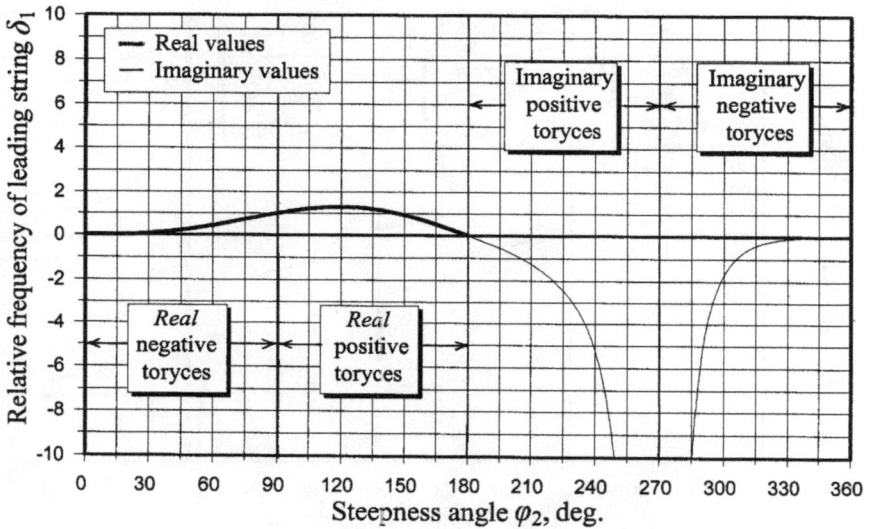

φ_2	$360/0^0$	90^0	180^0	270^0
δ_1	$-0i/+0$	$+1$	$+0/-0i$	$-\infty i$

Figure 7.10. Relative frequency of the leading string δ_1 as a function of the steepness angle of the trailing string φ_2.

7.11 Frequency of Trailing String

The relative frequency of the trailing string δ_2 is given by the equation:

$$\delta_2 = 1 - \cos s\varphi_2 \quad (0^0 < \varphi_2 < 360^0) \tag{7.11-1}$$

Figure 7.11 shows a plot of the above equation.

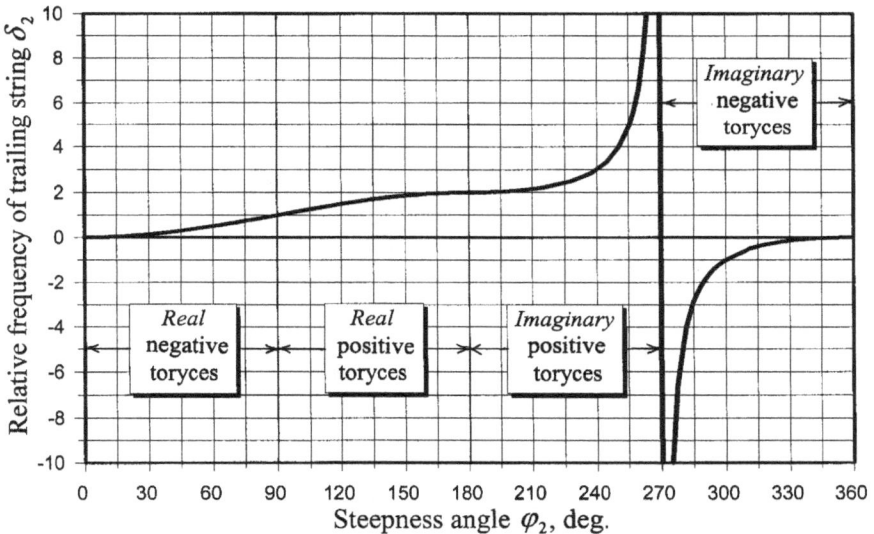

φ_2	$360/0^0$	90^0	180^0	270^0
δ_2	$-0/+0$	$+1$	$+2$	$+\infty/-\infty$

Figure 7.11. Relative frequency of the trailing string δ_2 as a function of the steepness angle of the trailing string φ_2.

7.12 Instant Velocities of Trailing String

To maintain space-time integrity of the toryx while its trailing string propagates at a constant relative spiral velocity $\beta_2 = 1$, the relative translational and rotational velocities, β_{2t} and β_{2r} must vary in a very specific way.

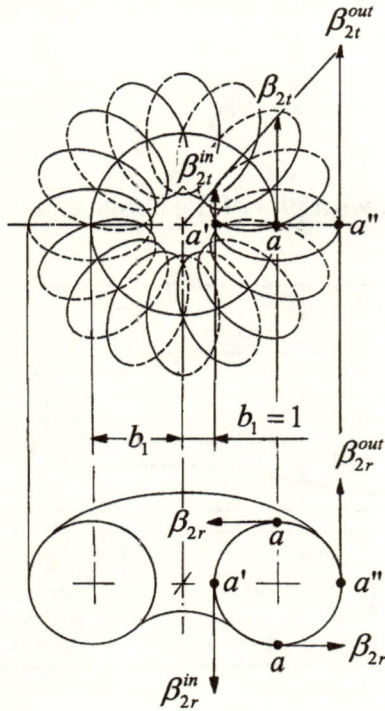

Figure 7.12.1. Relative peripheral velocities of the trailing string.

As shown in Fig. 7.12.1, the translational velocity at each point of the trailing string must be proportional to the distance of said point from the toryx center. At the same time, the rotational velocity of the trailing string must vary to satisfy the fundamental equation (5.6-3), according to which the trailing string must propagate at a constant relative spiral string velocity $\beta_2 = 1$.

Table 7.12.1 shows equations for the relative peripheral translational and rotational velocities of the trailing string. Within the ranges of b_1 (7.12-5) and (7.12-6), the inner and outer peripheral translational velocities β_{2t}^{in} and β_{2t}^{out} of the trailing string exceed velocity of light.

Figure 7.12.2 shows the variation of the relative instant translational and rotational velocities, β_{2t}^{i} and β_{2r}^{i} for the case when $b_1 = 2$. Notably, between 0.25 and 0.75 portions of the cycle of the trailing string, β_{2t}^i exceeds velocity of light, while β_{2r}^i is expressed with imaginary numbers.

Table 7.12.1. Relative peripheral velocities of trailing string.

Velocities at the point a'		Velocities at the point a''	
$\beta_{2t}^{in} = \dfrac{\sqrt{2b_1 - 1}}{b_1^2}$	(7.12-1)	$\beta_{2t}^{out} = \dfrac{(2b_1 - 1)^{3/2}}{b_1^2}$	(7.12-2)
$\beta_{2r}^{in} = \dfrac{\sqrt{b_1^4 - 2b_1 + 1}}{b_1^2}$	(7.12-3)	$\beta_{2r}^{out} = \dfrac{\sqrt{b_1^4 - (2b_1 - 1)^3}}{b_1^2}$	(7.12-4)
$0.544 < b_1 < 1.0$	(7.12-5)	$1.0 < b_1 < 6.222$	(7.12-6)

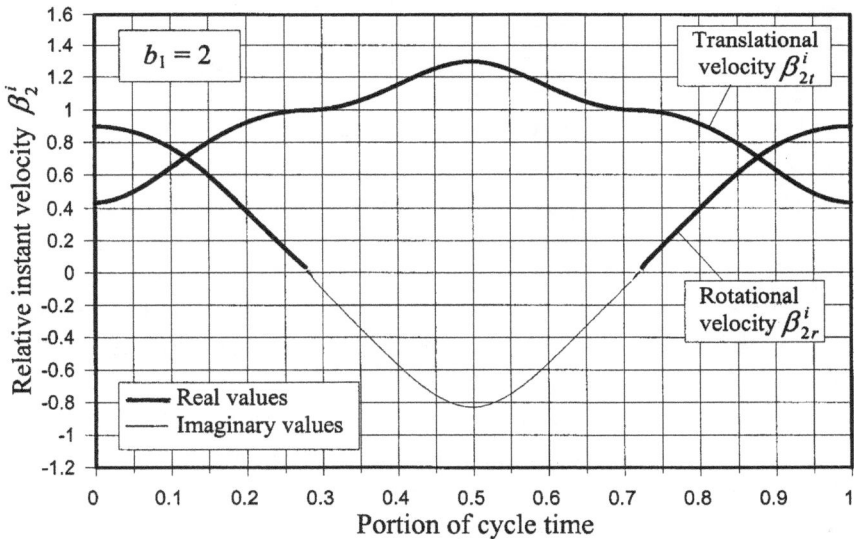

Figure 7.12.2. Variation of instant velocities β_{2t}^i and β_{2r}^i of the trailing string within one cycle of the trailing string.

7.13 Toryx Golden Polarization Factor

The toryx golden polarization factor G_t is equal to the product of the toryx vorticity V and the toryx reality R with an opposite sign. From Eqs. (7.1-1) and (7.2-1) this factor is defined by the equation:

$$G_t = -VR = \frac{(b_1 - 1)\sqrt{2b_1 - 1}}{b_1} \qquad (7.13\text{-}1)$$

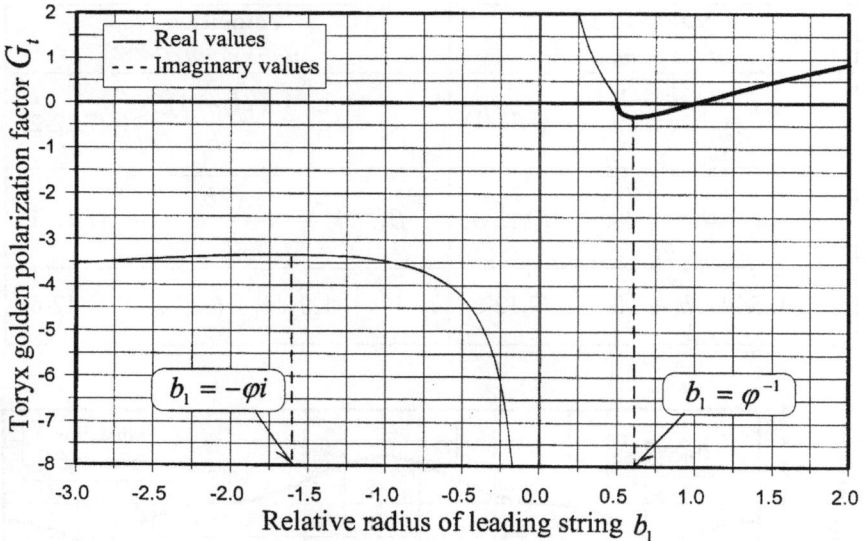

Figure 7.13. Toryx golden polarization factor G_t as a function of the relative radius of the leading string b_1.

Figure 7.13 shows a plot of Eq. (7.13-1) in which the toryx golden polarization factor G_t is expressed as a function of the relative radius of the leading string b_1.

Notably, the plot of G_t has two extreme values of G_t at which b_1 is directly related to the golden ratio φ. The first extreme value of G_t corresponds to $b_1 = -\varphi i$; here the value of G_t is maximum. The second extreme value of G_t corresponds to $b_1 = \varphi^{-1}$; here the value of G_t is minimum.

Your Notes

8

INVERSION STATES OF TORYX

A branch of sciences that studies inversion of curved spaces is called *topology*. One of the most famous topologists of the 19th century was the German mathematician and astronomer Ferdinand Möbius (1790-1868). To illustrate the idea of inverted surface, Möbius described a topological shape that became known as the *Möbius strip* in a paper discovered only after his death. Take a piece of paper, rotate one end through 180 degrees and connect the ends together (Fig. 8).

Outverted
side

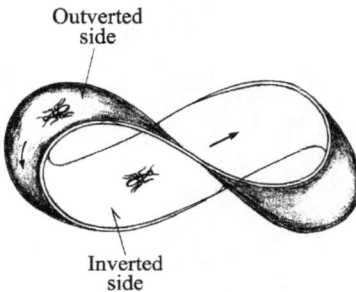

Figure 8.
An ant walking along
the Möbius strip.

Inverted
side

An ant walking on the inside part of the Möbius strip will eventually end up on the outer part of the strip. Thus, the Möbius strip makes a smooth transition from the inverted surface to the surface turned inside out. For the ant, the Möbius strip is a two-dimensional surface. Let us now consider the inversions of the toryx leading string, trailing string and peripheral circle.

8.1 Inversion of Leading String

To illustrate the inversion of its circular leading string let us envision it in the form of an extremely thin and narrow circular ribbon with the relative radius b_1 (Fig. 8.1). Let us assume that the leading string propagates outward when b_1 decreases from positive infinity $(+\infty)$. In that case, the outer color of the ribbon is black, while its inner color is white. As the ribbon radius b_1 decreases while remaining positive, the ribbon begins to appear smaller, but still propagating outward until b_1 reduces to positive infinility $(+0)$. Then the leading string inverts making b_1 approaching negative infinity (-0). As the negative values of b_1 increase and extend to negative infinity $(-\infty)$, its outer color of the leading string appears white, while its inner color looks black.

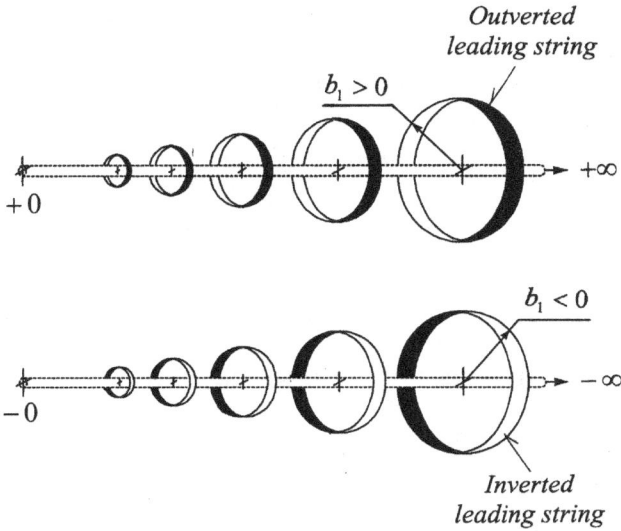

Fig. 8.1. Inversion of a leading string.

8.2 Inversion of Trailing String

To illustrate the inversion of the toryx trailing string let us envision it in the form of an extremely thin and narrow toroidal ribbon (Fig. 8.2).

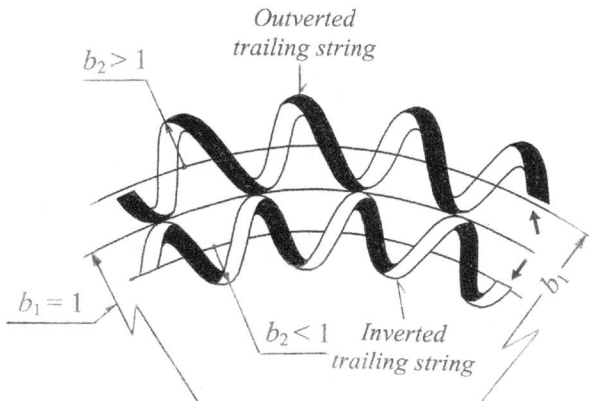

Fig. 8.2. Inversion of a trailing string.

We assume that the trailing string propagates outward when the relative radius of the leading string b_1 is greater than 1. In this case, the radius of the trailing string b_2 is positive and the outer color of the toroidal ribbon is black. When b_1 becomes infinitesimally close to 1, the radius of trailing string b_2 approaches positive infinility (+0). So, the toryx reduces to a circle seen in Figure 8.2 as a colorless edge of a circular ribbon with the radius $b_1 = 1$. When b_1 becomes less than 1, the sign of the radius b_2 changes from positive to negative and the ribbon becomes inverted. Consequently, its outer color becomes white while its inner color becomes black.

8.3 Inversion of Spherical Membrane

Visualize an extremely thin spherical membrane with the relative radius b_3 (Fig. 8.3).

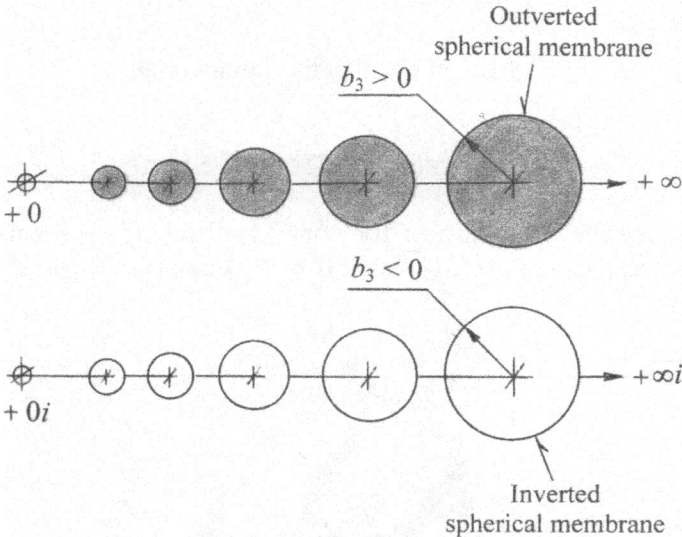

Fig. 8.3. Inversion of a spherical membrane.

Let us start from the outverted spherical membrane in which b_3 decreases from positive infinity (+∞). In that case, its outer color is assumed to be grey, while its inner color is white. As b_3 decreases while remaining positive, the spherical membrane begins to appear smaller, and it eventually reduces, so b_3 approaches real positive infinility $(+0)$. Then the spherical membrane inverts, so b_3 approaches imaginary positive infinility $(+0i)$. As the imaginary positive values of b_3 increase and extend to imaginary positive infinity $(+\infty i)$, the outer color of the spherical membrane appears white, while its inner color looks grey.

8.4 Inversion of Toryces

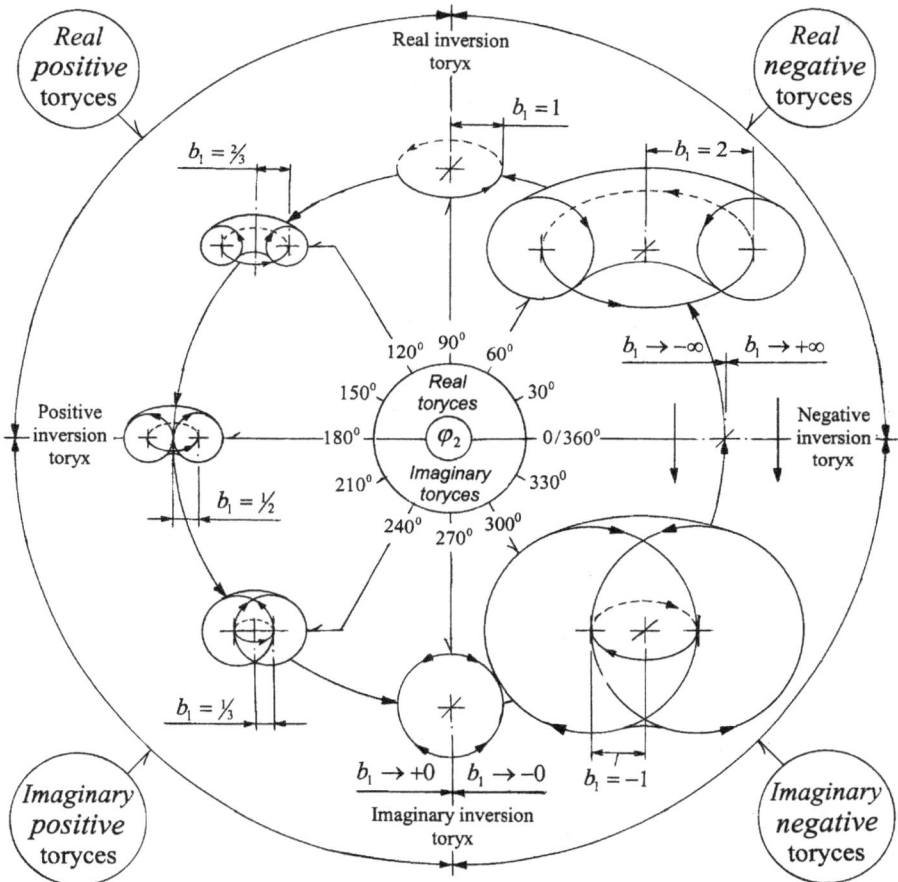

Figure 8.4. Metamorphoses of the toryx leading and trailing strings as a function of the steepness angle of the trailing string φ_2.

Figure 8.4 shows metamorphoses of the toryx leading string with the relative radius b_1 and the trailing strings with the relative radius b_2 as the steepness angle of the toryx trailing string φ_2 increases from 0^0 to 360^0.

Four kinds of *inversion toryces* are located at the boundaries of the circular diagram between the four main groups of toryces.

- Negative inversion toryx ($\varphi_2 \to +0^0/360^0$)- At this point, $b_1 \to \pm\infty$ and $b_2 \to \pm\infty$. Consequently, the toryx appears as two parallel lines separated by the distance equal to the diameter of real inversion string. At $\varphi_2 = +0^0/360^0$, both leading and trailing strings become inverted.

- Real inversion toryx ($\varphi_2 \to 90^0$) – At this point, $b_1 \to +1$, $b_2 \to \pm 0$. Consequently, the toryx appears as a circle with the relative radius $b_1 \to +1$. At $\varphi_2 = 90^0$, only the trailing string becomes inverted.

- Positive inversion toryx ($\varphi_2 \to 180^0$) – At this point, $b_1 \to +\frac{1}{2}$ and $b_2 \to -\frac{1}{2}$. Consequently, the toryx appears as an extreme case of a *spindle torus* with the inner parts of its windings touching one another. At $\varphi_2 = 180^0$, only the wavelength of the toryx trailing string becomes inverted.

- Imaginary inversion toryx ($\varphi_2 \to 270^0$) – At this point, $b_1 \to \pm 0$, $b_2 \to -1$. Consequently, and the toryx appears as a circle with the relative radius approaching -1. The circle is located at the plane perpendicular to the plane of the real inversion string. At $\varphi_2 = 270^0$, only the leading string becomes inverted.

8.5 Transformations of Toryces within Each Group

Located between the inversion toryces on the circular diagram of Figure 8.4 are the toryces that belong to their four main groups.

Figures 8.5.1 – 8.5.4 show the transformations of the toryces within each main group.

Real negative toryces (Fig. 8.5.1) - These toryces belong to the top right quadrant of the circular diagram shown in Figure 8.4. The trailing strings of these toryces are wound counter-clockwise outside of the real inversion string. As φ_2 increases, both b_1 and b_2 decrease, so that the trailing string appears like a conventional torus.

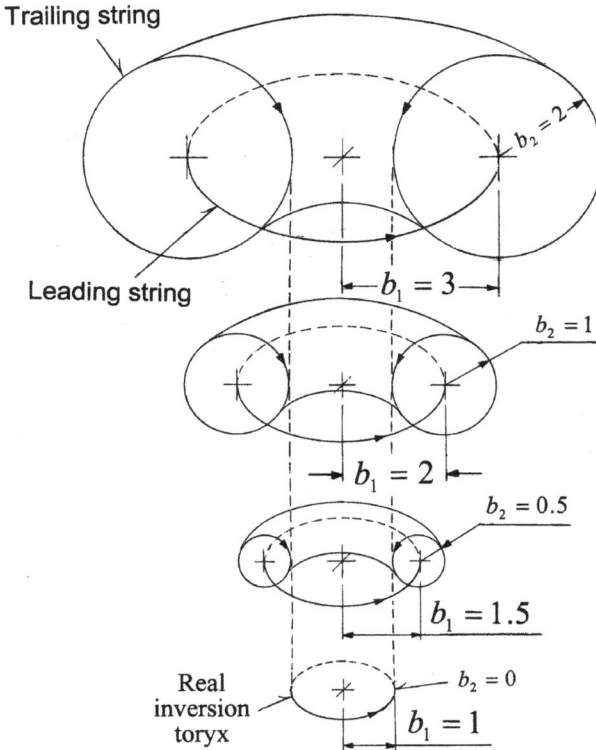

Range	φ_2	b	b_1	b_2	$\eta_2/2\pi$	w_2	β_{2t}	β_{2r}
From	0^0	$+\infty$	$+\infty$	$+\infty$	$+\infty$	$+\infty$	$+0$	1.0
To	90^0	1.0	1.0	$+0$	1.0	1.0	1.0	$+0$

Figure 8.5.1. Transformation of the real negative toryx.

Real positive toryces (Fig. 8.5.2) - These toryces belong to the top left quadrant of the circular diagram shown in Figure 8.4. Within this range the trailing string is inverted, so that its windings are now wound clockwise inside the real inversion toryx. As φ_2 increases, b_1 decreases while the negative value of b_2 increases. Consequently, the toryx appears as an inverted toroidal spiral.

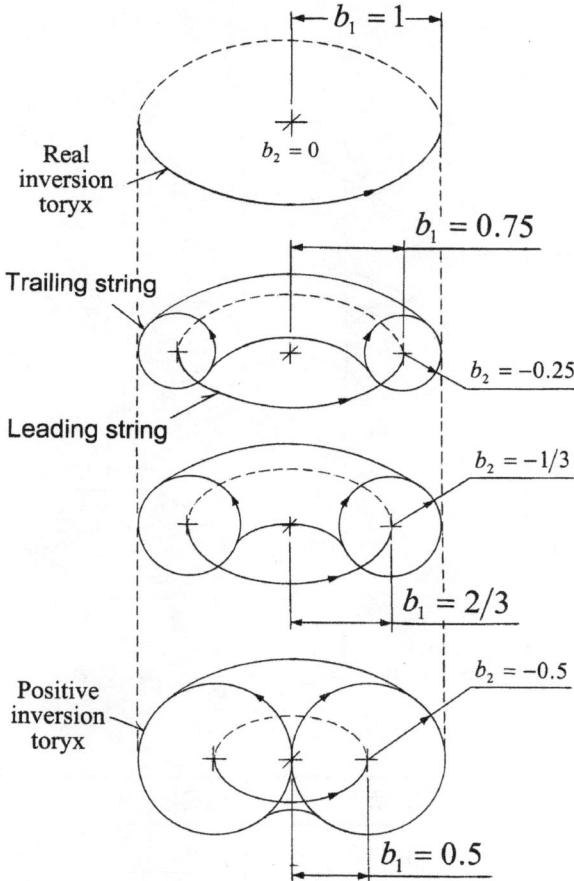

Range	φ_2	b	b_1	b_2	$\eta_2/2\pi$	w_2	β_{2t}	β_{2r}
From	90^0	1.0	1.0	-0	1.0	1.0	1.0	-0
To	180^0	$+0$	$\frac{1}{2}$	$-\frac{1}{2}$	$+0$	$+\infty$	$+0$	-1.0

Figure 8.5.2. Transformation of the real positive toryx.

Imaginary positive toryces (Fig. 8.5.3) - These toryces belong to the bottom left quadrant of the circular diagram shown in Figure 8.4. As φ_2 increases, b_1 decreases while negative values of b_2 increase. Within this range, the opposite parts of windings of the trailing string intersect with one another like in a *spindle torus*.

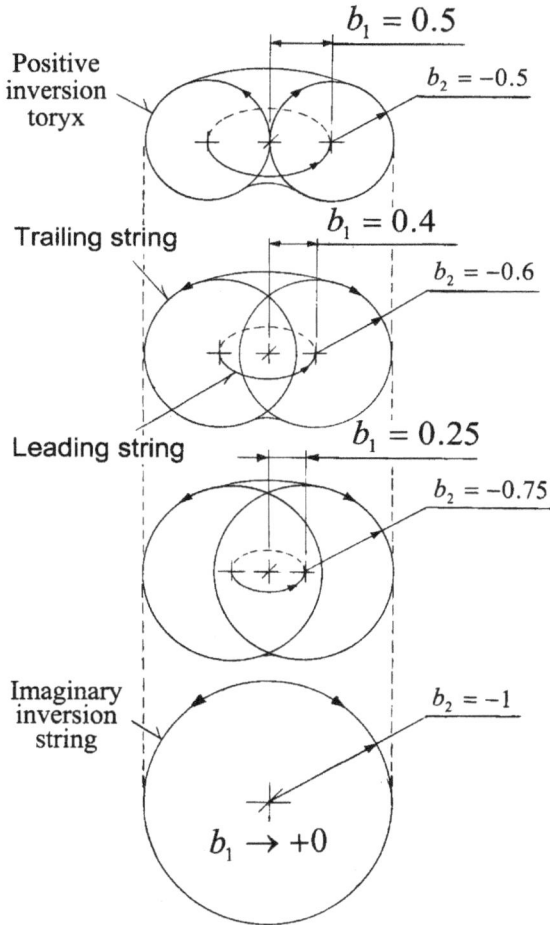

Range	φ_2	b	b_1	b_2	$\eta_2/2\pi$	w_2	β_{2t}	β_{2r}
From	180^0	-0	$\frac{1}{2}$	$-\frac{1}{2}$	$-0i$	$-\infty i$	$-0i$	-1.0
To	270^0	-1.0	$+0$	-1.0	$-i$	$-0i$	$-\infty i$	$-\infty$

Figure 8.5.3. Transformation of the imaginary positive toryx.

Imaginary negative toryces (Fig. 8.5.4) - These toryces belong to the bottom right quadrant of the circular diagram shown in Figure 8.4. Here the leading string becomes inverted. As the negative value of b_1 increases, the negative values of b_2 also increase. Within this range the toryx windings are located outside of the imaginary inversion toryx.

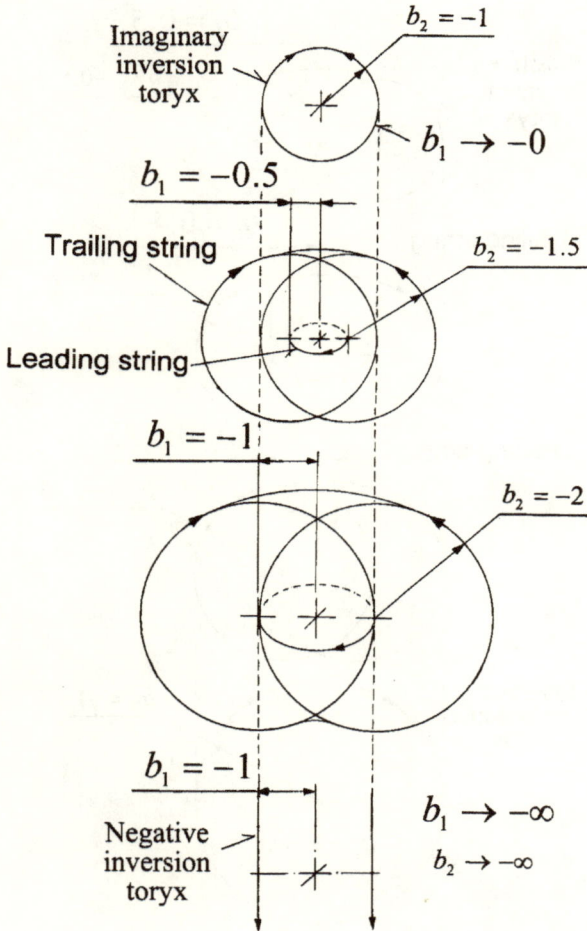

Range	φ_2	b	b_1	b_2	$\eta_2/2\pi$	w_2	β_{2t}	β_{2r}
From	270^0	-1.0	-0	-1.0	$-i$	$+0i$	$-\infty i$	$+\infty$
To	360^0	$-\infty$	$-\infty$	$-\infty$	$-\infty i$	$+\infty i$	$-0i$	1.0

Figure 8.5.4. Transformation of the imaginary negative toryces.

8.6 Summary of Toryx Transformations

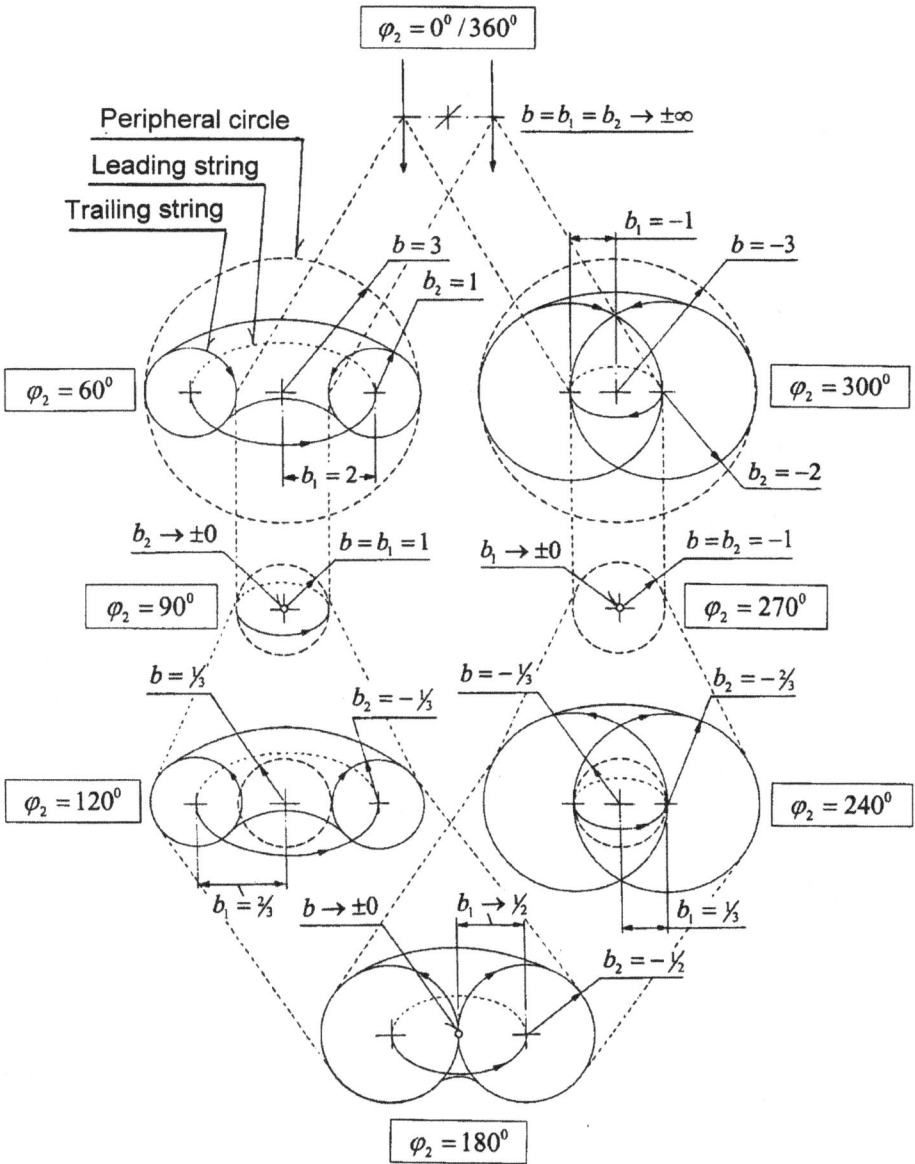

Figure 8.6. Transformations of the toryx peripheral circle, leading string and trailing string as a function of the steepness angle of the trailing string φ_2.

Figure 8.6 provides a complete overview of transformation of toryces by showing together the metamorphoses of the toryx peripheral circle, leading string and trailing string with the respective relative radii b, b_1 and b_2. Shown in this figure are the transformations of the toryx peripheral circle, leading string and trailing string as the steepness angle of the trailing string φ_2 increases from 0^0 to 360^0.

- $\varphi_2 = +0^0$ – At this point, all three radii b, b_1 and b_2 approach positive infinity $(+\infty)$.
- $+0^0 < \varphi_2 < 90^0$ - Within this range, b, b_1 and b_2 decrease as φ_2 increases.
- $\varphi_2 \rightarrow 90^0$ – At this point, $b = b_1 = 1$ and b_2 approaches positive infinility $(+0)$; the *trailing strings becomes inverted* and b_2 becomes negative.
- $90^0 < \varphi_2 < 180^0$ - Within this range, b_1 and b continue to decrease, while negative values of b_2 increase starting from negative infinity (-0).
- $\varphi_2 \rightarrow 180^0$ – At this point, $b_1 = \frac{1}{2}$, $b_2 = -\frac{1}{2}$ and b approaches positive infinity $(+0)$; the *peripheral circle becomes inverted* and b becomes negative.
- $180^0 < \varphi_2 < 270^0$ - Within this range, b_1 continues to decrease, negative values of b_2 continue to increase, while negative values of b increase starting from negative infinility (-0).
- $\varphi_2 \rightarrow 270^0$ – At this point, $b = b_2 = -1$, while b_1 approaches positive infinility $(+0)$; the *leading string becomes inverted* and b_1 becomes negative.
- $270^0 < \varphi_2 < 360^0$ - Within this range, negative values of b_1 increase starting from negative infinility (-0), while negative values of b and b_2 continue to increase.
- $\varphi_2 \rightarrow 360^0$ – At this point, all three radii b, b_1 and b_2 approach negative infinity $(-\infty)$.

The inversions of the toryx leading string, trailing string and peripheral circle are summarized in Table 8.6:

- The inversions of the toryx leading string, trailing string and peripheral circle occur when their radii approach either infinility (± 0) or infinity $(\pm \infty)$.

- Relative radius of the toryx trailing string b_2 approaches infinility (± 0) when $\varphi_2 \to 90^0$.
- Relative radius of the toryx peripheral circle b approaches infinility (± 0) when $\varphi_2 \to 180^0$.
- Relative radius of the toryx leading string b_1 approaches infinility (± 0) when $\varphi_2 \to 270^0$.
- Relative radii of the peripheral circle b, the leading string b_1 and the trailing string b_2 approach infinity $(\pm \infty)$ when $\varphi_2 \to 0^0/360^0$.
- Relative radii of the toryx leading string b_1 and the toryx peripheral circle b approach positive unity $(+1)$ when $\varphi_2 \to 90^0$.
- Relative radii of the toryx trailing string b_2 and the toryx peripheral circle b approach negative unity (-1) when $\varphi_2 \to 270^0$.

Table 8.6. Inversion points of the toryx leading string and trailing strings and its peripheral circle.

Toryx component	Relative radius	Steepness angle
Peripheral circle	$b \to \pm\infty$	$\varphi_2 \to 0^0/360^0$
	$b \to \pm 0$	$\varphi_2 \to 180^0$
Leading string	$b_1 \to \pm\infty$	$\varphi_2 \to 0^0/360^0$
	$b_1 \to \pm 0$	$\varphi_2 \to 270^0$
Trailing string	$b_2 \to \pm\infty$	$\varphi_2 \to 0^0/360^0$
	$b_2 \to \pm 0$	$\varphi_2 \to 90^0$

9

FORMATION OF MATTER PARTICLES

Structure of elementary particles and their classification was a principle subject of the quark model developed during the second half of the 20th century as briefly described in Chapter 1. The UST describes this subject differently.

9.1 Classification of Particles

The UST divides particles into two groups: *elementary* and *composite*. The elementary particles are called *trons*. Each tron is made up of two polarized toryces, while each composite particle is composed from two or more trons. Neutron and proton are the examples of composite particles.

According to the UST, there are only four elementary particles called *electrons, positrons, ethertrons* and *singulatrons*. Figure 9.1.1 shows a general presentation of formation of four elementary particles by the unification of polarized toryces.

- The ethertrons $a_{m,n,q}^{0}$ are vorticity-polarized real trons; they are made up of the real negative toryces $A_{m,n,q}^{-}$ and the real positive toryces $A_{m,n,q}^{+}$.

- The singulatrons $\breve{a}_{m,n,q}^{0}$ are vorticity-polarized imaginary trons; are made up of the imaginary negative toryces $\breve{A}_{m,n,q}^{-}$ and the imaginary positive toryces $\breve{A}_{m,n,q}^{+}$.

- The electrons $e_{m,n,q}^{-}$ are reality-polarized negative trons; they are made up of the real negative toryces $E_{m,n,q}^{-}$ and the imaginary negative toryces $\breve{E}_{m,n,q}^{-}$.

- The positrons $e_{m,n,q}^{+}$ are reality-polarized positive trons; they are made up of the real positive toryces $E_{m,n,q}^{+}$ and the imaginary positive toryces $\breve{E}_{m,n,q}^{+}$.

We use the same letters for symbols of trons and their constituent toryces; the lower-case letters are used for trons and capital letters for toryces. The superscripts in the symbols of toryces and trons indicate the signs of their vorticities V and, in some cases, the magnitudes of these vorticities. The subscripts indicate the toryx quantum states m, n and q that are explained later in this Chapter.

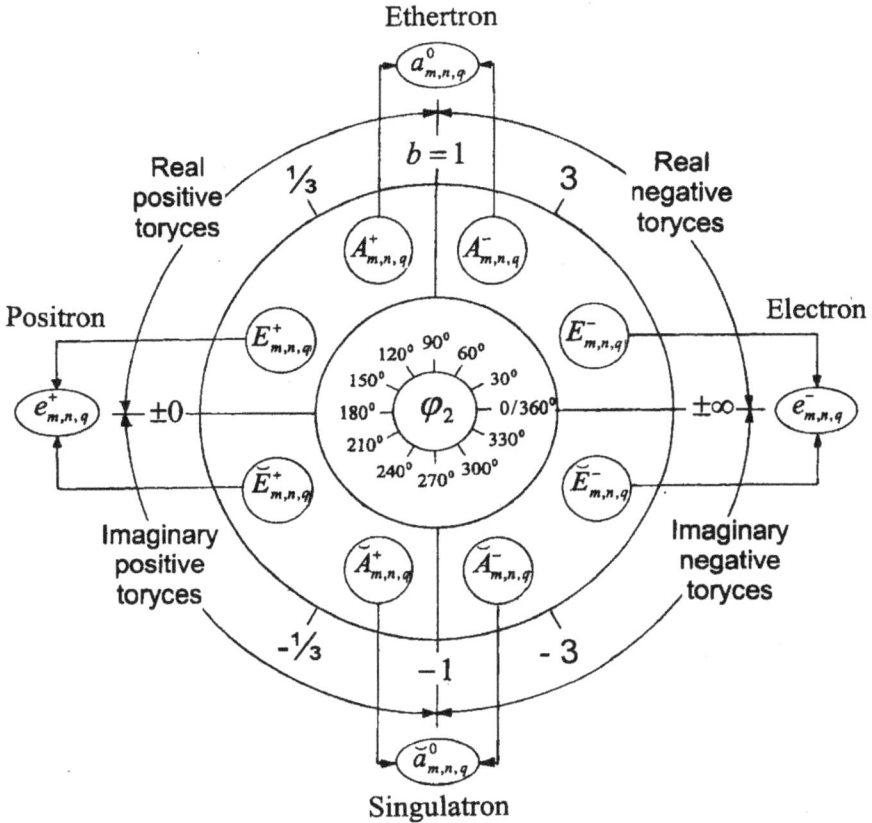

Figure 9.1.1. General presentation of formation of four elementary matter particles (trons) from polarized toryces.

Constituent toryces of trons exist interchageably with one another and the tron vorticity V_t is equal to the arithmetic average of the vorticities of its constituent toryces:

$$V_t = \frac{V' + V''}{2} \tag{9.1-1}$$

The relative radii of leading strings of constituent toryces, b_1' and b_1'', are related to one another by the equation:

$$b_1' = \frac{b_1''}{2b_1''(1+V_t)-1} \tag{9.1-2}$$

Eq. (9.1-2) reduces to the following forms for some particular cases:

$$V_t = 0: \quad b_1' = \frac{b_1''}{2b_1''-1} \tag{9.1-3}$$

$$V_t = -1: \quad b_1' = -b_1'' \tag{9.1-4}$$

$$V_t = +1: \quad b_1' = \frac{b_1''}{4b_1''-1} \tag{9.1-5}$$

$$V_t = -0.5: \quad b_1' = \frac{b_1''}{b_1''-1} \tag{9.1-6}$$

$$V_t = +0.5: \quad b_1' = \frac{b_1''}{3b_1''-1} \tag{9.1-7}$$

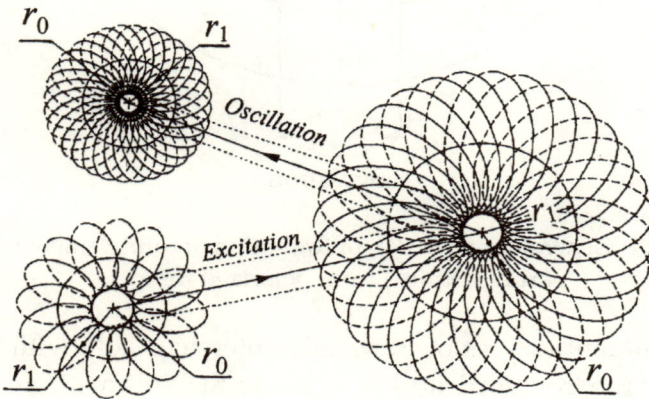

Figure 9.1.2. Excitation and oscillation of a toryx.

Constituent toryces of trons change their dimensions in quantum steps by two processes called *excitation* and *oscillation*. As shown in Figure 9.1.2, during the excitation of a toryx the radius of the toryx leading string r_1 increases, while its eye radius r_0 remains constant. During the oscillation of the toryx its radii, r_1 and r_0, decrease proportionally to the *toryx oscillation factor* Q_q.

9.2 Spacetime Conservation Law

According to the UST, the *toryx spacetime polarization* P_t is equal to:

$$P_t = Q_q V R^2 = -Q_q \frac{(b_1 - 1)(2b_1 - 1)}{b_1} \tag{9.2-1}$$

To assure stability of a system made up of many toryces, it is necessary to add an additional limitation of the toryx degrees of freedom defined by the proposed *Spacetime Conservation Law*.

In a stable enclosed system containing N toryces the total toryx spacetime polarization $P_{t\Sigma}$ approaches infinility (± 0) as given by the equation:

$$P_{t\Sigma} = \sum_{i=1}^{N} P_{ti} = \left(Q_q V R^2 \right)_i \rightarrow \pm 0 \tag{9.2-2}$$

The toryces obeying the Spacetime Conservation Law are called the *basic toryces*. These toryces form the *basic trons*. It follows from Eq. (9.2-1) and Figs. 7.1 and 7.2 that in the basic trons the relationships between the radii of peripheral circles of their constituent basic toryces are as follows.

For the reality-polarized electrons:

$$\breve{b}^- = -b^- \tag{9.2-3}$$

For the reality-polarized positrons:

$$\breve{b}^+ = -b^+ \tag{9.2-4}$$

For the vorticity-polarized ethertrons:

$$b^+ = \frac{1}{b^-} \qquad \text{(9.2-5)}$$

For the vorticity-polarized singulatrons:

$$\breve{b}^+ = \frac{1}{\breve{b}^-} \qquad \text{(9.2-6)}$$

Also, in a stable system containing both the real and imaginary matched basic toryces, the number of the real toryces must exceed the number of the imaginary toryces by the *toryx reality ratio T* that is equal to:

$$T = \left(\frac{b_1}{b_1 - 1}\right)^2 = \left(\frac{b+1}{b-1}\right)^2 \qquad \text{(9.2-7)}$$

where b_1 and b are respectively the relative radii of leading strings and peripheral circles of the real toryces.

9.3 Quantization Equations for Excited Toryces

To derive the quantization equations for excited toryces the UST takes into account the toryx fine structure and also the well-known experimental data, including:

- Measured charge, mass and magnetic moments of the nucleons and hydrogen atom.
- Neutron must decay into a free proton, an electron and an electron antineutrino.
- Measured frequencies of the photons emitted by hydrogen atom.

The quantization equations for excited toryces are expressed as a function of the *toryx quantization parameter z* given by the equation:

$$z = 2(n\Lambda)^m + n \qquad \text{(9.3-1)}$$

where:

$m = 0, 1, 2, \ldots$, toryx exponential excitation quantum state

$n = 0, 1, 2, \ldots$, toryx linear excitation quantum state.

The toryces in which the toryx exponential excitation quantum state $m = 0$ are called the *harmonic toryces*, because the frequencies of their trailing strings are related to one another by simple harmonic ratios. For these toryces, Eq. (9.3-1) reduces to:

$$z = 2 + n$$

We consider below quantization equations for three kinds of toryces: basic, resonant and oscillated.

Basic toryces – The basic toryces obey the Spacetime Conservation Law. Table 9.3.1 shows quantization equations for the relative radii of leading strings and peripheral circles of basic toryces forming the basic trons.

Table 9.3.1. Quantization equations for the relative radii of leading strings and peripheral circles of basic toryces.

Basic tron	Toryx	Relative radius of leading string		Relative radius of peripheral circle	
Electron $e_{m,n,q}^{-}$	$E_{m,n,q}^{-}$	$b_1^{-} = z$	(9.3-2)	$b^{-} = 2z - 1$	(9.3-10)
	$\breve{E}_{m,n,q}^{-}$	$\breve{b}_1^{-} = 1 - z$	(9.3-3)	$\breve{b}^{-} = 1 - 2z$	(9.3-11)
Positron $e_{m,n,q}^{+}$	$E_{m,n,q}^{+}$	$b_1^{+} = \dfrac{z}{2z - 1}$	(9.3-4)	$b^{+} = \dfrac{1}{2z - 1}$	(9.3-12)
	$\breve{E}_{m,n,q}^{+}$	$\breve{b}_1^{+} = \dfrac{1 - z}{1 - 2z}$	(9.3-5)	$\breve{b}^{+} = \dfrac{1}{1 - 2z}$	(9.3-13)
Ethertron $a_{m,n,q}^{0}$	$A_{m,n,q}^{-}$	$b_1^{-} = \dfrac{z}{z - 1}$	(9.3-6)	$b^{-} = \dfrac{z + 1}{z - 1}$	(9.3-14)
	$A_{m,n,q}^{+}$	$b_1^{+} = \dfrac{z}{z + 1}$	(9.3-7)	$b^{+} = \dfrac{z - 1}{z + 1}$	(9.3-15)
Singulatron $\breve{a}_{m,n,q}^{0}$	$\breve{A}_{m,n,q}^{-}$	$\breve{b}_1^{-} = \dfrac{1}{1 - z}$	(9.3-8)	$\breve{b}^{-} = \dfrac{1 + z}{1 - z}$	(9.3-16)
	$\breve{A}_{m,n,q}^{+}$	$\breve{b}_1^{+} = \dfrac{1}{1 + z}$	(9.3-9)	$\breve{b}^{+} = \dfrac{1 - z}{1 + z}$	(9.3-17)

The quantization equations shown in Table 9.3.1 were derived by using the following equations:

- Eq. (9.3-2) is based on Eq. (9.3-1).
- Eq. (9.3-10) is derived from Eq. (9.3-2) by using Eq. (5.6-11).
- Eq. (9.3-11) is derived from Eq. (9.3-10) by using Eq. (9.2-3).
- Eq. (9.3-12) is derived from Eq. (9.3-10) by using Eq. (9.2-5).
- Eq. (9.3-13) is derived from Eq. (9.3-11) by using Eq. (9.2-6).
- Eq. (9.3-3) is derived from Eq. (9.3-11) by using Eq. (5.6-11).
- Eq. (9.3-4) is derived from Eq. (9.3-12) by using Eq. (5.6-11).
- Eq. (9.3-5) is derived from Eq. (9.3-13) by using Eq. (5.6-11).
- Eq. (9.3-6) is derived from Eq. (9.3-2) by using Eq. (9.1-6).
- Eq. (9.3-14) is derived from Eq. (9.3-6) by using Eq. (5.6-11).
- Eq. (9.3-15) is derived from Eq. (9.3-14) by using Eq. (9.2-5).
- Eq. (9.3-16) is derived from Eq. (9.3-14) by using Eq. (9.2-3).
- Eq. (9.3-17) is derived from Eq. (9.3-16) by using Eq. (9.2-6).
- Eq. (9.3-7) is derived from Eq. (9.3-15) by using Eq. (5.6-11).
- Eq. (9.3-8) is derived from Eq. (9.3-16) by using Eq. (5.6-11).
- Eq. (9.3-9) is derived from Eq. (9.3-17) by using Eq. (5.6-11).

Resonant toryces – In the resonant toryces forming elementary matter particles, the frequencies of their trailing strings have the same amplitudes but opposite signs. Table 9.3.2 shows equations for resonant toryces making up a resonant electron $e\bar{r}_{0,n,q}$.

Table 9.3.2. Quantization equations for the relative radii of leading strings and peripheral circles of resonant excited toryces.

Resonant tron	Toryx	Relative radius of leading string	Relative radius of peripheral circle
Electron $e\bar{r}_{m,n,q}$	$\bar{E}_{m,n,q}$	$\bar{b}_1^- = z$ (9.3-2)	$b^- = 2z - 1$ (9.3-10)
	$\breve{E}\bar{r}_{m,n,q}$	$\breve{b}_1^- = -z$ (9.3-18)	$\breve{b}^- = -(2z+1)$ (9.3-19)

Oscillated toryces – During oscillation of a toryx, its radii, r_0 and r_1, change proportionally to the *toryx oscillation factor* Q_q. The toryx oscillation factor Q_q is assumed to be expressed by the equation:

$$Q_0 = 1; \quad Q_q = 3\left(\frac{\Lambda}{2(q-1)}\right)^{q-1} \quad (q = 1, 2,..) \quad (9.3\text{-}20)$$

where

$q = 0, 1, 2, \ldots$, toryx oscillation quantum state.

The values of Q_q calculated from Eq. (9.3-20) are shown in Figure 9.3 and Table 9.3.3. At $q = q_m$ and $\Lambda = 137$, the toryx oscillation factor Q_q reaches its maximum value Q_{qm} defined by the equations:

$$q_m = 1 + \frac{\Lambda}{2e} = 26.199742$$

$$(9.3\text{-}21)$$

$$Q_{qm} = 3e^{\Lambda/2e} = 2.637728 \times 10^{11}$$

When $q > q_m$, the toryx oscillation factor Q_q sharply decreases and at $q = 70.589986$ its magnitude reduces to 1. After that as q continues to increase and approaches infinity, Q_q decreases and approaches infinility.

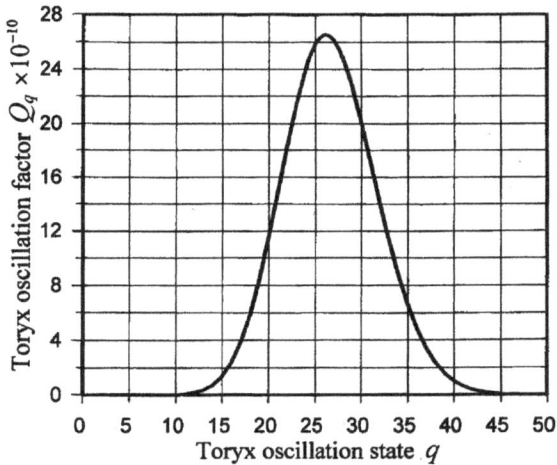

Figure 9.3. Toryx oscillation factor Q_q as a function of the toryx oscillation state q.

The relationship between the toryx parameters and the toryx oscillation factor Q_q is given by the equation:

$$Q_q = \frac{r_{i0}}{r_{iq}} = \frac{r_{10}}{r_{1q}} = \frac{f_{iq}}{f_{i0}} \qquad (9.3\text{-}22)$$

Table 9.3.3. Toryx oscillation factor Q_q as a function of the toryx oscillation quantum state q.

q	0	1	2	3	4	5
Q_q	1.000	3.000	205.500	3511.1875	35713.236	258014.18

In Eq. (9.3-22), the parameters with the subscripts "0" and "q" correspond respectively to $q = 0$ and to $q > 0$.

10

PHYSICAL PROPERTIES OF MICRO-TORYCES

The UST is based on an assumption that both space and time are universal terms. The history of humanity tells us that thousands years ago people learned how to measure distances between various objects by comparing these distances, for instance, with lengths of their feet. Thanks to the periodicity of sunsets they were also able to measure a passage of time by counting the number of days. The same discoveries could be made by intelligent beings that might exist in any part of the universe.

It would, however, be a completely different story when it comes to describing physical properties of matter. Physical terms, such as mass, charge, force, energy, magnetic induction, etc. are purely man-made inventions based on their theories. It is very likely, however, that other intelligent beings developed completely different physical theories of nature and, consequently, use completely different physical terms. In the UST, physical properties of micro-toryces are expressed as a function of spacetime parameters comprehendible by any intelligent beings in the universe.

Another important feature of the UST is that toryces are applicable to the entities of both micro- and macro-worlds. They are respectively called *micro-toryces* and *macro-toryces*. There is one principal difference between micro- and macro-toryces. Micro-toryces describe properties of prime elements of nature making up elementary particles, while macro-toryces describe properties of fields associated with much larger entities. Spacetime properties of both micro- and macro-toryces are described by the same spacetime equations, except for equations describing radii of their real inversion toryces. To make a distinction between these radii, we use the symbols r_i for micro-toryces and the symbol r_j for macro-toryces.

10.1 Radius of Real Inversion Toryx

The radius of the real inversion toryx r_i is one of major parameters allowing us to establish a correlation between physical and spacetime properties of micro-toryces.

A real inversion toryx is shown at the top of Figure 8.4. It corresponds to the case when the relative radius of leading string b_1

= 1. Thus, in application to a micro-toryx, the radius of the real inversion toryx r_i is equal to the toryx eye radius r_0.

$$r_i = r_0 \qquad (10.1\text{-}1)$$

It possible to express the radius r_i in physical terms by comparing the relative velocity of the toryx leading string β_1 based on the Spacetime Law of Planetary Motion described by Eq. (5.8-1) with relative orbital velocity of atomic electron based on classical law of planetary motion.

The law of planetary motion for an atomic electron based on classical physics can be derived by equating two forces:

- The attraction electric force F_e between the positive electric charge Ze and the negative electric charge e separated by the distance r_1

- The centrifugal forces F_c applied to the electron with the mass m_e orbiting a nucleon with the orbital velocity V_1 (Fig. 10.1.).

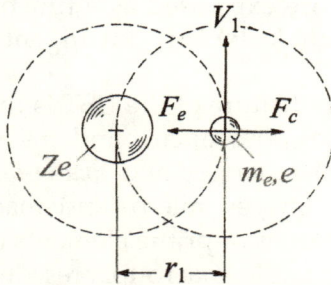

Figure 10.1. Atomic planetary system with two charged particles.

Based on Coulomb's law of electric force and an equation for centrifugal force, we obtain that at the equilibrium state:

$$F_e = F_c = \frac{Ze^2}{4\pi \varepsilon_0 r_1^2} = \frac{m_e V_1^2}{r_1} \qquad (10.1\text{-}2)$$

From Eqs. (5.7-1), (10.1-1) and (10.1-2) we obtain for the case when $b_1 \gg 1$ that the radius of the real inversion toryx r_i is equal to:

$$r_i = r_0 = \frac{r_e}{2} = \frac{Ze^2}{8\pi \varepsilon_0 m_e c^2} \qquad (10.1\text{-}3)$$

where
 Z = atomic number
 r_e = classical electron radius (Z = 1)
 e = elementary charge
 ε_0 = electric constant.

10.2 Toryx Charge, Gravitational & Inertial Masses

In the UST the toryx charge, gravitational and inertial masses are related to the ratios between the radii of their leading string, trailing string and peripheral circle as shown below.

Toryx charge - The toryx relative charge e_t/e is equal to the ratio of the radii of the toryx trailing and leading strings:

$$\frac{e_t}{e} = -\frac{r_2}{r_1} = -\frac{b_2}{b_1} = -\frac{b_1 - 1}{b_1} \tag{10.2-1}$$

Toryx gravitational mass - The toryx relative gravitational mass m_{tg}/m_e is equal to the product of the toryx oscillation factor Q_q and absolute value of the ratio of the radii of the toryx trailing and leading strings.

$$\frac{m_{tg}}{m_e} = Q_q \left| \frac{r_2}{r_1} \right| = Q_q \left| \frac{b_2}{b_1} \right| = Q_q \left| \frac{b_1 - 1}{b_1} \right| \tag{10.2-2}$$

Toryx inertial mass – The toryx relative inertial mass m_{ti}/m_e is equal to the product of the $2Q_q$ and the ratio of the radii of the toryx trailing and peripheral circle:

$$\frac{m_{ti}}{m_e} = 2Q_q \frac{r_2}{r} = 2Q_q \frac{b_2}{b} = 2Q_q \frac{b_1 - 1}{2b_1 - 1} \tag{10.2-3}$$

The UST yields a completely different set of relativistic equations than those used in Einstein's special theory of relativity. Main features of the toryx relativistic equations shown in Table 10.2 are:

• Not only masses of toryces but also their charges are dependent on velocities of their leading strings.

- In real toryces the absolute values of their relative charges and masses decrease with an increase of the relative spiral velocities of their leading strings β_1.
- Conversely, in imaginary toryces the absolute values of their relative charges and masses increase as the relative spiral velocities of their leading string β_1 increase.

Table 10.2 Relativistic equations for the toryx relative charge and gravitational mass.

Toryx parameters	Equations
Relative charge	$\dfrac{e_t}{e} = \sqrt{1 - \beta_1^2}$ (10.2-4)
Relative gravitational mass	$\dfrac{m_{tg}}{m_e} = Q_q\left\lvert\sqrt{1 - \beta_1^2}\right\rvert$ (10.2-5)

The toryx base frequency f_0 and the toryx base period T_0 are equal to:

$$f_0 = \frac{4\varepsilon_0 m_e c^3}{Ze^2} \tag{10.2-6}$$

$$T_0 = \frac{Ze^2}{4\varepsilon_0 m_e c^3} \tag{10.2-7}$$

10.3 Toryx Mechanical Properties

Described below are three toryx mechanical properties: density, elasticity and angular momentum. We defined these properties in a similar way as they would be defined by classical mechanics.

Toryx density – The toryx density ρ_t is equal to the ratio of the toryx gravitational mass m_{tg} to the volume occupied by its trailing string U_2, thus:

$$\rho_t = \frac{m_{tg}}{U_2} \tag{10.3-1}$$

From Eqs. (10.2-2) and (10.3-1), the toryx density ρ_t is equal to:

$$\rho_t = \frac{m_e Q_q}{2\pi^2 r_i^3} \left| \frac{b_1 - 1}{b_1} \right| \frac{1}{b_1(b_1 - 1)^2} \qquad (10.3\text{-}2)$$

From Eq. (10.3-2), the toryx relative density ρ_{tr} is equal to (see Fig. 10.3.1):

$$\rho_{tr} = \rho_t \frac{2\pi^2 r_i^3}{m_e Q_q} = \left| \frac{b_1 - 1}{b_1} \right| \frac{1}{b_1(b_1 - 1)^2} \qquad (10.3\text{-}3)$$

Notably, the toryx density approaches infinility (± 0) when φ_2 approaches 0^0 and 360^0, and it approaches infinity $(\pm\infty)$ when φ_2 approaches 90^0 and 270^0.

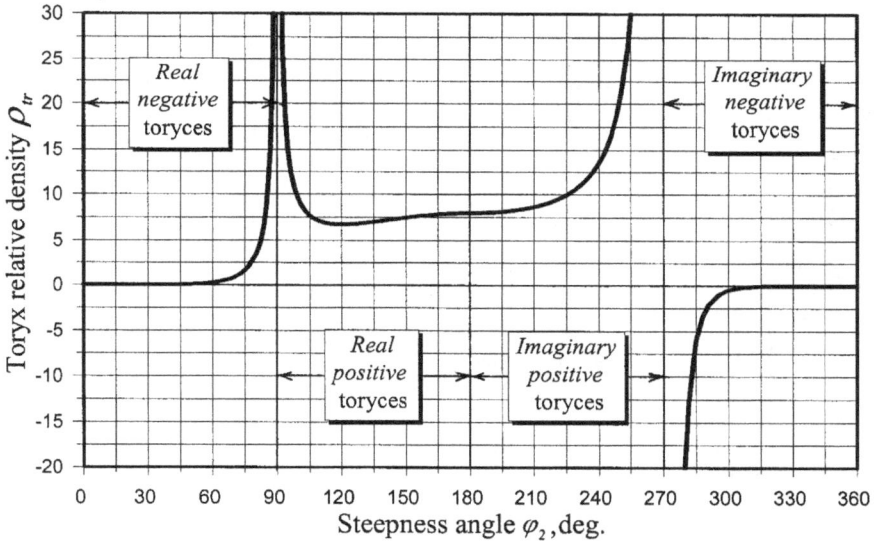

φ_2	$360^0 / 0^0$	90^0	180^0	270^0
ρ_{tr}	$-0/+0$	$+\infty$	8.0	$+\infty/-\infty$

Figure 10.3.1. Toryx relative density ρ_{tr} as a function of the steepness angle of trailing string φ_2.

Toryx elasticity – In classical mechanics, elastic properties, or rigidity, of compressible media contained in a certain volume are defined by the *bulk modulus of elasticity*. It is equal to the ratio of change in pressure inside the media to the resulting fractional change in the volume of the media. Let compression waves, like sound waves, travel through the toryx trailing string with velocity equal to the spiral velocity of the toryx leading string V_1. In that case, the toryx bulk modulus B_t will be equal to:

$$B_t = \rho_t V_1^2 = \frac{Q_q m_e c^2}{2\pi^2 r_i^3} \left|\frac{b_1 - 1}{b_1}\right| \frac{2b_1 - 1}{b_1^3 (b_1 - 1)^2} \qquad (10.3\text{-}4)$$

From Eqs. (5.8-1), (10.3-1) and (10.3-4), the toryx relative bulk modulus of elasticity B_{tr} is equal to:

$$B_{tr} = B_t \frac{2\pi^2 r_i^3}{Q_q m_e c^2} = \left|\frac{b_1 - 1}{b_1}\right| \frac{2b_1 - 1}{b_1^3 (b_1 - 1)^2} \qquad (10.3\text{-}5)$$

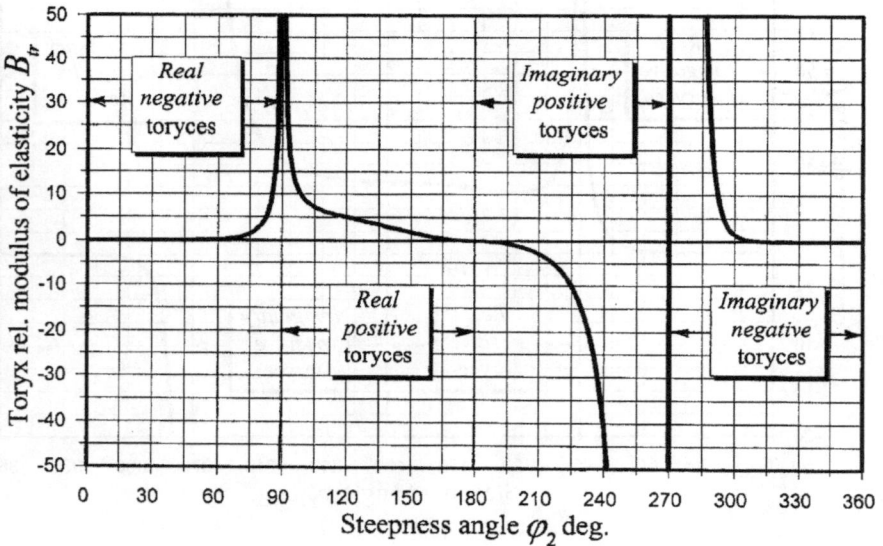

φ_2	$360^0 / 0^0$	90^0	180^0	270^0
B_{tr}	0	$+\infty$	$+0/-0$	$-\infty/+\infty$

Figure 10.3.2. Toryx relative bulk modulus of elasticity B_{tr} as a function of the steepness angle of trailing string φ_2.

The toryx relative bulk modulus of elasticity B_{tr} as a function of the steepness angle of trailing string φ_2 is equal to (see Fig. 10.3.2):

$$B_{tr} = B_t \frac{2\pi^2 r_i^3}{Q_q m_e c^2} = \left|\frac{b_1 - 1}{b_1}\right| \frac{2b_1 - 1}{b_1^3 (b_1 - 1)^2} \qquad (10.3\text{-}6)$$

Notably, the toryx relative bulk modulus of elasticity B_{tr} approaches infinility (± 0) when φ_2 approaches 0^0 and 360^0, and it approaches infinity ($\pm\infty$) when φ_2 approaches 90^0 and 270^0.

Toryx orbital angular momentum – The toryx orbital angular momentum L_t is defined by the classical mechanics equation:

$$L_t = m_{ti} V_1 r_1 \qquad (10.3\text{-}7)$$

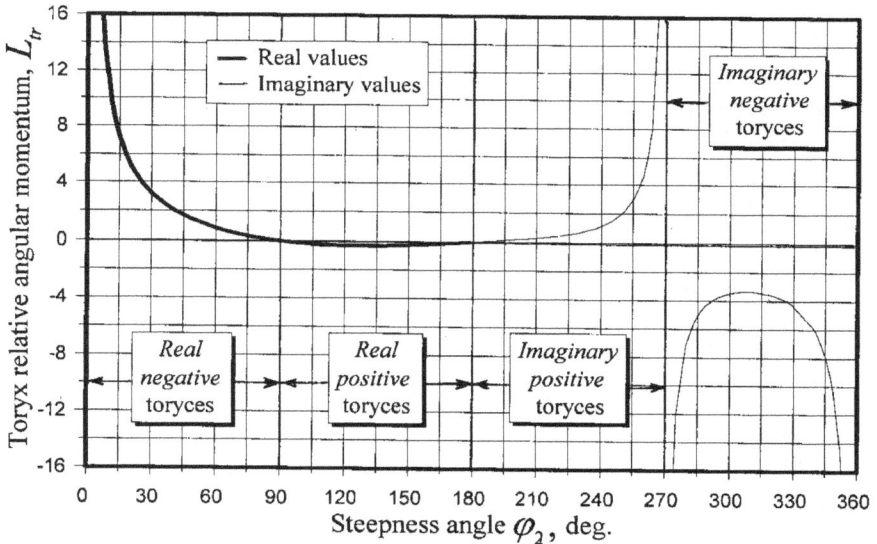

φ_2	$360^0 / 0^0$	90^0	180^0	270^0
L_{tr}	$-\infty i / +\infty$	$+0/-0$	$-0/+0i$	$+\infty i/-\infty i$

Figure 10.3.3. Toryx relative orbital angular momentum L_{tr} as a function of the steepness angle of trailing string φ_2.

From Eqs. (5.7-1a), (5.7-7a), (10.2-3) and (10.3-7), relative values of the toryx angular momentum are equal to (see Fig. 10.3.3):

$$\text{Real toryces: } L_{tr} = \frac{L_t}{Q_q m_e c r_i} = \pm \frac{(b_1 - 1)\sqrt{2b_1 - 1}}{b_1} \qquad (10.3\text{-}8)$$

$$\text{Imaginary toryces: } \breve{L}_{tr} = \frac{L_t}{Q_q m_e c r_i} = \pm \frac{(b_1 - 1)\sqrt{2b_1 - 1}}{ib_1} \qquad (10.3\text{-}9)$$

10.4 Toryx Energies

Toryx kinetic energy - The toryx kinetic energy K is defined by the classical mechanics equation:

$$K = \frac{m_{ti} V_1^2}{2} \qquad (10.4\text{-}1)$$

From Eqs. (5.6-8a), (10.2-3) and (10.4-1), the toryx kinetic energy K_t is equal to:

$$K_t = Q_q m_e c^2 \frac{(b_1 - 1)}{b_1^2} \qquad (10.4\text{-}2)$$

Toryx potential energy – Similarly to the Bohr's quantum model of hydrogen atom, the toryx potential energy U_t is equal to:

$$U_t = -\frac{e e_t}{4\pi\varepsilon_0 r_1} \qquad (10.4\text{-}3)$$

From Eqs. (5.6-1a), (10.1-3), (10.2-1) and (10.4-3) when $Z = 1$, the toryx potential energy U_t is equal to:

$$U_t = -2 Q_q m_e c^2 \frac{(b_1 - 1)}{b_1^2} \qquad (10.4\text{-}4)$$

Toryx total energy – The toryx total energy is equal to the sum of the toryx kinetic and potential energies. Thus, we obtain from Eqs. (10.4.2 and 10.4.4) that the toryx total relative energy $E_t / m_e c^2$ is equal to:

$$E_t / (Q_q m_e c^2) = -\frac{(b_1 - 1)}{b_1^2} \qquad (10.4\text{-}5)$$

Toryx field & matter energy – The toryx field & matter energy E_{fm} is equal to the sum of the toryx field energy E_f and the toryx matter energy E_m:

$$E_{fm} = E_f + E_m = Q_q m_e c^2 \qquad (10.4\text{-}6)$$

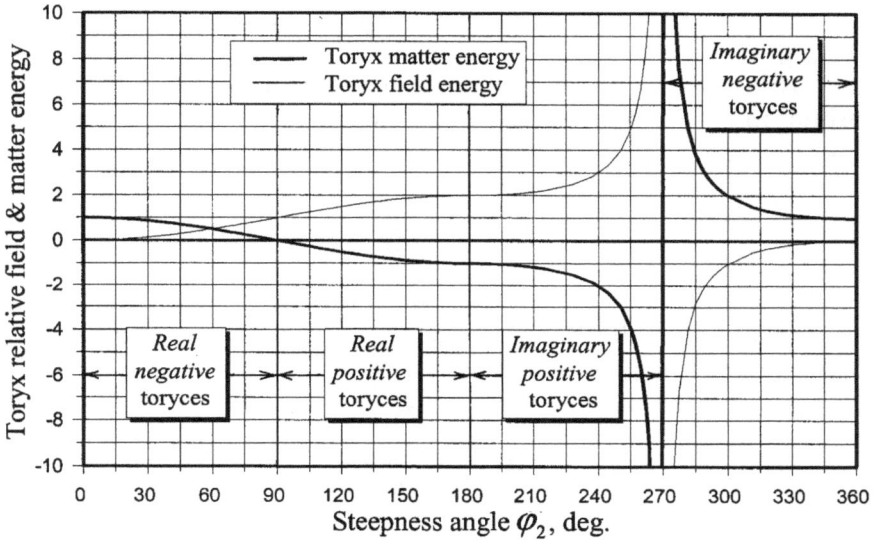

φ_2	$360^0/0^0$	90^0	180^0	270^0
$E_f/(Q_q m_e c^2)$	$-0/+0$	$+1$	$+2$	$+\infty/-\infty$
$E_m/(Q_q m_e c^2)$	$+1.0$	$+0/-0$	-1.0	$-\infty/+\infty$

Figure 10.4.1. Toryx relative field energy $E_f/(Q_q m_e c^2)$ and matter energy $E_m/(Q_q m_e c^2)$ as a function of the steepness angle of trailing string φ_2.

Toryx matter energy – The toryx matter energy E_m is defined by the equation:

$$E_m = Q_q m_e c^2 \frac{b_1 - 1}{b_1} \qquad (10.4\text{-}7)$$

Toryx field energy – From Eqs (10.4-6) and (10.4-7), the toryx field energy E_f is associated with toryx field. It is defined by the equation (see Fig. 10.4.1):

$$E_f = \frac{Q_q m_e c^2}{b_1} \qquad (10.4\text{-}8)$$

10.5 Toryx Orbital Magnetic Moment

According to the classical theory of electromagnetism the toryx orbital magnetic moment μ_t can be expressed by the equation (see Fig. 10.5.1):

$$\mu_t = \frac{e_t V_1 r_1}{2 Q_q} \qquad (10.5\text{-}1)$$

We express the toryx magnetic moment in relative terms in respect to either the Bohr magneton μ_B or the nuclear magneton μ_{N}. The Bohr magneton μ_B is given by the equation:

$$\mu_B = \frac{e^3}{8\pi\alpha\varepsilon_0 m_e c} \qquad (10.5\text{-}2)$$

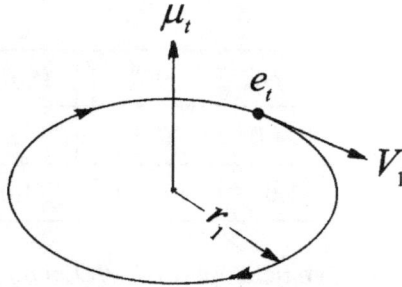

Figure 10.5.1. Toryx magnetic moment μ_t.

From Eqs. (5.7-1a), (5.7-8a), (10.2-1), (10.5-1) and (10.5-2) the toryx relative magnetic moment in respect to the Bohr magneton μ_t/μ_B is equal to (see Fig. 10.5.2):

$$\text{Real toryces:} \quad \frac{\mu_t}{\mu_B} = \pm\alpha \frac{(b_1 - 1)\sqrt{2b_1 - 1}}{2 Q_q b_1} \qquad (10.5\text{-}3)$$

$$\text{Imaginary toryces:} \quad \frac{\breve{\mu}_t}{\mu_B} = \pm \alpha i \frac{(\breve{b}_1 - 1)\sqrt{2\breve{b}_1 - 1}}{2Q_q \breve{b}_1} \quad (10.5\text{-}4)$$

From Eqs. (7.13-1), (10.3-8), (10.3-9), (10.5-3) and (10.5-4), the toryx relative magnetic moment in respect to the Bohr magneton μ_t/μ_B is proportional to the toryx golden polarization factor G_t and to the toryx orbital angular momentum L_t.

The toryx relative magnetic moment in respect to the nuclear magneton μ_t/μ_N is related to the toryx relative magnetic moment in respect to the Bohr magneton μ_t/μ_B by the equation:

$$\frac{\mu_t}{\mu_N} = \frac{\mu_t}{\mu_B} \frac{m_p}{m_e} \quad (10.5\text{-}5)$$

where m_p is proton rest mass.

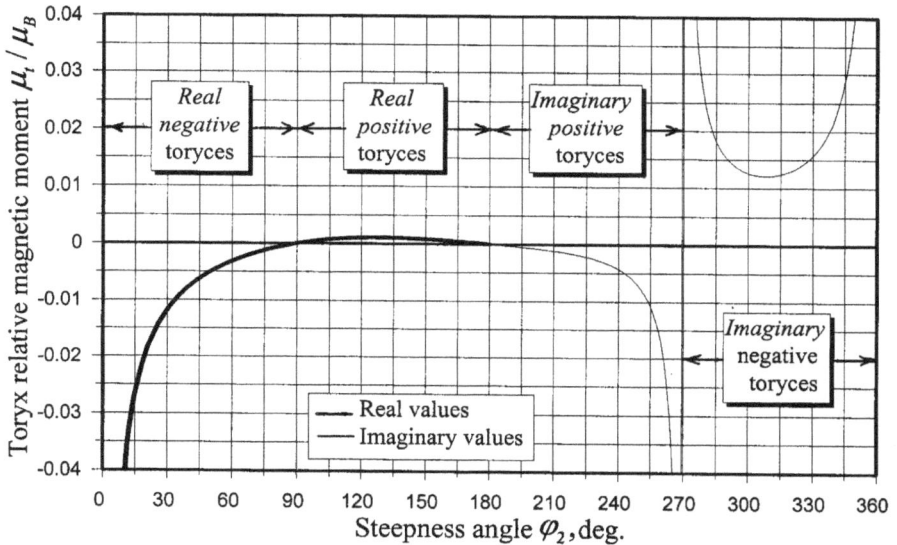

φ_2	$360°/0°$	$90°$	$180°$	$270°$
μ_t/μ_B	$+\infty i/-\infty$	$-0/+0$	$+0/-0i$	$-\infty i/+\infty i$

Figure 10.5.2. Toryx relative magnetic moment in respect to Bohr magneton μ_t/μ_B as a function of the steepness angle of trailing string φ_2.

Your notes

11

FROM QUANTUM VACUUM TO ATOMS

11.1 Micro-Genesis

According to the UST, the formation of nucleons and atoms from quantum vacuum occurs in five major phases:

1. Formation of harmonic trons
2. Formation of nucleon crystal structures
3. Formation of oscillated and excited trons
4. Formation of nucleon cores
5. Formation of nucleons and atoms.

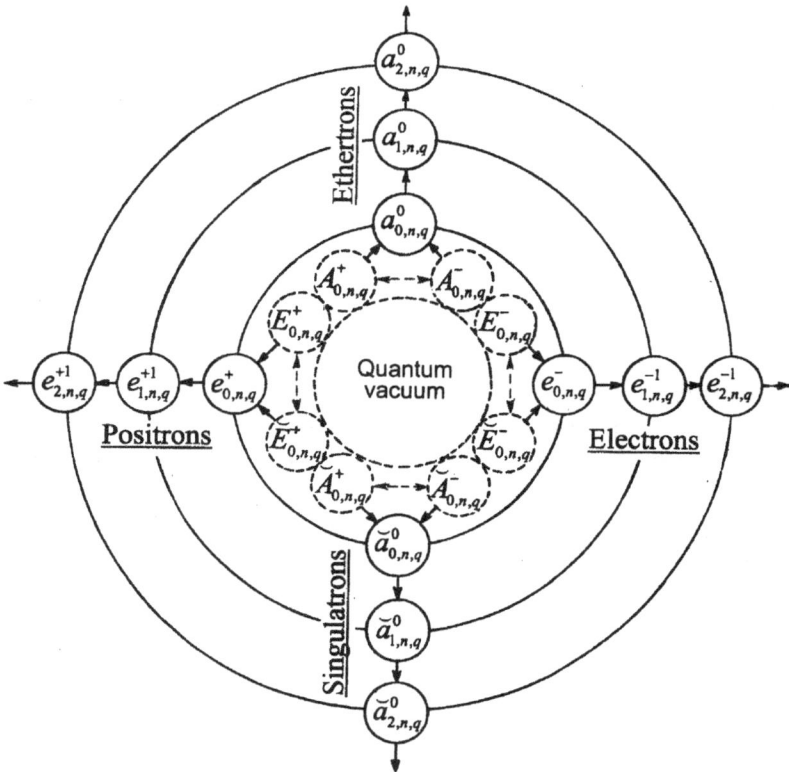

Figure 11.1. Formation of harmonic and excited trons.

Figure 11.1 shows schematically the formation of harmonic and excited trons. The interchangeable interactions between the oppositely-polarized harmonic toryces are shown with straight horizontal and vertical dotted lines.

The atoms are made up of assemblies of harmonic and excited trons. In the harmonic trons, the exponential excitation quantum states $m = 0$. In the excited trons, the exponential excitation quantum states m depends on the *spacetime levels L* as shown in Table 11.1.

Table 11.1. Spacetime levels of atoms.

Spacetime levels of atoms L	Exponential excitation quantum states m Of atomic components			
	Ethertron	Singulatron	Electron	Positron
L0 (harmonic matter)	$m = 0$	$m = 0$	$m = 0$	$m = 0$
L1	$m = 0$	$m = 0$	$m = 1$	$m = 1$
L2 (ordinary matter)	$m = 1$	$m = 1$	$m = 2$	$m = 2$
L3	$m = 2$	$m = 2$	$m = 3$	$m = 3$
Lx	$m = x - 1$	$m = x - 1$	$m = x$	$m = x$

11.2 Formation of Harmonic Trons

The UST treats harmonic trons as particular cases of *virtual particles* produced spontaneously in *quantum vacuum* by quantum fluctuations. In quantum theory, the creation of virtual particles from quantum vacuum is governed by the Heisenberg's uncertainty principle. This principle states that a virtual particle with the energy ΔE may pop in and out of existence and will exist during the time Δt if the following condition is met:

$$\Delta E \, \Delta t \leq \frac{h}{2\pi} \qquad (11.2\text{-}1)$$

where h is the Planck constant expressed by the equation:

$$h = \frac{e^2}{2\alpha c \varepsilon_0} \qquad (11.2\text{-}2)$$

According to the UST, the pairs of polarized virtual particles can be transformed into the harmonic trons by the *natural selection process*. For this process to take place, the following two conditions must be met:

1. The relative radii of the leading strings of the constituent toryces b_1 must be within the ranges defined by the *Spacetime Uncertainty Principle*.
2. The particles must be made up of polarized toryces which spacetime properties are based on three fundamental spacetime postulates outlined in Chapter 5.

The Spacetime Uncertainty Principle is derived from Eqs. (11.2-1) and (11.2-2) by replacing the energy ΔE with the toryx matter energy E_m expressed by Eq. (10.4-7) and the time Δt with the period of the toryx leading string T_1 expressed by Eq. (5.6-10a). Consequently, we obtain:

$$\text{Real toryces:} \quad \frac{b_1(b_1-1)}{\sqrt{2b_1-1}} \leq \frac{1}{\pi\alpha Q_q} \tag{11.2-3}$$

$$\text{Imaginary toryces:} \quad \frac{ib_1(b_1-1)}{\sqrt{2b_1-1}} \leq \frac{1}{\pi\alpha Q_q} \tag{11.2-4}$$

For the non-oscillated harmonic toryces ($Q_q = 1$), we obtain from Eqs. (11.2-3) and (11.2-4):

$$0.5000164241 < b_1 < 16.1229570320 \tag{11.2-5}$$

$$0.4999835759 > b_1 > -15.1229570320 \tag{11.2-6}$$

Figures 11.2-1 – 11.2-5 and Tables 11.2.1 – 11.2.5 show the composition and properties of the harmonic trons and their constituent toryces.

Basic harmonic electron – The basic harmonic electron $\downarrow e_{0,0,0}^{-1}$ is made up of two components: two basic harmonic real negative toryx $\downarrow E_{0,0,0}^{-\frac{1}{2}}$ and one half of the basic harmonic imaginary negative toryx $\downarrow \breve{E}_{0,0,0}^{-2}$ as shown by the equation:

$$\downarrow e_{0,0,0}^{-1} = 2 \downarrow E_{0,0,0}^{-\frac{1}{2}} + \frac{1}{2} \downarrow \breve{E}_{0,0,0}^{-2}$$

Figure 11.2.1 and Table 11.2.1 show the composition, properties, cross-section and dimensions of the basic harmonic electron $\downarrow e_{0,0,0}^{-1}$.

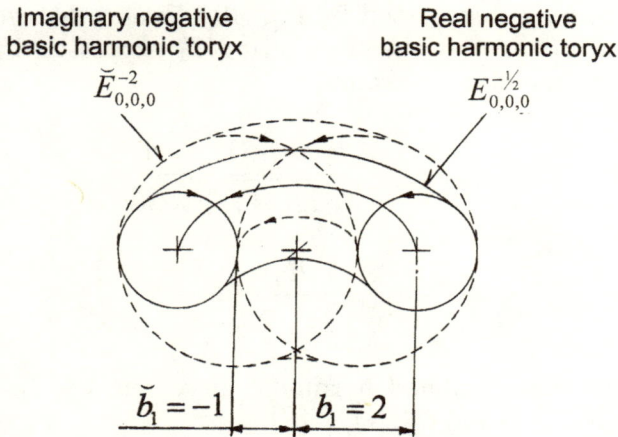

Figure 11.2.1. Cross-section and dimensions of the basic harmonic electron $\downarrow e_{0,0,0}^{-1}$ and its constituent toryces.

Table 11.2.1. Composition and properties of the basic harmonic electron $\downarrow e_{0,0,0}^{-1}$ and its constituent toryces.

Toryx	b_1	e_t / e	μ_t / μ_N	m_{tg} / m_e	P_t	N
$\downarrow E_{0,0,0}^{-\frac{1}{2}}$	2.0	-0.50	-5.80196032	0.5000	- 1.5	2
$\downarrow \breve{E}_{0,0,0}^{-2}$	-1.0	-2.00	-23.20784127	2.0000	+ 6.00	0.5
Basic electron $\downarrow e_{0,0,0}^{-1}$		-1.00	-11.60392064	1.0000	0.00	1

Basic harmonic positron – The basic harmonic positron $\uparrow e_{0,0,0}^{+1}$ is made up of two components: two basic harmonic real positive toryx $\uparrow E_{0,0,0}^{+\frac{1}{2}}$ and one half of the basic harmonic imaginary positive toryx $\uparrow \breve{E}_{0,0,0}^{+2}$ as shown by the equation:

$$\uparrow e_{0,0,0}^{+1} = 2 \uparrow E_{0,0,0}^{+\frac{1}{2}} + \frac{1}{2} \uparrow \breve{E}_{0,0,0}^{+2}$$

Figure 11.2.2 and Table 11.2.2 show the composition, properties, cross-section and dimensions of the basic harmonic positron $\uparrow e_{0,0,0}^{+1}$.

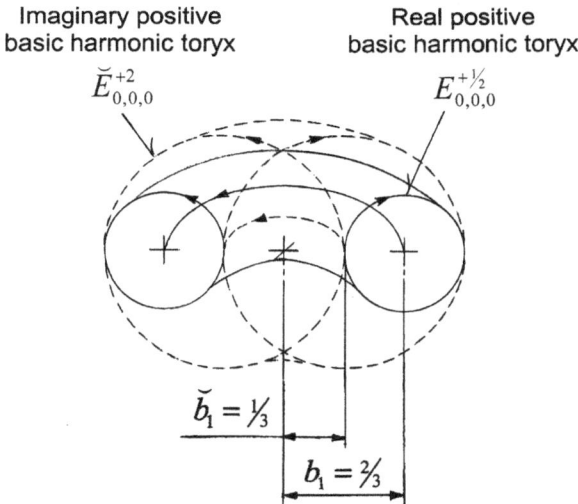

Imaginary positive basic harmonic toryx $\breve{E}_{0,0,0}^{+2}$

Real positive basic harmonic toryx $E_{0,0,0}^{+\frac{1}{2}}$

$\breve{b}_1 = \frac{1}{3}$

$b_1 = \frac{2}{3}$

Figure 11.2.2. Cross-section and dimensions of the basic harmonic positron $\uparrow e_{0,0,0}^{+1}$ and its constituent toryces.

Table 11.2.2. Composition and properties of the basic harmonic positron $\uparrow e_{0,0,0}^{+1}$ and its constituent toryces.

Toryx	b_1	e_t/e	μ_t/μ_N	m_{tg}/m_e	P_t	N
$\uparrow E_{0,0,0}^{+\frac{1}{2}}$	$\frac{2}{3}$	+ 0.50	+1.93398677	0.5000	$+\frac{1}{6}$	2
$\uparrow \breve{E}_{0,0,0}^{+2}$	$\frac{1}{3}$	+ 2.00	+7.73594709	2.0000	$-\frac{2}{3}$	0.5
Basic positron $\uparrow e_{0,0,0}^{+1}$		+1.00	+3.86797355	1.000	0.00	1

Resonant oscillated harmonic electron - Figure 11.2.3 and Table 11.2.3 show the composition, properties, cross-section and dimensions of the resonant oscillated harmonic electron $\downarrow er^{-1}_{0,(0/1),1}$ in which the toryx oscillation quantum state $q = 1$. It follows from Eqs. (9.3-20) and (9.3-22) that for these electrons all dimensions and magnetic moments of their constituent oscillated harmonic toryces are reduced by the toryx oscillation factor $Q_q = 3$. At the same time, their gravitational masses are increased by the same factor.

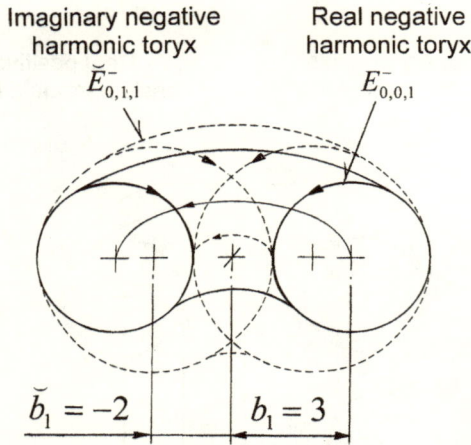

Imaginary negative harmonic toryx $\breve{E}^-_{0,1,1}$

Real negative harmonic toryx $E^-_{0,0,1}$

$\breve{b}_1 = -2$ $b_1 = 3$

Figure 11.2.3. Cross-section and dimensions of the resonant oscillated harmonic electron $\downarrow er^{-1}_{0,(0/1),1}$ and its constituent toryces.

Table 11.2.3. Composition and properties of the resonant oscillated harmonic electron $\downarrow er^{-1}_{0,(0/1),1}$ and its constituent toryces.

Toryx	b_1	e_i/e	μ_i/μ_N	m_{tg}/m_e	P_t	N
$\downarrow E^{-1/2}_{0,0,1}$	2.0	-0.50	-1.93398677	1.5000	- 4.5	1
$\downarrow \breve{E}^{-3/2}_{0,1,1}$	-2.0	-1.50	-7.49029856	4.5000	+ 22.5	1
Resonant electron $\downarrow er^{-1}_{0,(0/1),1}$		-1.00	-4.71214267	3.0000	9.0	1

Basic harmonic ethertron - The basic harmonic ethertron $a_{0,0,0}^{0}$ is made up of two components: one basic harmonic real negative toryx $\downarrow A_{0,0,0}^{-\frac{1}{2}}$ and one basic harmonic real positive toryx $\uparrow A_{0,0,0}^{+\frac{1}{2}}$ as shown by the equation:

$$a_{0,0,0}^{0} = \downarrow A_{0,0,0}^{-\frac{1}{2}} + \uparrow A_{0,0,0}^{+\frac{1}{2}}$$

Figure 11.2.4 and Table 11.2.4 show the composition, properties, cross-section and dimensions of the basic harmonic ethertron $a_{0,0,0}^{0}$.

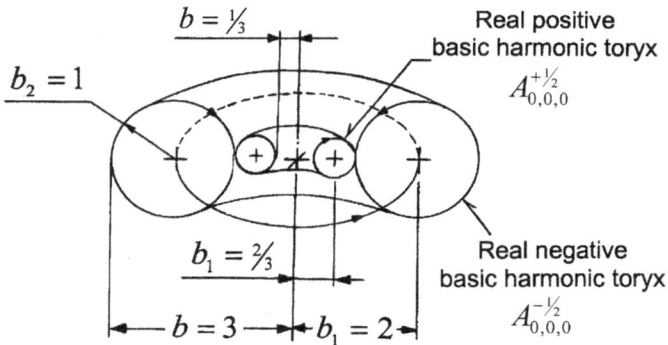

Figure 11.2.4. Cross-section and dimensions of the basic harmonic ethertron $a_{0,0,0}^{0}$ and its constituent toryces.

Table 11.2.4. Composition and properties of the basic harmonic ethertron $a_{0,0,0}^{0}$ and its constituent toryces.

Toryx	b_1	e_t / e	μ_t / μ_N	m_{tg} / m_e	P_t	N
$\downarrow A_{0,0,0}^{-\frac{1}{2}}$	2.0	-0.50	-5.80196032	0.500000	$-\frac{3}{2}$	1
$\uparrow A_{0,0,0}^{+\frac{1}{2}}$	$\frac{2}{3}$	+0.50	+1.93398677	0.500000	$+\frac{1}{6}$	1
Ethertron $a_{0,0,0}^{0}$	0.00	-1.93398677	0.500000	$-\frac{2}{3}$	1

Basic harmonic singulatron - The basic harmonic singulatron $\breve{a}_{0,0,0}^{0}$ is made up of two components: one basic harmonic imaginary negative toryx $\downarrow \breve{A}_{0,0,0}^{-2}$ and one basic harmonic imaginary positive toryx $\uparrow \breve{A}_{0,0,0}^{+2}$ as shown by the equation:

$$\breve{a}_{0,n,q}^{0} = \downarrow \breve{A}_{0,n,q}^{-2} + \uparrow \breve{A}_{0,n,q}^{+2}$$

Figure 11.2.5 and Table 11.2.5 show the composition, properties, cross-section and dimensions of the basic harmonic singulatron $\breve{a}_{0,0,0}^{0}$.

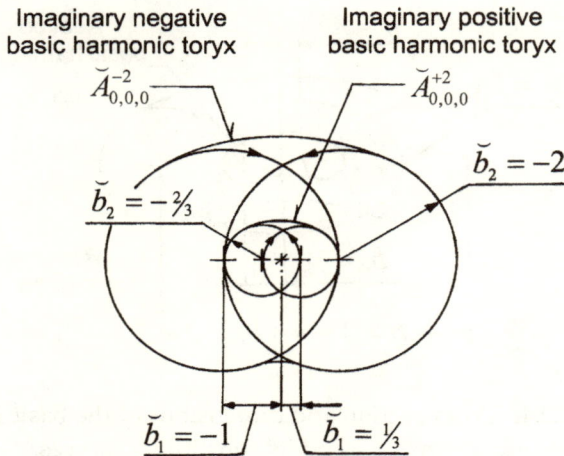

Figure 11.2.5 Cross-section and dimensions of the basic harmonic singulatron $\breve{a}_{0,n,q}^{0}$ and its constituent toryces.

Table 11.2.5. Composition and properties of the basic harmonic singulatron $\breve{a}_{0,0,0}^{0}$ and its constituent toryces.

Toryx	b_1	e_t/e	μ_t/μ_N	m_{tg}/m_e	P_t	N
$\downarrow \breve{A}_{0,0,0}^{-2}$	-1.0	-2.0	-23.20784127	2.000000	6.0	1
$\uparrow \breve{A}_{0,0,0}^{+2}$	$\frac{1}{3}$	+2.0	+7.73594709	2.000000	$-\frac{2}{3}$	1
Singulatron $\breve{a}_{0,0,0}^{0}$	**0.00**		**-7.73594709**	**2.000000**	$\frac{8}{3}$	1

11.3 Nucleon Crystal Structure

In nucleons, the degrees of freedom of their constituent trons are reduced by locating them at the centers and vertices of a nucleon bipyramid hexagonal crystal structure called the *nucleon crystal* (Fig. 11.3.1).

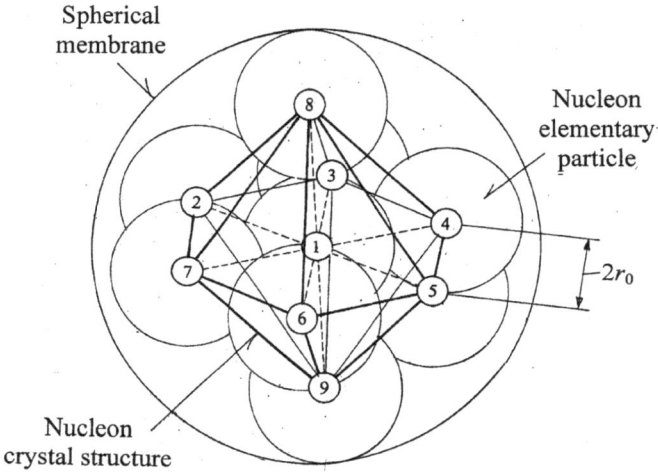

Figure 11.3.1. Nucleon crystal.

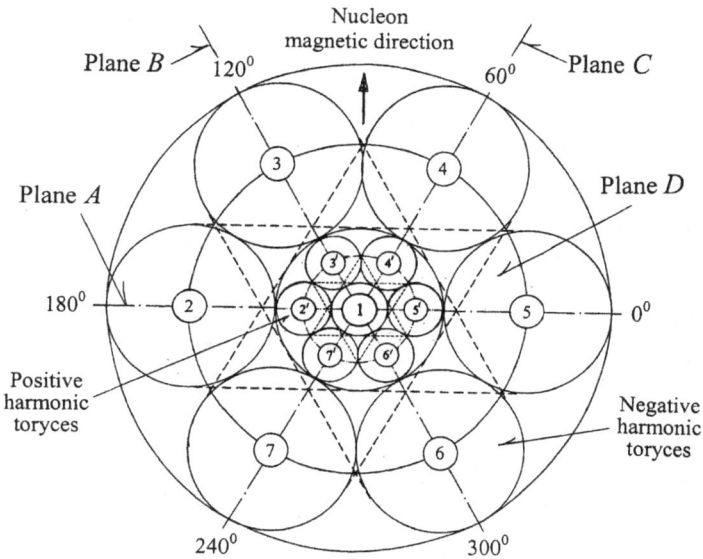

Figure 11.3.2. Plan view cross-section of the nucleon crystal.

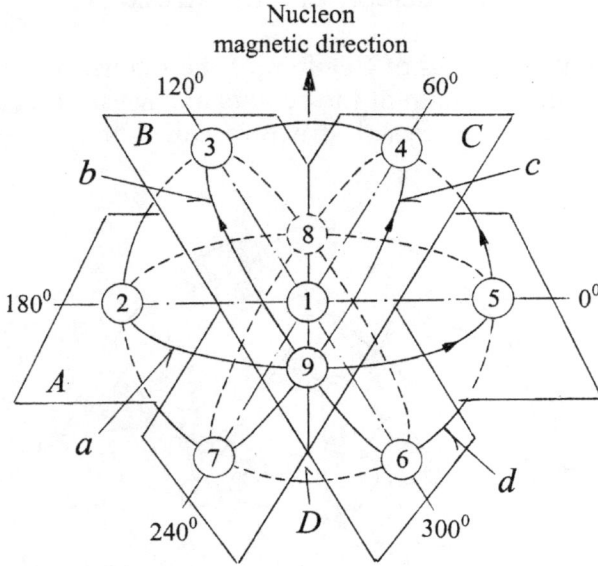

Figure 11.3.3. Formation of the nucleon crystal structure.

Table 11.3.1. Compositions and properties of the nucleon crystal $\downarrow cn^0_{0,0,0}$.

Trons	Locations Fig. 11.3.3	μ / μ_N	m_g / m_e	P_t	N
Singulatron $\downarrow \breve{a}^0_{0,0,0}$	Vertex 1	-7.73594709	2.0000	$+\frac{8}{3}$	1
Ethertron $\downarrow a^0_{0,0,0}$	Plane A	-1.93398677	0.5000	$-\frac{2}{3}$	1
Ethertron $\uparrow a^0_{0,0,0}$	Plane B	+ 0.96699339	0.5000	$-\frac{2}{3}$	1
Ethertron $\downarrow a^0_{0,0,0}$	Plane C	+ 0.96699339	0.5000	$-\frac{2}{3}$	1
Ethertron $\downarrow a^0_{0,0,0}$	Plane D	0.0	0.5000	$-\frac{2}{3}$	1
Nucleon crystal $\downarrow cn^0_{0,0,0}$		-3.86797355	2.0000	0.0	1

As shown in Fig. 11.3.2, there are two parts in the crystal, outer and inner. The inner part is three times smaller than the nucleon outer part and it resides inside the outer part. Each part of the nucleon crystal has the center 1 and eight vertices 2 through 9. The vertices of the outer part are formed at the intersections of leading strings a, b, c and d of the real negative harmonic toryces $A_{0,0,0}^{-7/2}$ as shown in Fig. 11.3.3, while the vertices of the inner part are formed at the intersections of leading strings of the real positive harmonic toryces $A_{0,0,0}^{+7/2}$. The leading strings a, b, c and d of the harmonic toryces are located in three different planes A, B, C and D, with the centers of the leading strings located at the crystal center 1.

Table 11.3.1 shows the compositions and properties of the nucleon crystal. Notably, the directions of the tron magnetic fields are perpendicular to the related planes A, B, C and D.

11.4 Excited Trons

According to the UST, the excited particles are created from the harmonic particles. The harmonic toryces, however, do not have sufficient inner energy to propel themselves into the stable excitation quantum states. Fortunately, there is a way of accumulating the needed energy.

The Spacetime Uncertainty Principle sets no limits on the number of harmonic trons that can be created within a certain limited space. Under a certain pressure the squeezed trons form oscillated harmonic trons which stored energy is then used to propel themselves into the stable excitation quantum states.

Figures 11.4-1 – 11.4-4 and Tables 11.4.1 – 11.4.4 show the composition and properties of the basic excited trons and their constituent toryces.

Basic excited electrons - The basic excited electron $\downarrow e^{-1}_{m,n,q}$ of the spacetime level L is made up of two components: one basic excited real negative toryx $\downarrow E^{-\frac{1}{2}}_{m,n,q}$ and one basic excited imaginary negative toryx $\downarrow \breve{E}^{-2}_{m,n,q}$ as shown by the equation:

$$\downarrow e^{-1}_{m,n,q} = \downarrow E^{-\frac{1}{2}}_{m,n,q} + \downarrow \breve{E}^{-2}_{m,n,q}$$

Figure 11.4.1 and Table 11.4.1 show the composition, properties, cross-section and dimensions of the basic excited electron $\downarrow e^{-1}_{2,1,0}$ of the spacetime level $L2$ (ordinary matter).

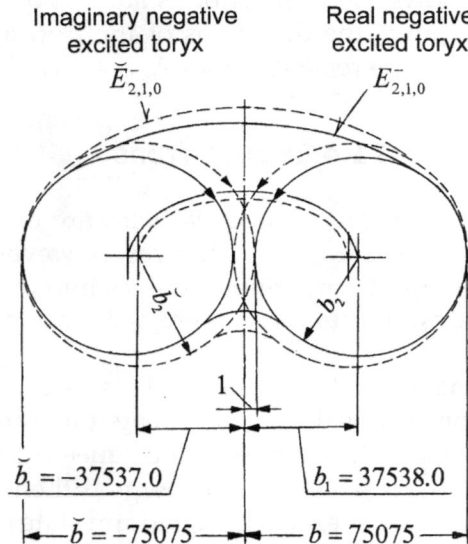

Figure 11.4.1. Cross-section and dimensions of the basic excited electron $\downarrow e^{-1}_{2,1,0}$ of of ordinary matter for $m = 2$, $n = 1$ and $q = 0$.

Table 11.4.1. Composition and properties of the basic excited electron $\downarrow e^{-1}_{2,1,0}$ of ordinary matter for $m = 2$, $n = 1$ and $q = 0$.

Toryx	b_1	e_t / e	μ_t / μ_N	m_{tg} / m_e	P_t	N
$\downarrow E^{-1}_{2,1,0}$	37538.0	-0.99997336	-1835.609190	0.99997336	-75075	1.000027
$\downarrow \breve{E}^{-1}_{2,1,0}$	-37537.0	-1.00002664	-1835.706994	1.00002664	+75075	0.999973
Electron $\downarrow e^{-1}_{2,1,0}$		- 1.0000000	-1835.658092	1.00000000	0.0	1

Basic excited positron - The basic excited positron $\uparrow e^{+1}_{m,n,q}$ of the spacetime level L is made up of two components: one basic excited real positive toryx $\uparrow E^{+\frac{1}{2}}_{m,n,q}$ and one basic excited imaginary positive toryx $\uparrow \breve{E}^{+2}_{m,n,q}$ as shown by the equation:

$$\uparrow e^{+1}_{m,n,q} = \uparrow E^{+\frac{1}{2}}_{m,n,q} + \uparrow \breve{E}^{+2}_{m,n,q}$$

Figure 11.4.2 and Table 11.4.2 show the composition, properties, cross-section and dimensions of the basic excited positron $\uparrow e^{+1}_{2,1,0}$ of the spacetime level $L2$ (ordinary matter).

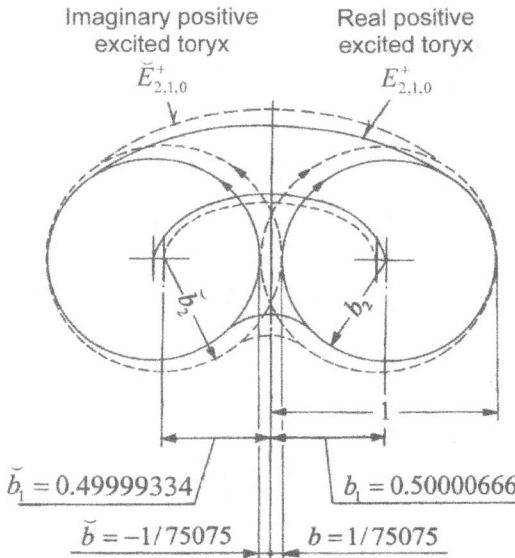

Figure 11.4.2. Cross-section and dimensions of the basic excited positron $\uparrow e^{+1}_{2,1,0}$ of ordinary matter for $m = 2$, $n = 1$ and $q = 0$.

Table 11.4.2. Composition and properties of the basic excited positron $e^{+1}_{2,1,0}$ of ordinary matter for $m = 2$, $n = 1$ and $q = 0$.

Toryx	b_1	e_t/e	μ_t/μ_N	m_{tg}/m_e	P_t	N
$\uparrow E^{+1}_{2,1,0}$	0.50000666	+0.99997336	+0.024450	0.99997336	$+1.3\cdot10^{-5}$	1.000027
$\uparrow \breve{E}^{+1}_{2,1,0}$	0.49999334	+1.00002664	+0.024452	1.00002664	$-1.3\cdot10^{-5}$	0.999973
Positron $\uparrow e^{+1}_{2,1,0}$		+1.0000000	+0.024451	1.00000000	0.0	1

Basic excited ethertron - The basic excited ethertron $a^0_{m,n,q}$ of the spacetime level L is made up of two components: one basic excited real negative toryx $\downarrow A^-_{m,n,q}$ and one basic excited real positive toryx $\uparrow A^+_{m,n,q}$ as shown by the equation:

$$a^0_{m,n,q} = \downarrow A^-_{m,n,q} + \uparrow A^+_{m,n,q}$$

Figure 11.4.3 and Table 11.4.3 show the composition, properties, cross-section and dimensions of the basic excited ethertron $a^0_{1,1,0}$ of the spacetime level $L2$ (ordinary matter).

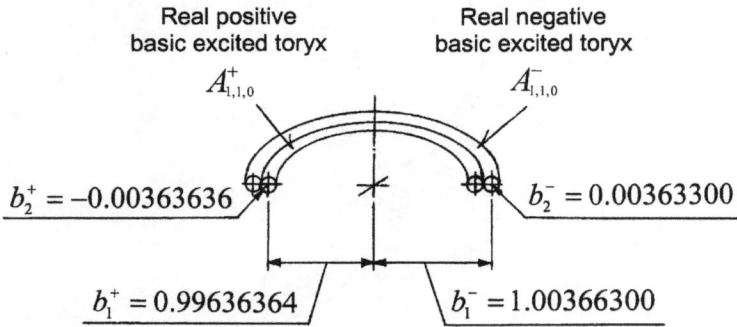

Real positive basic excited toryx $A^+_{1,1,0}$

Real negative basic excited toryx $A^-_{1,1,0}$

$b^+_2 = -0.00363636$

$b^-_2 = 0.00363300$

$b^+_1 = 0.99636364$

$b^-_1 = 1.00366300$

Figure 11.4.3. Cross-section and dimensions of the basic excited ethertron $a^0_{1,1,0}$ of ordinary matter for $m = 1$, $n = 1$ and $q = 0$.

Table 11.4.3. Composition and properties of the basic excited ethertron $a^0_{1,1,0}$ of ordinary matter for $m = 1$, $n = 1$ and $q = 0$.

Toryx	b_1	e_t/e	μ_t/μ_N	m_{tg}/m_e	P_t	N
$A^-_{1,1,0}$	1.003663	- 0.00365	-0.02454023	0.00364964	- 0.003676	1
$A^+_{1,1,0}$	0.996364	+ 0.00365	+0.02436175	0.00364964	+ 0.003623	1
Ethertron $a^0_{1,1,0}$	0.00	-0.00008924	0.00364964	- 0.000053	1

Basic excited singulatron - The basic excited singulatron $\breve{a}^0_{m,n,q}$ of the spacetime level L is made up of two components: one basic excited imaginary negative toryx $\downarrow \breve{A}^-_{m,n,q}$ and one basic excited imaginary positive toryx $\uparrow \breve{A}^+_{m,n,q}$ as shown by the equation:

$$\breve{a}^0_{m,n,q} = \downarrow \breve{A}^-_{m,n,q} + \uparrow \breve{A}^+_{m,n,q}$$

Figure 11.2.5 and Table 11.2.5 show the composition, properties, cross-section and dimensions of the basic excited singulatron $\breve{a}^0_{1,1,0}$ of the spacetime level $L2$ (ordinary matter).

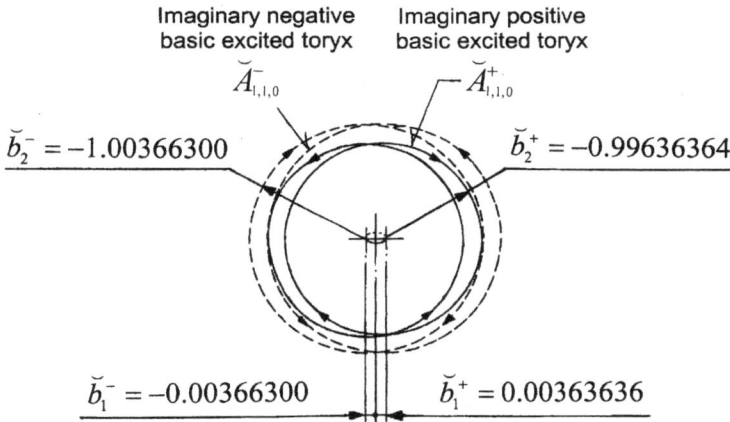

Figure 11.4.4. Cross-section and dimensions of the basic excited singulatron $\breve{a}^0_{1,1,0}$ of ordinary matter for $m = 1$, $n = 1$ and $q = 0$.

Table 11.4.4. Properties of the basic excited singulatron $\breve{a}^0_{1,1,0}$ of ordinary matter for $m = 1$, $n = 1$ and $q = 0$.

Toryx	b_1	e_t / e	μ_t / μ_N	m_{tg} / m_e	P_t	N
$\breve{A}^-_{1,1,0}$	- 0.0036630	-274.0	-1842.382113	274.000	+276.007	1
$\breve{A}^+_{1,1,0}$	0.0036364	+274.0	+1828.982971	274.000	-276.007	1
Singulatron $\breve{a}^0_{1,1,0}$	0.00		-6.69957132	274.000	+2.0	1

11.5 Oscillated Excited Trons

The oscillated excited electrons and positrons represent leptons of the Standard Model.

Table 11.5. Comparison of calculated properties of oscillated excited electrons of ordinary matter with measured properties of respective leptons.

Oscillated excited electrons		Physical properties		
Name	Symbol	μ/μ_B	m_g/m_e	e/e_0
Electron	$e_{2,1,0}^{-1}$	**-0.99973052**	**1.000000000**	**-1.00**
Measured values:		-1.00115965	1.000000000	-1.00
Calc./measured ratio:		0.9986	1.0000	1.00
Name	Symbol	μ/μ_B	m_g/m_e	e/e_0
Negative 3e-tron	$e_{2,1,1}^{-1}$	**-0.33324351**	**3.000000000**	**-1.00**
Measured values:		?	?	-1.00
Calc./measured ratio:		?	?	1.00
Name	Symbol	μ/μ_B	m_g/m_e	e/e_0
Negative μ-tron	$e_{2,1,2}^{-1}$	**-4.864869×10^{-3}**	**205.5000000**	**-1.00**
Measured values		-4.841970×10^{-3}	206.7682841	-1.00
Calc./measured ratio		1.0047	0.9939	1.00
Name	Symbol	μ/μ_B	m_g/m_e	e/e_0
Negative τ-tron	$e_{2,1,3}^{-1}$	**-2.840799×10^{-4}**	**3519.187500**	**-1.00**
Measured values		?	3477.482833	-1.00
Calc./measured ratio		?	1.0120	1.00
Name	Symbol	μ/μ_B	m_g/m_e	e/e_0
Negative x-tron	$e_{2,1,4}^{-1}$	**-2.799328×10^{-5}**	**35713.23612**	**-1.00**
Measured values		?	?	-1.00
Calc./measured ratio		?	?	1.00

As follows from Table 11.5, the UST predicts new leptons: the negative 3e-tron that is three times heavier than electron and negative trons, beginning from the negative x-tron, that are heavier than the negative τ-tron.

11.6 Nucleon Core

The nucleon core, or shortly the *nucore*, provides a bulk of the nucleon mass. The singulatrons of the nucore coexist interchangeably with the ethertrons. To obey the Spacetime Conservation Law, each singulatron must be matched with T ethertrons.

In the nucore $nc_{m,n,q}^{0}$ the magnetic moments of its trons are arranged according to the equation:

$$nc_{m,n,q}^{0} = 3(\downarrow \breve{a}_{m,n,q}^{0} + T \downarrow a_{m,n,q}^{0}) + 4(\uparrow \breve{a}_{m,n,q}^{0} + T \uparrow a_{m,n,q}^{0})$$

Table 11.6.1. Components and properties of the nucore $nc_{1,1,0}^{0}$ of ordinary matter for $m = 1$, $n = 1$ and $q = 0$.

Trons	Vertices Fig. 11.3.2	μ / μ_N	m_g / m_e	P_t	N
Singulatron $\downarrow \breve{a}_{1,1,0}^{0}$	1, 2, 5	- 6.69957132	274.00	+2.0	3
Ethertron $\downarrow a_{1,1,0}^{0}$	1, 2, 5	- 0.00008924	0.00364964	-2.66×10^{-5}	$3T$
Singulatron $\uparrow \breve{a}_{1,1,0}^{0}$	3, 4, 6, 7	+ 6.69957132	274.00	+2.0	4
Ethertron $\uparrow a_{1,1,0}^{0}$	3, 4, 6, 7	+ 0.00008924	0.00364964	-2.66×10^{-5}	$4T$
Nucore $nc_{1,1,0}^{0}$		+ 6.69957132	1918.0000	0.0	1

Table 11.6.1 shows the compositions and properties of the nucore $nc_{1,1,0}^{0}$ of ordinary matter. This table also shows locations of singulatrons and ethertrons making up the nucores. According to Eq. (9.2-7), to comply with the Spacetime Conservation Law, each excited singulatron must be matched with $T = 75076$ ethertrons.

11.7 Nucleons

Described below are the compositions and properties of proton and two kinds of neutrons. All nucleons have the same crystal structure shown in Figure 11.7.1. Shown with circles are the vacant vertices occupied by elementary particles of adjacent nucleons during formation of complex nuclei.

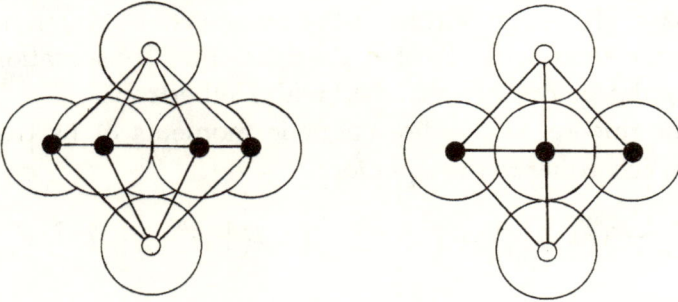

Figure 11.7.1. Front view (left) and side view (right) of a nucleon crystal structure.

Proton – The proton $\uparrow p_L^{+1}$ of the matter level L is made up of three components: one nucleon crystal $\downarrow cn_{m,n,q}^0$, one nucore $\uparrow nc_{m,n,q}^0$ and one excited positron $\downarrow e_{m,n,q}^{+1}$ as shown by the equation:

$$\uparrow p_L^{+1} = \downarrow cn_{0,n,q}^0 + \uparrow nc_{m,n,q}^0 + \downarrow e_{m,n,q}^{+1}$$

Table 11.7.1 shows the composition and properties of the proton of ordinary matter. The center of the excited positron is located at the crystal center 1 (Fig. 11.3.1).

Notably, the direction of magnetic field of excited positron is reversed compared to its direction shown in Table 11.4.2. The proton is stable, because its spacetime polarization $P_t = 0$.

Table 11.7.1. Components and properties of proton
of ordinary matter.

Components	Table	μ/μ_N	m_{tg}/m_e	P_t	N
Nucleon crystal $\downarrow cn^0_{0,n,0}$	11.3.1	− 3.86797355	2.00	0.0	1
Nucore $\uparrow nc^0_{1,1,0}$	11.6.1	+ 6.69957132	1918.00	0.0	1
Basic excited positron $\downarrow e^{+1}_{2,1,0}$	11.4.2	− 0.02445099	1.00	0.0	1
Proton $\uparrow p^{+1}_{L2}$		**+ 2.80714679**	**1921.00**	**0.0**	**1**
Measured values		+ 2.79284736	1836.15267	-	-
Calculated/measured ratio		1.0051	1.0462	-	-

Basic neutron – The basic neutron $\downarrow n^0_L$ of the spacetime level L is made up of two components: one proton $\uparrow p^0_L$ and one basic harmonic electron $\downarrow e^{-1}_{0,n,q}$ as shown by the equation:

$$\downarrow n^0_L = \uparrow p^{+1}_L + \downarrow e^{-1}_{0,n,q}$$

Table 11.7.2 shows the composition and properties of the basic neutron of ordinary matter. The center of the basic harmonic electron is located at the crystal center 1 (Fig. 11.3.1).

Table 11.7.2. Components and properties of basic neutron
of ordinary matter.

Components	Table	μ/μ_N	m_{tg}/m_e	P_t	N
Proton $\uparrow p^{+1}_{L2}$	11.7.1	+ 2.80714679	1921.00	0.0	1
Basic harmonic electron $\downarrow e^{-1}_{0,1,0}$	11.2.1	− 11.60392064	1.0000	0.0	1
Basic neutron $\downarrow n^0_{L2}$		**− 8.79677385**	**1922.00**	**0.0**	**1**

The basic neutron is stable, because its spacetime polarization $P_t = 0$.

Resonant oscillated neutron – The resonant oscillated neutron $\downarrow nr_L^0$ of the spacetime level L is made up of two components: one proton $\downarrow p_L^0$ and one resonant oscillated electron $\downarrow er_{0,n,q}^{-1}$ as shown by the equation:

$$\downarrow n_L^0 = \uparrow p_L^{+1} + \downarrow er_{0,n,q}^{-1}$$

Table 11.7.3 shows the composition and properties of the resonant oscillated neutron of ordinary matter. The center of the resonant oscillated electron is located at the crystal center 1 (Fig. 11.3.1).

Table 11.7.3. Components and properties of resonant oscillated neutron of ordinary matter.

Components	Table	μ / μ_N	m_{tg} / m_e	P_t	N
Proton p_{L2}^{+1}	11.7.1	+ 2.80714679	1921.00	0.0	1
Resonant electron $er_{0,1,1}^{-1}$	11.2.3	- 4.71214267	3.0000	-9.0	1
Resonant neutron nr_{L2}^{0}		- 1.90499588	1924.00	-9.0	1
Measured values		- 1.91304272	1838.68366	-	-
Calculated/measured ratio		0.9958	1.0464	-	-

The resonant oscillated neutron is unstable, because its spacetime polarization $P_t > 0$.

11.8 Hydrogen & Helium

Described below are the compositions of hydrogen atom, helium atom and their isotopes. When forming nuclei containing more than one nucleon, the adjacent nucleons interlock by filling vacant vertices of their crystal structures.

Hydrogen atom – The hydrogen atom $\downarrow H_L^0$ of the spacetime level L is made up of two components: one proton $p_L^0 \uparrow$ and one basic excited electron $\downarrow e_{m,n,q}^{-1}$ as shown by the equation:

$$\downarrow H_L^{+1} = \downarrow p_L^{+1} + \downarrow e_{m,n,q}^{-1}$$

Table 11.8.1 shows the composition and properties of the hydrogen atom of ordinary matter. The center of the basic excited electron is located at the crystal center 1 (Fig. 11.3.1).

The hydrogen atom is stable, because its spacetime polarization $P_t = 0$.

Table 11.8.1. Components and properties of hydrogen atom of ordinary matter.

Components	Table	μ / μ_N	m_{tg} / m_e	P_t	N
Proton $\uparrow p_{L2}^{+1}$	11.7.1	+ 2.80714679	1921.00	0.0	1
Basic excited electron $\downarrow e_{2,1,0}^{-1}$	11.4.1	- 1835.658092	1.0000	0.0	1
Hydrogen $\downarrow_1^1 H_{L2}^0$		- 1832.850945	1922.00	**0.0**	**1**
Measured values		-	1836.154060	-	-
Calculated/measured ratio		-	1.0468	-	-

Crystal structure of the hydrogen nucleon is the same as the proton crystal structure shown in Fig. 11.7.1.

Deuterium - The deuterium $\downarrow_1^2 H_L^0$ of the spacetime level L is made up of one proton $\uparrow p_L^{+1}$, one basic neutron $\downarrow n_L^0$ and one basic excited electron $\downarrow e_{m,n,q}^{-1}$ according to the equation:

$$\downarrow_1^2 H_L^0 = \uparrow p_L^{+1} + \downarrow n_L^0 + \downarrow e_{m,n,q}^{-1}$$

Figure 11.8.1 shows crystal structure of the deuterium nucleus.

Table 11.8.2 shows the composition and properties of the deuterium of ordinary matter. The deuterium is stable because its spacetime polarization $P_t = 0$.

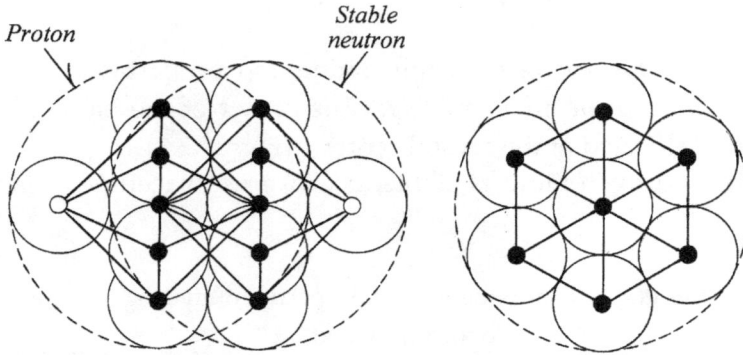

Figure 11.8.1. Crystal structure of the deuterium nucleus: front view (left) and side view (right).

Table 11.8.2. Components and properties of deuterium of ordinary matter.

Components	Table	μ/μ_N	m_{tg}/m_e	P_t	N
Proton $\uparrow\, p_{L2}^{+1}$	11.7.1	+ 2.80714679	1921.00	0.0	1
Basic neutron $\downarrow\, n_{L2}^{0}$	11.7.2	− 8.79677385	1922.00	0.0	1
Basic excited electron $\downarrow\, e_{2,1,0}^{-1}$	11.4.1	− 1835.658092	1.0000	0.0	1
Deuterium $\downarrow_1^2 H_{L2}^0$		**− 1841.647719**	**3844.00**	**0.0**	**1**
Measured values		-	3671.485770	-	-
Calculated/measured ratio		-	1.0470	-	-

Tritium - The tritium $\downarrow_1^3 H_L^0$ of the spacetime level L is made up of one proton $\uparrow\, p_L^{+1}$, one basic neutron $\downarrow\, n_L^0$, one resonant oscillated neutron $\downarrow\, nr_L^0$ and one basic excited electron $\downarrow\, e_{m,n,q}^{-1}$ according to the equation:

$$\downarrow_1^3 H_L^0 = \uparrow\, p_L^{+1} + \downarrow\, n_L^0 + \downarrow\, nr_L^0 + \downarrow\, e_{m,n,q}^{-1}$$

Figure 11.8.2 shows crystal structure of the tritium nucleus. Table 11.8.3 shows the composition and properties of the tritium

of ordinary matter. The tritium is unstable because its spacetime polarization $P_t > 0$.

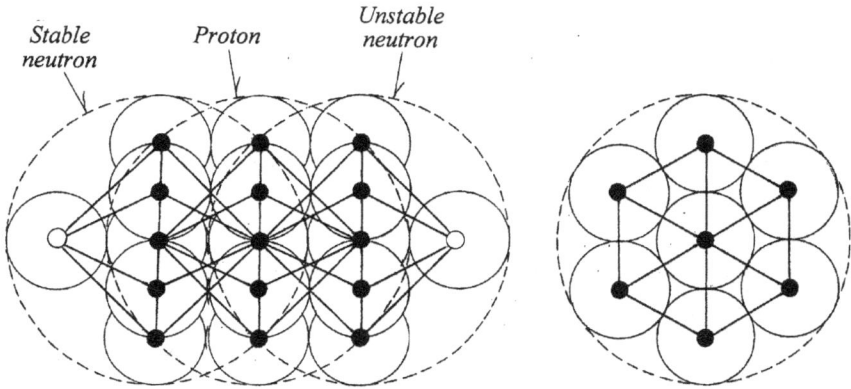

Figure 11.8.2. Crystal structure of the tritium nucleus: front view (left) and side view (right)

Table 11.8.3. Components and properties of tritium of ordinary matter.

Components	Table	μ / μ_N	m_{tg} / m_e	P_t	N
Proton $\uparrow p_{L2}^{+1}$	11.7.1	+ 2.80714679	1921.00	0.0	1
Basic neutron $\downarrow n_{L2}^{0}$	11.7.2	- 8.79677385	1922.00	0.0	1
Resonant neutron $\downarrow nr_{L2}^{0}$	11.7.3	- 1.90499588	1924.00	-18.0	1
Basic excited electron $\downarrow e_{2,1,0}^{-1}$	11.4.1	- 1835.658092	1.0000	0.0	1
Tritium $\downarrow {}_1^3 H_{L2}^0$		**-1843.552720**	**5768.00**	**18.0**	**1**
Measured values		-	5497.925226	-	-
Calculated/measured ratio		-	1.0491	-	-

Helium-3 - The helium-3 $\downarrow {}_2^3 He_L^0$ of the spacetime level L is made up of two protons $\uparrow p_L^{+1}$, one basic neutron $\downarrow n_L^0$ and two basic excited electrons $\updownarrow e_{m,n,q}^{-1}$ according to the equation:

$$\downarrow {}_{2}^{3}He_{L}^{0} = 2 \uparrow p_{L}^{+1} + \downarrow n_{L}^{0} + 2 \updownarrow e_{m,n,q}^{-1}$$

Figure 11.8.3 shows structure of the helium-3 nucleus crystal structure. Table 11.8.4 shows the composition and properties of the helium-3 of ordinary matter. The helium-3 is stable because its spacetime polarization $P_t = 0$.

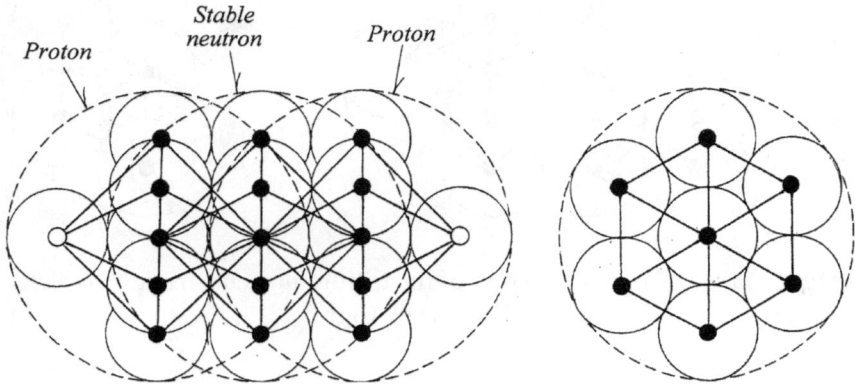

Figure 11.8.3. Crystal structure of the helium-3 nucleus: front view (left) and side view (right).

Table 11.8.4. Components and properties of the helium-3 of ordinary matter.

Components	Table	μ / μ_N	m_{tg} / m_e	P_t	N
Proton $\uparrow p_{L2}^{+1}$	11.7.1	+ 2.80714679	1921.00	0.0	2
Basic neutron $\downarrow n_{L2}^{0}$	11.7.2	- 8.79677385	1922.00	0.0	1
Basic excited electron $\downarrow e_{2,1,0}^{-1}$	11.4.1	- 1835.658092	1.0000	0.0	1
Basic excited electron $\uparrow e_{2,1,0}^{-1}$	11.4.1	+ 1835.658092	1.0000	0.0	1
Helium-3 $\downarrow {}_{2}^{3}H_{L2}^{0}$		- 3.182485	5766.00	0.0	1
Measured values		-	5497.888769	-	-
Calculated/measured ratio		-	1.0488	-	-

Helium-4 - The helium-4 $\,_2^4H_L^0$ of the spacetime level L is made up of two protons $\updownarrow p_L^{+1}$, two basic neutrons $\updownarrow n_L^0$ and two basic excited electrons $\updownarrow e_{m,n,q}^{-1}$ according to the equation:

$$_2^4He_L^0 = 2 \updownarrow p_L^{+1} + \updownarrow 2n_L^0 + 2 \updownarrow e_{m,n,q}^{-1}$$

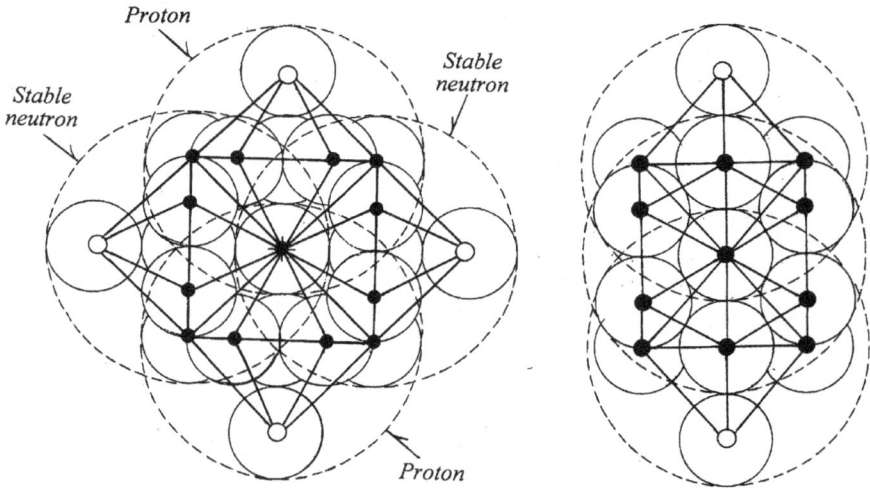

Figure 11.8.4 Crystal structure of the helium-4 nucleus: front view (left) and side view (right).

Table 11.8.5. Components and properties of the helium-4 of ordinary matter.

Components	Table	μ / μ_N	m_{tg} / m_e	P_t	N
Proton $\updownarrow p_{L2}^{+1}$	11.7.1	$\pm$ 2.80714679	1921.00	0.0	2
Basic neutron $\updownarrow n_{L2}^0$	11.7.2	$\mp$ 8.79677385	1922.00	0.0	2
Basic excited electron $\updownarrow e_{2,1,0}^{-1}$	11.4.1	$\mp$ 1835.658092	1.0000	0.0	2
Helium-4 $_2^4H_{L2}^0$		**0.00**	**7688.00**	**0.0**	**1**
Measured values		-	7296.303079	-	-
Calculated/measured ratio		-	1.0537	-	-

Figure 11.8.4 shows structure of the helium nucleus crystal structure. Table 11.8.5 shows the composition and properties of the helium-4 of ordinary matter. The helium-4 is stable because its spacetime polarization $P_t = 0$.

11.9 Effect of Spacetime Levels

Since the toryx exponential quantum state m increases with increase of the spacetime level L, the toryx excitation quantization parameter z increases too, as follows from Eq. (9.3-1). This produces substantial changes in the spacetime and physical properties of the nucleons and hydrogen atom.

Table 11.9.1. Composition and properties of the stable protons $\uparrow p_L^{+1}$ of several spacetime levels L.

Spacetime level	Proton	μ / μ_N	m_{tg} / m_e	Mass ratio
$L1$	$\uparrow p_{L1}^{+1}$	$+ 3.58152247$	17.0	1/113
$L2$ (ordinary matter)	$\uparrow p_{L2}^{+1}$	$+ 2.80714679$	1921.0	1.00
$L3$	$\uparrow p_{L3}^{+1}$	$+ 2.82946437$	262769.0	136.8

Table 11.9.2. Relative properties of the hydrogen atoms $\downarrow H_L^0$ in the ground state $n = 1$ of several spacetime levels L.

Spacetime levels L	Hydrogen atoms	Electron orbital radius ratio	Orbital magnetic moment ratio	Mass ratio	Atomic density ratio
$L1$	$\downarrow H_{L1}^0$	1/137.25	1/11.7	1/106.8	24200
$L2$ (ordinary matter)	$\downarrow H_{L2}^0$	1.0	1.0	1.0	1.0
$L3$	$\downarrow H_{L3}^0$	137.0	11.7	136.7	1/18800

Table 11.9.1 shows the properties of the stable protons of the spacetime levels $L1$, $L2$ and $L3$. Both composition and properties of the protons are greatly dependent on the spacetime level L. Notably, the proton mass of the spacetime level $L1$ is about 113 times less than the proton mass of the spacetime level $L2$, while

the proton mass of the spacetime level *L3* is about 137 times greater than the proton mass of the spacetime level *L2*.

Table 11.9.2 shows relative properties of the hydrogen atoms ↓ H_L^0 in the ground state *n* = 1 of the spacetime levels *L1* and *L3* in respect to these properties of the spacetime level *L2*.

- In the spacetime level *L1*, the nucleons are about 107 times lighter, the orbital radii of their atomic electrons are about 137 times smaller and, consequently, the atomic densities are about 24200 times greater than in the spacetime level *L2* (ordinary matter).
- In the spacetime level *L3*, the nucleons are about 137 times heavier, the orbital radii of their atomic electrons are about 137 times larger and, consequently, the atomic densities are about 18800 times smaller than in the spacetime level *L2* (ordinary matter).

Similar dependences of atomic properties on the spacetime levels are expected for all atoms. This opens a path for the creation of new materials with unique properties, including lighter materials with greater strength.

Since the atomic density decreases with the increase in the spacetime level, the UST proposes that the observed expansion of the universe could be caused by the expansion of spacetimes from lower to higher levels *L*.

11.10 Ether

The UST defines the ether as a spacetime entity in which an excited singulatron is surrounded by *T* ethertrons as shown in Figure 11.10. For the basic ethertrons and singulatrons, the number of the ethertons *T* is defined by Eq. (9.2-7).

We find from the equations describing physical properties of toryces (see Chapter 10) that, in the extreme case when the relative radius of the singulatron toryx leading string $b_1 \to 0$, the relative masses of the ethertrons are extremely small, while their bulk modulus of elasticity (rigidity) is very big. The rotational velocities of the trailing strings of their toryces approach the velocity of light *c*.

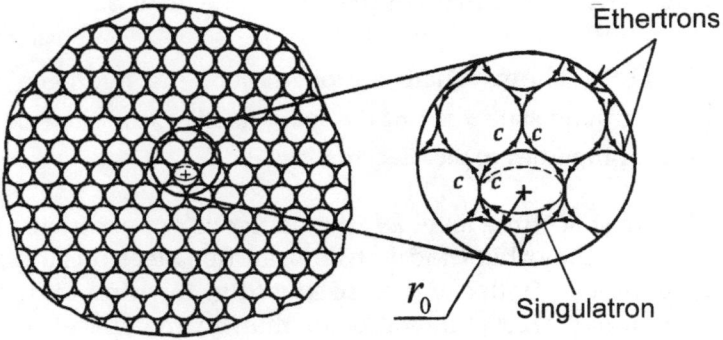

Figure 11.10. Structure of ether.

As shown in Figure 11.10, the directions of these velocities in the adjacent ethertrons are opposite to one another, making it very difficult to detect the ethertrons.

PART 3

Elementary Radiation Particles

12

BASIC CONCEPT OF HELYX

According to the UST, helyces are prime elements of radiation particles. Their structures and properties are based on the same basic principles as the structures and properties of toryces.

12.1 Helyx Basic Structure

Helyx is geometrically the 5-D dual-level helicola described in Chapter 2 and shown in Fig. 2.9. The first level of the helyx is in the form of a double-helical spiral shown in Figs. 12.1.1 and 12.1.2. There is an easy way to visualize the double-helical spiral. Consider two points, a and b, rotating around a pivot point m with the rotational velocity $\tilde{V}_{1r}$, while the pivot point m propagates along a straight line O_1O_1 with the translational velocity $\tilde{V}_{1t}$. Each moving point, a and b, forms a helical branch winding around the straight line O_1O_1. These two branches form a double helix called the *helyx leading string A_1*.

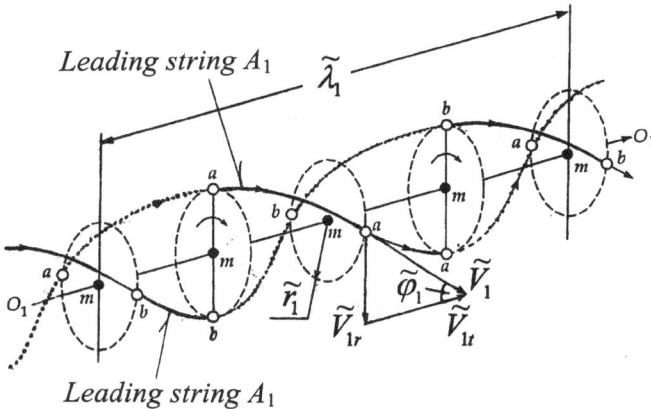

Figure 12.1.1. Structure of the helyx leading string.

The second level of the helyx is made up of two double helices forming the helyx *trailing string A_2* as shown in Fig. 12.1.2. Four branches of the helyx trailing string A_2 are wrapped around two branches of the helyx leading string A_1. For the sake of simplicity, Figure 12.1.2 shows a helyx with only one branch of the helyx leading string A_1 that is accompanied by two branches of the helyx trailing string A_2.

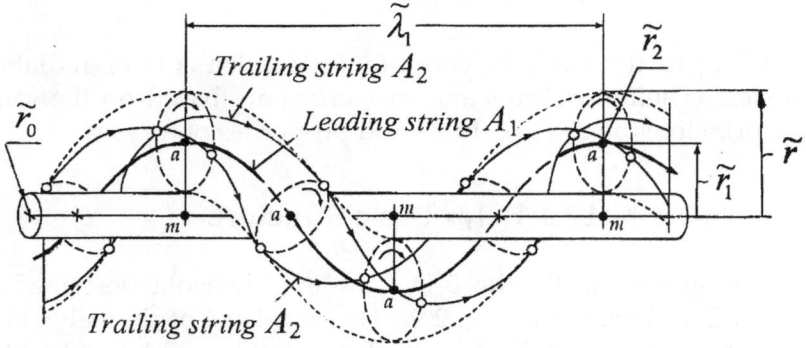

Figure 12.1.2. One of two double-helical leading strings A_1 accompanied by two branches of the trailing string A_2.

12.2 Helyx Spacetime Parameters in Absolute Units

Similarly to the toryx, the relationships between spacetime parameters of the helyx are also described by the Pythagorean Theorem.

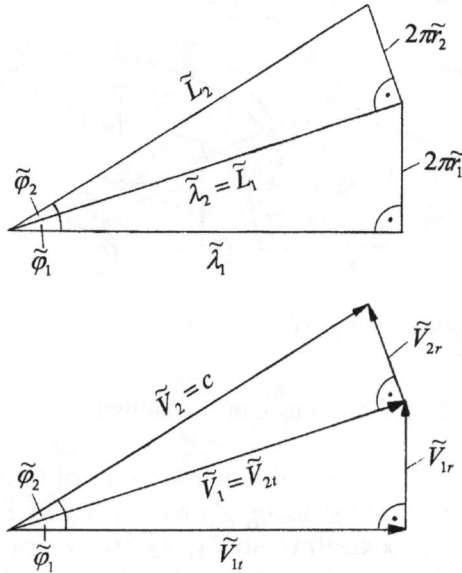

Figure 12.2. Hodograph of the helyx absolute spacetime parameters expressed in relation to the middle point of the trailing string.

In the diagrams shown in Figure 12.2 the spacetime parameters of the helyx leading and trailing strings form sides of right triangles. Therefore, we can readily establish the relationships between these parameters by using the Pythagorean Theorem. Helical trailing strings of the helyx propagate along their helical paths in synchrony with the helyx leading strings, so, the translational velocity of the trailing string $\tilde{V}_{2t}$ is equal to the spiral velocity of the helyx leading string $\tilde{V}_1$.

Notably, the symbols used for defining the helyx parameters are the same as those used for the toryx, except for the "wave" mark (tilde) over the symbols of the helyx parameters. Below is a list of the helyx spacetime parameters expressed in absolute values.

$\tilde{f}_0$ = helyx base frequency
$\tilde{f}_1$ = frequency of the helyx leading string
$\tilde{f}_2$ = frequency of the helyx trailing string
$\tilde{L}_1$ = spiral length of one winding of the helyx leading string
$\tilde{L}_2$ = spiral length of one winding of the helyx trailing string
$\tilde{r}$ = helyx peripheral radius
$\tilde{r}_0$ = helyx eye radius
$\tilde{r}_1$ = radius of the helyx leading string
$\tilde{r}_2$ = radius of the helyx trailing string
$\tilde{T}_0$ = helyx base period
$\tilde{T}_1$ = period of the helyx leading string
$\tilde{T}_2$ = period of the helyx trailing string
$\tilde{V}_1$ = spiral velocity of the helyx leading string
$\tilde{V}_{1r}$ = rotational velocity of the helyx leading string
$\tilde{V}_{1t}$ = translational velocity of the helyx leading string
$\tilde{V}_2$ = spiral velocity of the helyx trailing string
$\tilde{V}_{2r}$ = rotational velocity of the helyx trailing string
$\tilde{V}_{2t}$ = translational velocity of the helyx trailing string
$\tilde{w}_1$ = the number of windings of the helyx leading string
$\tilde{w}_2$ = the number of windings of the helyx trailing string
$\tilde{\lambda}_1$ = wavelength of the helyx leading string
$\tilde{\lambda}_2$ = wavelength of the helyx trailing string
$\tilde{\varphi}_1$ = apex angle of the helyx leading string
$\tilde{\varphi}_2$ = apex angle of the helyx trailing string.

12.3 Helyx Spacetime Postulates in Absolute Units

The helyx spacetime postulates is a set of four fundamental equations limiting the degrees of several helyx parameters, making possible to establish relationships between all spacetime parameters of the helyx. Here is how the helyx spacetime postulates look when the helyx spacetime parameters are expressed in absolute values.

Helyx Spacetime Postulates
(in absolute units of the helyx spacetime parameters)

- The wavelength of the trailing string $\widetilde{\lambda}_2$ is equal to the length of one winding of the leading string $\widetilde{L}_1$:

$$\widetilde{\lambda}_2 = \widetilde{L}_1 \quad (-\infty < \widetilde{r}_1 < +\infty) \qquad (12.3\text{-}1)$$

- The helyx eye radius $\widetilde{r}_0$ is equal to a real positive constant:

$$\widetilde{r}_0 = \widetilde{r}_1 - \widetilde{r}_2 = const. \quad (-\infty < \widetilde{r}_1 < +\infty) \qquad (12.3\text{-}2)$$

- The spiral velocity of the trailing string $\widetilde{V}_2$ is constant and equals to the velocity of light c at each point of its spiral path. Its components, the translational velocity $\widetilde{V}_{2t}$ and the rotational velocity $\widetilde{V}_{2r}$, relate to the spiral velocity $\widetilde{V}_2$ by the Pythagorean Theorem:

$$\widetilde{V}_2 = \sqrt{\widetilde{V}_{2t}^2 + \widetilde{V}_{2r}^2} = c = const. \quad (-\infty < \widetilde{r}_1 < +\infty) \quad (12.3\text{-}3)$$

- The cosine of apex angle of trailing string $\cos s\widetilde{\varphi}_2$ is equal to the ratio of the radii of trailing and leading strings:

$$\cos s\widetilde{\varphi}_2 = \frac{\widetilde{r}_2}{\widetilde{r}_1} \quad (-\infty < \widetilde{r}_1 < +\infty) \qquad (12.3\text{-}4)$$

One can clearly see some similarity between the spacetime postulates (12.3-1) – (12.3-3) for the helyx and the spacetime postulates (5.3-1) – (5.3-3) for the toryx. Similarly to the toryx, it follows from equation (12.3-2) that when the radius of the helyx leading string $\tilde{r}_1$ is equal to the radius of the helyx eye $\tilde{r}_0$ the radius of the helyx trailing string $\tilde{r}_2$ reduces to zero. Consequently, the helyx transforms into a circular *real inversion helyx* with the radius $\tilde{r}_0$ and the helyx base frequency $\tilde{f}_0$.

Notably the helyx eye radius $\tilde{r}_0$ is equal to the toryx eye radius r_0, while the helyx base frequency $\tilde{f}_0$ is equal to the toryx base frequency f_0.

$$\tilde{r}_0 = r_0 \tag{12.3-5}$$

$$\tilde{f}_0 = f_0 = \frac{c}{2\pi\tilde{r}_0} = \frac{c}{2\pi r_0} \tag{12.3-6}$$

12.4 Helyx Spacetime Postulates (in relative values)

The helyx spacetime postulates can be simplified by expressing the helyx spacetime parameters in relative values in respect to the helyx eye radius r_0 and the velocity of light c.

$\tilde{b} = \tilde{r}/\tilde{r}_0$ = relative peripheral radius of the helyx

$\tilde{b}_0$ = helyx relative eye radius

$\tilde{b}_1 = \tilde{r}_1/\tilde{r}_0$ = relative radius of the helyx leading string

$\tilde{b}_2 = \tilde{r}_2/\tilde{r}_0$ = relative radius of the helyx trailing string

$\tilde{l}_1 = L_1/\tilde{r}_0$ = relative length of the helyx leading string

$\tilde{l}_2 = L_2/\tilde{r}_0$ = relative length of the helyx trailing string

$\tilde{t}_1 = \tilde{T}_1/\tilde{T}_0$ = relative period of the helyx leading string

$\tilde{t}_2 = \tilde{T}_2/\tilde{T}_0$ = relative period of the helyx trailing string

$\tilde{\beta}_1 = \tilde{V}_1/c$ = relative spiral velocity of the helyx leading string

$\tilde{\beta}_{1t} = \tilde{V}_{1t}/c$ = relative translational velocity of the helyx leading string

$\tilde{\beta}_{1r} = \tilde{V}_{1r}/c$ = relative rotational velocity of the helyx leading string

$\tilde{\beta}_2 = \tilde{V}_2/c$ = relative spiral velocity of the helyx trailing string

$\tilde{\beta}_{2t} = \tilde{V}_{2t}/c$ = relative translational velocity of helyx trailing string

$\tilde{\beta}_{2r} = \tilde{V}_{2r}/c$ = relative rotational velocity of helyx trailing string

$\tilde{\delta}_1 = \tilde{f}_1/\tilde{f}_0$ = relative frequency of the helyx leading string

$\tilde{\delta}_2 = \tilde{f}_2/\tilde{f}_0$ = relative frequency of the helyx trailing string

$\tilde{\eta}_1 = \lambda_1/r_0$ = wavelength of the helyx leading string

$\tilde{\eta}_2 = \tilde{\lambda}_2 / r_0 =$ wavelength of the helyx trailing string.

Helyx Spacetime Postulates
(in relative values of the helyx spacetime parameters)

- The relative wavelength of the helyx trailing string $\tilde{\eta}_2$ is equal to the relative length of one winding of the helyx leading string $\tilde{l}_2$:

$$\tilde{\eta}_2 = \tilde{l}_1 \quad (-\infty < \tilde{b}_1 < +\infty) \qquad (12.4\text{-}1)$$

- The helyx relative eye radius $\tilde{b}_0$ is equal to 1:

$$\tilde{b}_0 = \tilde{b}_1 - \tilde{b}_2 = 1 \quad (-\infty < \tilde{b}_1 < +\infty) \qquad (12.4\text{-}2)$$

- The relative spiral velocity of the trailing string $\tilde{\beta}_2$ is equal to 1 at each point of its spiral path; it relates to the translational velocity $\tilde{\beta}_{2t}^2$ and the rotational velocity $\tilde{\beta}_{2r}^2$ by the Pythagorean Theorem:

$$\tilde{\beta}_2 = \sqrt{\tilde{\beta}_{2t}^2 + \tilde{\beta}_{2r}^2} = 1 \quad (-\infty < \tilde{b}_1 < +\infty) \qquad (12.4\text{-}3)$$

- The cosine of apex angle of trailing string $\cos s\tilde{\varphi}_2$ is equal to the ratio of the relative radii of trailing and leading strings:

$$\cos s\tilde{\varphi}_2 = \frac{\tilde{b}_2}{\tilde{b}_1} \quad (-\infty < \tilde{r}_1 < +\infty) \qquad (12.4\text{-}4)$$

12.5 Derivative Spacetime Equations of Helyces

Based on the structure of the basic toryx and also on equations of the toryx spacetime postulates, it is possible to derive equations for all other toryx space and spacetime time parameters shown in Table 12.5.1.

In the hodograph shown in Figure 12.5.1, the spacetime parameters of the helyx leading and trailing strings form sides of right triangles.

Table 12.5.1. Spacetime parameters of the helyx leading and trailing strings as a function of the relative radius of the helyx leading string $\widetilde{b}_1$.

Relative parameter	Leading string Eq.(a)	Trailing string Eq. (b)
Radius Eq.(12.5-1)	$\widetilde{b}_1 = \dfrac{\widetilde{r}_1}{\widetilde{r}_0}$	$\widetilde{b}_2 = \widetilde{b}_1 - 1$
Apex angle Eq.(12.5-2)	$\sin s\widetilde{\varphi}_1 = \dfrac{\widetilde{b}_1\sqrt{2\widetilde{b}_1 - 1}}{(\widetilde{b}_1 - 1)^2}$	$\cos s\widetilde{\varphi}_2 = \dfrac{\widetilde{b}_1 - 1}{\widetilde{b}_1}$
Wavelength Eq.(12.5-3)	$\widetilde{\eta}_1 = \dfrac{\widetilde{\lambda}_1}{2\pi r_0} = \sqrt{\dfrac{(\widetilde{b}_1 - 1)^4 - \widetilde{b}_1^{\,2}(2\widetilde{b}_1 - 1)}{2\widetilde{b}_1 - 1}}$	$\widetilde{\eta}_2 = \dfrac{\widetilde{\lambda}_2}{2\pi r_0} = \dfrac{(\widetilde{b}_1 - 1)^2}{\sqrt{2\widetilde{b}_1 - 1}}$
Length of one winding Eq.(12.5-4)	$\widetilde{l}_1 = \dfrac{\widetilde{L}_1}{2\pi r_0} = \dfrac{(\widetilde{b}_1 - 1)^2}{\sqrt{2\widetilde{b}_1 - 1}}$	$\widetilde{l}_2 = \dfrac{\widetilde{L}_2}{2\pi r_0} = \dfrac{\widetilde{b}_1(\widetilde{b}_1 - 1)}{\sqrt{2\widetilde{b}_1 - 1}}$
Transl. velocity Eq.(12.5-5)	$\widetilde{\beta}_{1t} = \dfrac{\widetilde{V}_{1t}}{c} = \dfrac{\sqrt{(\widetilde{b}_1 - 1)^4 - \widetilde{b}_1^{\,2}(2\widetilde{b}_1 - 1)}}{\widetilde{b}_1(\widetilde{b}_1 - 1)}$	$\widetilde{\beta}_{2t} = \dfrac{\widetilde{V}_{2t}}{c} = \dfrac{\widetilde{b}_1 - 1}{\widetilde{b}_1}$
Rotational velocity Eq.(12.5-6)	$\widetilde{\beta}_{1r} = \dfrac{\widetilde{V}_{1r}}{c} = \dfrac{\sqrt{2\widetilde{b}_1 - 1}}{\widetilde{b}_1 - 1}$	$\widetilde{\beta}_{2r} = \dfrac{\widetilde{V}_{2r}}{c} = \dfrac{\sqrt{2\widetilde{b}_1 - 1}}{\widetilde{b}_1}$
Spiral velocity Eq.(12.5-7)	$\widetilde{\beta}_1 = \dfrac{\widetilde{V}_1}{c} = \dfrac{\widetilde{b}_1 - 1}{\widetilde{b}_1}$	$\widetilde{\beta}_2 = \dfrac{\widetilde{V}_2}{c} = 1$
Frequency Eq.(12.5-8)	$\widetilde{\delta}_1 = \widetilde{\delta}_2 = \widetilde{\delta} = \dfrac{\widetilde{f}_1}{\widetilde{f}_0} = \dfrac{\widetilde{f}_2}{\widetilde{f}_0} = \dfrac{\sqrt{2\widetilde{b}_1 - 1}}{\widetilde{b}_1(\widetilde{b}_1 - 1)}$	
Period Eq.(12.5-9)	$\widetilde{\tau}_1 = \widetilde{\tau}_2 = \tau = \dfrac{\widetilde{T}_1}{\widetilde{T}_0} = \dfrac{\widetilde{T}_2}{\widetilde{T}_0} = \dfrac{\widetilde{b}_1(\widetilde{b}_1 - 1)}{\sqrt{2\widetilde{b}_1 - 1}}$	

From Eq. (12.5-8) we can express the relative radius of the leading string $\widetilde{b}_1$ as a function of the helyx relative frequency $\widetilde{\delta}$ for the two extreme cases.

$$\tilde{b}_1 = \left(\frac{2}{\tilde{\delta}^2}\right)^{1/3} \quad (\tilde{b}_1 \gg 1) \tag{12.5-10}$$

$$\tilde{b}_1 = -\frac{i}{\tilde{\delta}} \quad (\tilde{b}_1 \ll 1) \tag{12.5-11}$$

From equations shown in Table 12.5.1, we obtain:

$$\tilde{b} = 2\tilde{b}_1 - 1 \tag{12.5-12}$$

$$\tilde{l}_1 = \tilde{\eta}_2 \tag{12.5-13}$$

$$\tilde{\beta}_1 = \tilde{\beta}_{2t} \tag{12.5-14}$$

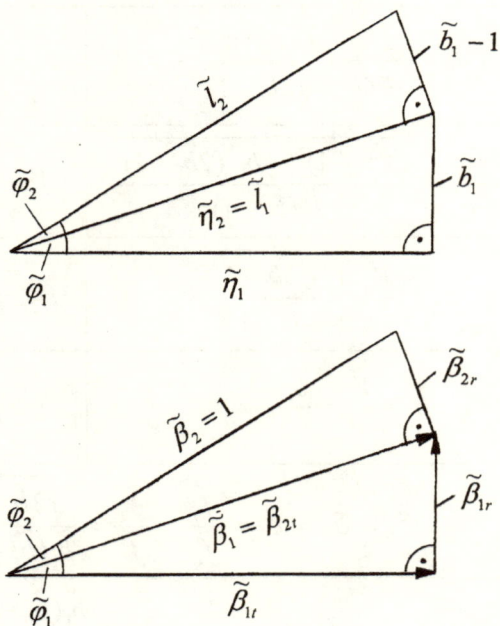

Figure 12.5.1. Hodograph of the helyx relative spacetime parameters expressed in relation to the middle point of the trailing string.

Figure 12.5.2 shows the transformations of relative velocities of the helyx trailing string at the middle point of trailing string.

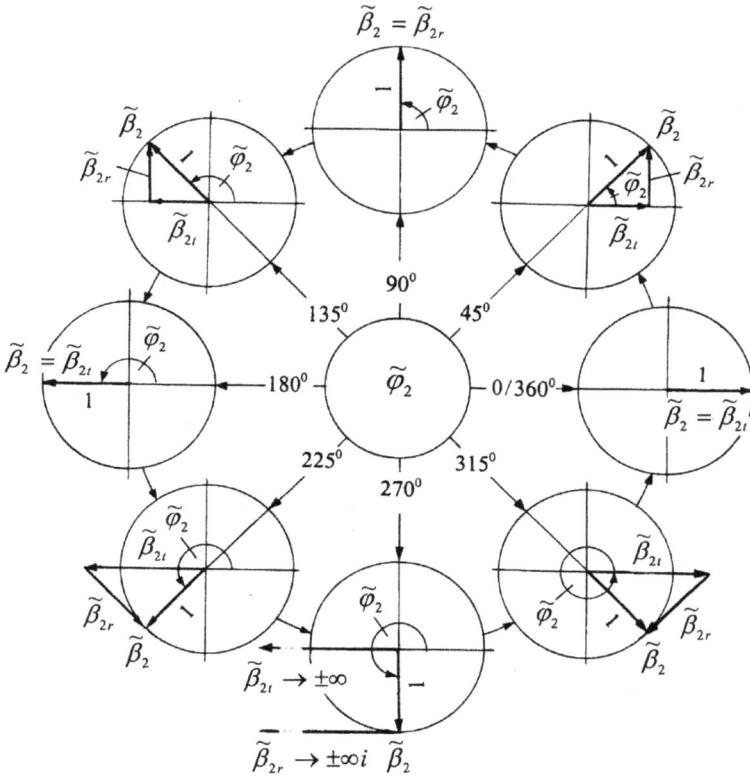

Figure 12.5.2. Transformations of the relative velocities of the helyx trailing string at the middle point of trailing string.

Similarly to the toryces, two integrated spacetime parameters of helyces define their main classification: the *helyx vorticity* $\tilde{V}$ and the *helyx reality* $\tilde{R}$.

12.6 Helyx Vorticity

The helyx vorticity $\tilde{V}$ is equal to the ratio of the radius of the trailing string $\tilde{r}_2$ to the radius of the leading string $\tilde{r}_1$ with the opposite sign.

$$\widetilde{V} = -\frac{\widetilde{r}_2}{\widetilde{r}_1} = -\frac{\widetilde{b}_2}{\widetilde{b}_1} = -\frac{\widetilde{b}_1 - 1}{\widetilde{b}_1} = -\frac{\widetilde{b} - 1}{\widetilde{b} + 1} = -\cos s\widetilde{\varphi}_2 \quad (12.6\text{-}1)$$

Figure 12.6 shows a circular diagram of the helyx vorticity $\widetilde{V}$ as a function of the apex angle of the trailing string $\widetilde{\varphi}_2$. Helyces with positive vorticity $\widetilde{V}$ are called *positive* and with negative vorticity $\widetilde{V}$ *negative*.

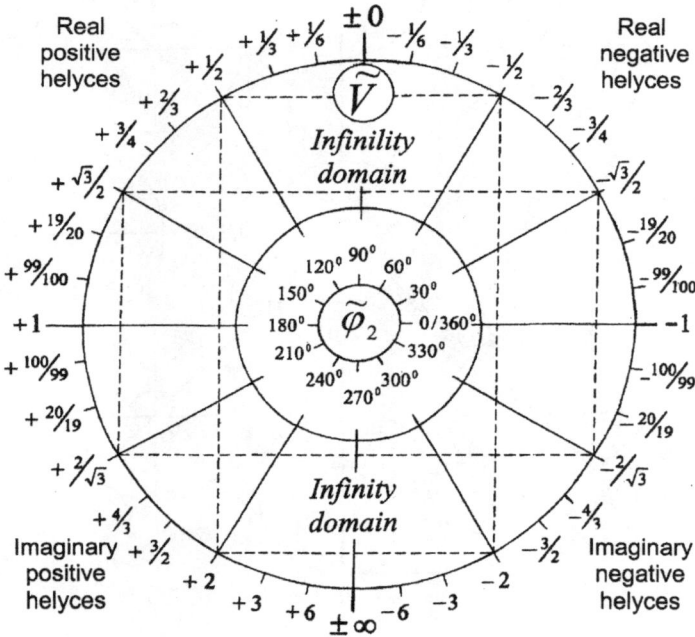

Figure 12.6. Helyx vorticity $\widetilde{V}$ as a function of the apex angle of the trailing string $\widetilde{\varphi}_2$.

The circular diagram of the helyx vorticity $\widetilde{V}$ is divided into two domains, infinity domain and infinility domain, each occupying equal sectors of the diagram. The infinility domain occupies two top quadrants; it contains the values of $\widetilde{V}$ extending clockwise from the positive unity $(+1)$ and passing through infinility (± 0) to the negative unity (-1). The infinity domain resides in two bottom quadrants; it contains the values of $\widetilde{V}$ extending counterclockwise from the positive unity $(+1)$ and passing through infinity $(\pm\infty)$ to the negative unity (-1).

Notably, positive infinility $(+0)$ merges with negative infinility (-0) at $\varphi_2 \to 90^\circ$, while negative infinity $(-\infty)$ merges with positive infinity $(+\infty)$ at $\varphi_2 \to 270^\circ$. There are two kinds of symmetrical polarization between the helyx vorticities $\widetilde{V}$ of helyces that belong to four quadrants of the circular diagram, inverse-symmetrical and reverse-symmetrical.

The helyces that belong to the top and bottom quadrants are inverse-symmetrically polarized by their vorticity $\widetilde{V}$. The magnitudes of their vorticities $\widetilde{V}$ are symmetrically inversed while the signs of $\widetilde{V}$ are the same. The helyces that belong to the right and left quadrants are reverse-symmetrically polarized by their vorticity $\widetilde{V}$, because the signs of their vorticities $\widetilde{V}$ are symmetrically reversed while the magnitudes of $\widetilde{V}$ are the same.

12.7 Helyx Reality

The helyx reality $\widetilde{R}$ is equal to the square root of the relative peripheral radius $\widetilde{b}$ of the helyx.

$$\widetilde{R} = \sqrt{\widetilde{b}} = \sqrt{2\widetilde{b}_1 - 1} = \sqrt{\frac{1 + \cos s\widetilde{\varphi}_2}{1 - \cos s\widetilde{\varphi}_2}} \qquad (12.7\text{-}1)$$

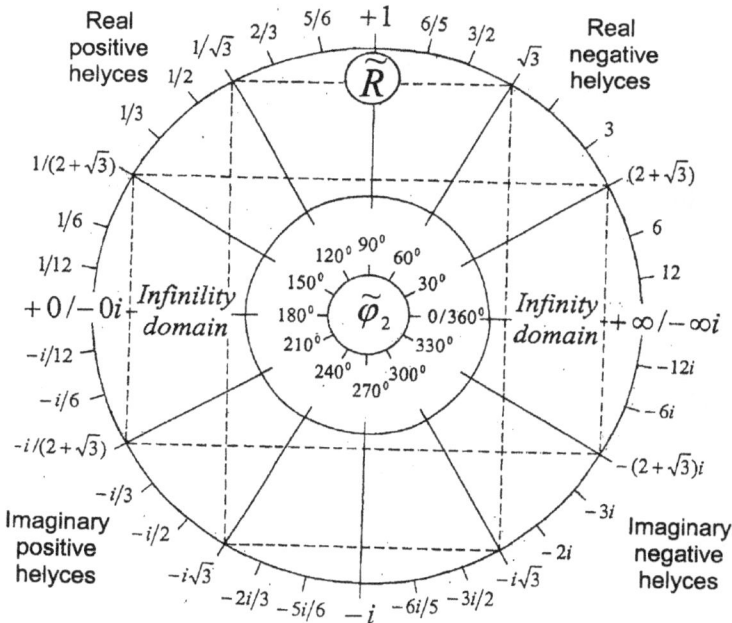

Figure 12.7. Helyx reality $\widetilde{R}$ as a function of the apex angle of the trailing string $\widetilde{\varphi}_2$.

Figure 12.7 shows a circular diagram of the helyx reality $\widetilde{R}$ as a function of the apex angle of the trailing string $\widetilde{\varphi}_2$. The circular diagram of the helyx reality $\widetilde{R}$ is divided into infinility and infinity domains occupying left and right quadrants respectively. There are two kinds of symmetrical polarization between the values of the helyx reality $\widetilde{R}$ of the helyces that belong to the four quadrants of the circular diagram, inverse-symmetrical and reverse-symmetrical.

The helyces that belong to the right and left quadrants are inverse-symmetrically polarized by their reality $\widetilde{R}$, because the magnitudes of their realities $\widetilde{R}$ are symmetrically inversed while the signs of $\widetilde{R}$ are the same. The helyces that belong to the top and bottom quadrants are reverse-symmetrically polarized by their reality $\widetilde{R}$. In these toryces the signs and realities of $\widetilde{R}$ are symmetrically reversed while the magnitudes of $\widetilde{R}$ stay the same.

Consequently, the helyces are divided into four main groups according to their vorticity $\widetilde{V}$ and reality $\widetilde{R}$ as shown in Table 12.7.

Table 12.7. Main classification of helyces.

Helyx name	$\widetilde{\varphi}_2$	$\widetilde{R}$	$\widetilde{V}$
Real negative	$0^0 - 90^0$	real	$(-)$
Real positive	$90^0 - 180^0$	real	$(+)$
Imaginary positive	$180^0 - 270^0$	imaginary	$(+)$
Imaginary negative	$270^0 - 360^0$	imaginary	$(-)$

12.8 Helyx Golden Polarization Factor

The helyx golden polarization factor $\widetilde{G}_h$ is equal to the product of the helyx vorticity $\widetilde{V}$ and the square root of the helyx reality $\widetilde{R}$ with an opposite sign as given by the equation:

$$\widetilde{G}_h = -\widetilde{V}\sqrt{\widetilde{R}} = \frac{(\widetilde{b}_1 - 1)\sqrt{2\widetilde{b}_1 - 1}}{\widetilde{b}_1} \tag{12.8-1}$$

Figure 12.8 shows a plot of Eq. (12.8-1) in which the helyx golden polarization factor $\widetilde{G}_h$ is expressed as a function of the relative radius of the leading string $\widetilde{b}_1$.

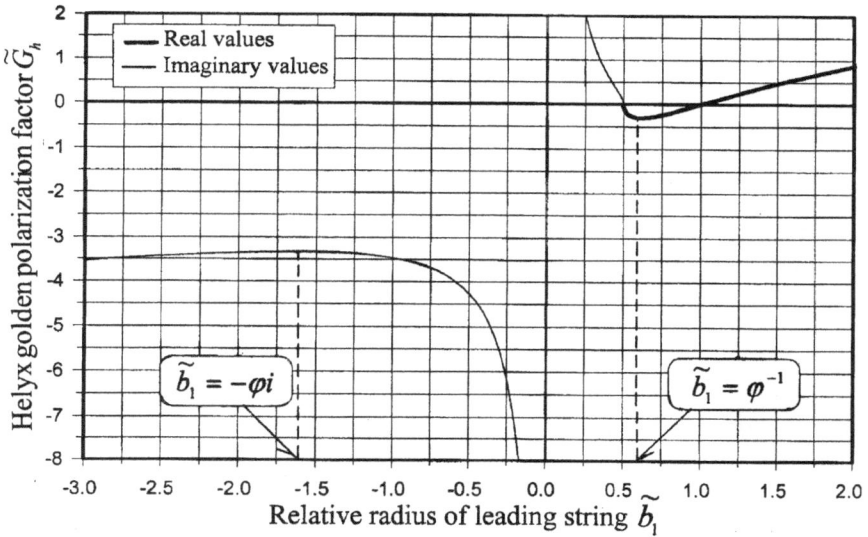

Figure 12.8. Helyx golden polarization factor $\widetilde{G}_h$.

Notably, the plot of $\widetilde{G}_h$ has two extreme values of $\widetilde{G}_h$ at which $\widetilde{b}_1$ is directly related to the golden ratio φ. The first extreme value of $\widetilde{G}_h$ corresponds to $\widetilde{b}_1 = -\varphi i$; here the value of $\widetilde{G}_h$ is maximum. The second extreme value of $\widetilde{G}_h$ corresponds to $\widetilde{b}_1 = \varphi^{-1}$; here the value of $\widetilde{G}_t$ is minimum.

Your Notes

13

TRENDS OF HELYX SPACETIME PARAMETERS

Similarly to the toryx, the helyx spacetime parameters change significantly as the radius of its leading string changes from positive to negative infinity. Also similar to the toryx is the math that describes the helyx spacetime parameters.

13.1 Helyx Spacetime Trigonometry

Trigonometry used for helyx is the same as the toryx spacetime trigonometry, except for one difference. In the toryx, trigonometric functions are related to the steepness angle of the toryx trailing string φ_2, while in the helyx they are related to the apex angle of the helyx trailing string $\widetilde{\varphi}_2$.

Thus, for the helyx, the relationship between the helyx spacetime trigonometric function $\cos s\widetilde{\varphi}_2$ and the relative radius of the helyx leading string $\widetilde{b}_1$ is given by the equation:

$$\cos s\widetilde{\varphi}_2 = \frac{\widetilde{b}_1 - 1}{\widetilde{b}_1} \qquad (0^0 < \widetilde{\varphi}_2 < 360^0) \qquad (13.1\text{-}1)$$

Table 13.1 shows the relationship between spacetime and conventional trigonometric functions in application to the helyx.

Table 13.1 Relationship between spacetime and conventional trigonometric functions in application to the helyx.

Spacetime trigonometry	Conventional function	
$(0^0 < \varphi_2 < 360^0)$	$(0^0 < \varphi_2 < 180^0)$	$(180^0 < \varphi_2 < 360^0)$
$\cos s\widetilde{\varphi}_2$	$\cos \varphi_2$	$\sec \varphi_2$
$\sin s\widetilde{\varphi}_2$	$\sin \varphi_2$	$i \tan \varphi_2$

13.2 Helyx Relative Radii

The relative radii of the helyx leading string $\widetilde{b}_1$, the trailing string $\widetilde{b}_2$ and its peripheral radius $\widetilde{b}$ are given by the equations:

$$\widetilde{b}_1 = \frac{1}{1 - \cos s\widetilde{\varphi}_2} \qquad (0^\circ < \widetilde{\varphi}_2 < 360^\circ) \qquad \text{(13.2-1)}$$

$$\widetilde{b}_2 = \frac{\cos s\widetilde{\varphi}_2}{1 - \cos s\widetilde{\varphi}_2} \qquad (0^\circ < \widetilde{\varphi}_2 < 360^\circ) \qquad \text{(13.2-2)}$$

$$\widetilde{b} = \frac{1 + \cos s\widetilde{\varphi}_2}{1 - \cos s\widetilde{\varphi}_2} \qquad (0^\circ < \widetilde{\varphi}_2 < 360^\circ) \qquad \text{(13.2-3)}$$

Figure 13.2 shows a plot of Eqs. (13.2-1) - (13.2-3).

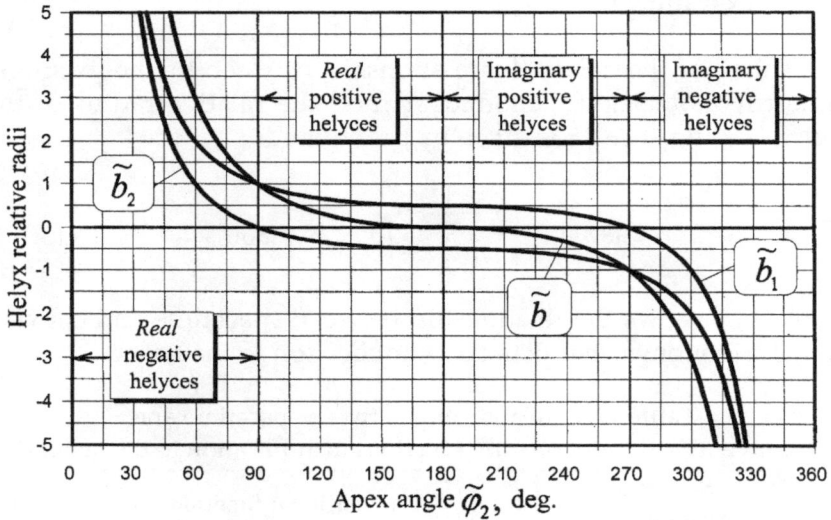

$\widetilde{\varphi}_2$	$360/0^0$	90^0	180^0	270^0
$\widetilde{b}_1$	$-\infty/+\infty$	1	$+\frac{1}{2}$	$+0/-0$
$\widetilde{b}_2$	$-\infty/+\infty$	$+0/-0$	$-\frac{1}{2}$	-1
$\widetilde{b}$	$-\infty/+\infty$	1	$+0/-0$	-1

Figure 13.2. Helyx relative radii as a function of the apex angle of the trailing string $\widetilde{\varphi}_2$.

13.3 Relative Wavelength of Leading String

The relative wavelength of the leading string $\tilde{\eta}_1$ is given by the equation:

$$\tilde{\eta}_1 = 2\pi \frac{\sqrt{\cos s^2\tilde{\varphi}_2 - \tan s^2\tilde{\varphi}_2}}{\tan s\tilde{\varphi}_2 - \sin s\tilde{\varphi}_2} \quad (0^0 < \varphi_2 < 360^0) \tag{13.3-1}$$

Figure 13.3 shows a plot of Eq. (13.3-1).

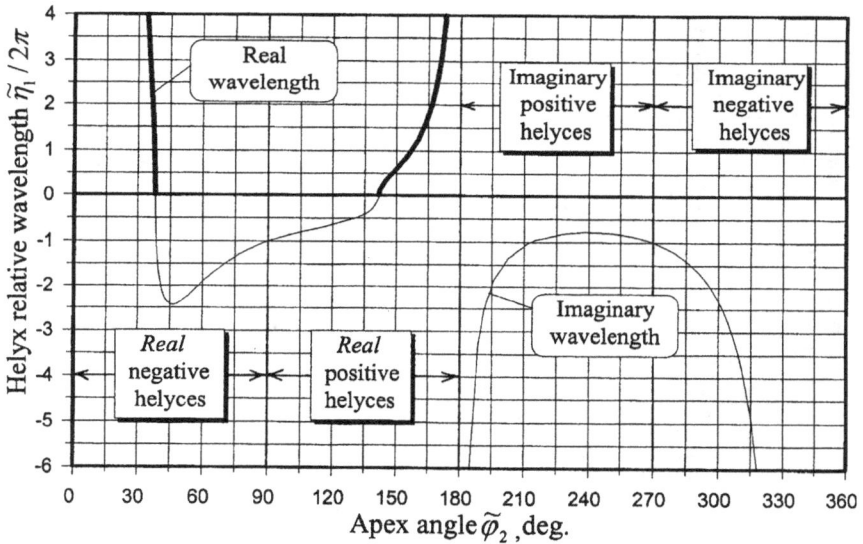

$\tilde{\varphi}_2$	$360/0^0$	90^0	180^0	270^0
$\tilde{\eta}_1 / 2\pi$	$-\infty i/+\infty$	$-i$	$+\infty/-\infty i$	$-i$

Figure 13.3. Relative wavelength of the leading string $\tilde{\eta}_1$ as a function of the apex angle of the trailing string $\tilde{\varphi}_2$.

13.4 Relative Wavelength of Trailing String

The relative wavelength of trailing string $\tilde{\eta}_2$ is given by the equation:

$$\tilde{\eta}_2 = \frac{2\pi \cos s\tilde{\varphi}_2}{(1 - \cos s\tilde{\varphi}_2)\tan s\tilde{\varphi}_2} \qquad (0^0 < \varphi_2 < 360^0) \qquad (13.4\text{-}1)$$

Figure 13.4 shows a plot of Eq. (13.4-1).

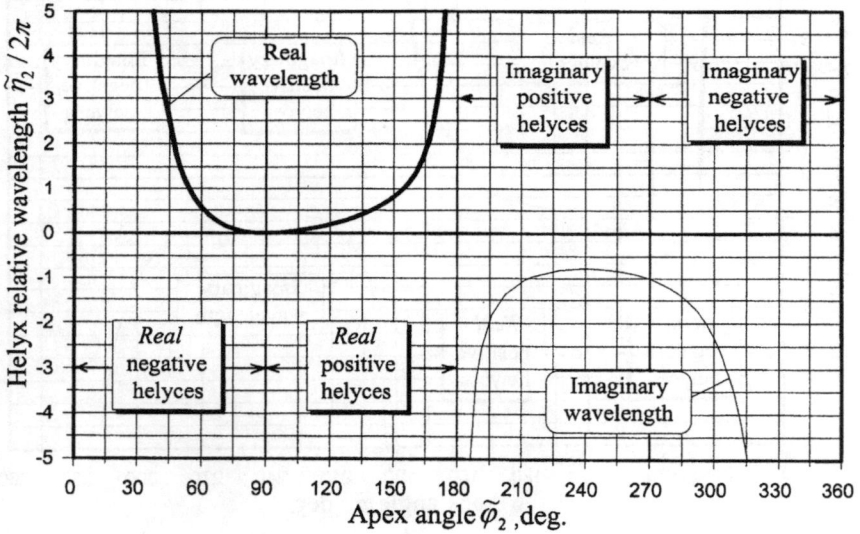

$\tilde{\varphi}_2$	$360/0^0$	90^0	180^0	270^0
$\tilde{\eta}_2 / 2\pi$	$-\infty i / +\infty$	$+0$	$+\infty / -\infty i$	$-i$

Figure 13.4. Relative wavelength of the trailing string $\tilde{\eta}_2$ as a function of the apex angle of the trailing string $\tilde{\varphi}_2$.

13.5 Relative Translational Velocity of Leading String

The relative translational velocity of the leading string $\widetilde{\beta}_{1t}$ is given by the equation:

$$\widetilde{\beta}_{1t} = \sqrt{\cos s^2\widetilde{\varphi}_2 - \tan s^2\widetilde{\varphi}_2} \qquad (0^0 < \varphi_2 < 360^0) \qquad (13.5\text{-}1)$$

Figure 13.5 shows a plot of Eq. (13.5-1).

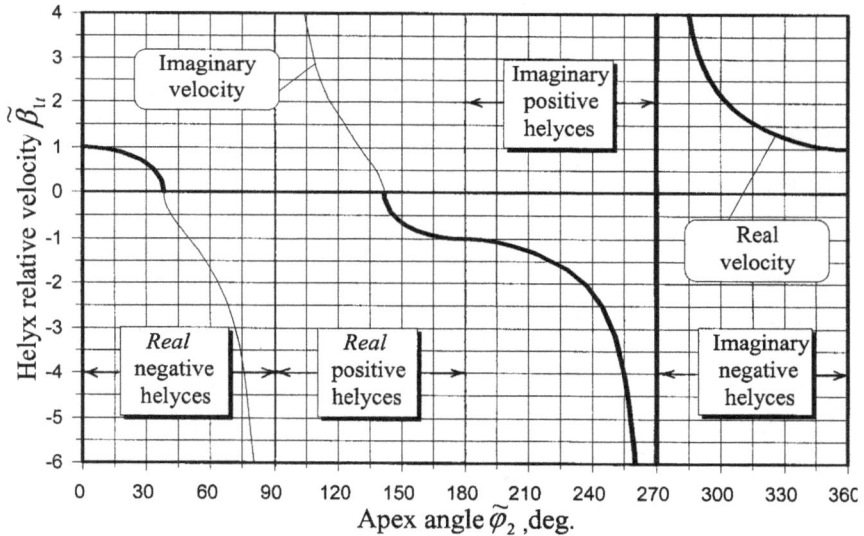

$\widetilde{\varphi}_2$	$360/0^0$	90^0	180^0	270^0
$\widetilde{\beta}_{1t}$	$+1$	$-\infty i/+\infty i$	-1	$-\infty/+\infty$

Figure 13.5. Relative translational velocity of the leading strings $\widetilde{\beta}_{1t}$ as a function of the apex angle of the trailing string $\widetilde{\varphi}_2$.

13.6 Relative Translational Velocity of Trailing String

The relative translational velocity of trailing string $\tilde{\beta}_{2t}$ is given by the equation:

$$\tilde{\beta}_{2t} = \cos s\tilde{\varphi}_2 \quad (0^0 < \varphi_2 < 360^0) \tag{13.6-1}$$

Figure 13.6 shows a plot of Eq. (13.6-1).

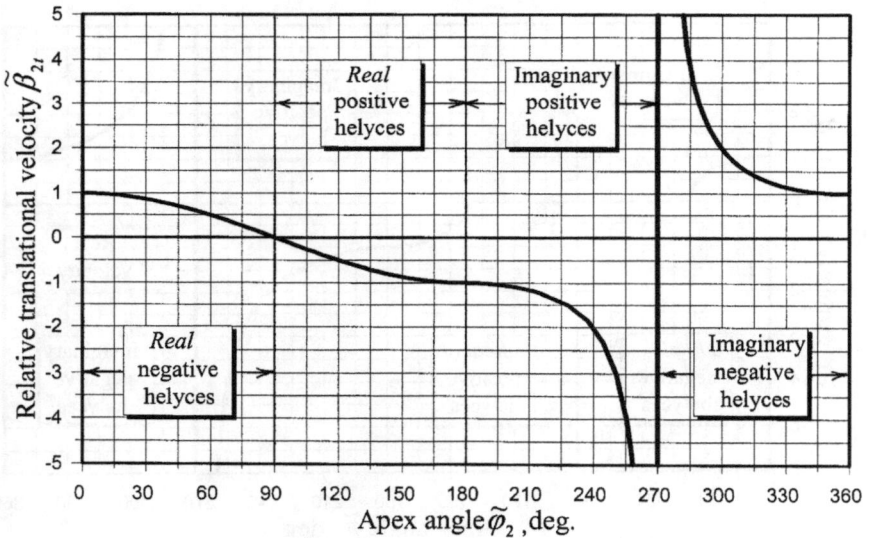

$\tilde{\varphi}_2$	$360/0^0$	90^0	180^0	270^0
$\tilde{\beta}_{2t}$	$+1$	$+0/-0$	-1	$-\infty/+\infty$

Figure 13.6. Relative translational velocity of the trailing strings $\tilde{\beta}_{2t}$ as a function of the apex angle of the trailing string $\tilde{\varphi}_2$.

13.7 Relative Rotational Velocity of Leading String

The relative rotational velocity of the trailing string $\widetilde{\beta}_{2r}$ is given by the equation:

$$\widetilde{\beta}_{1r} = \tan s\widetilde{\varphi}_2 \qquad (0^\circ < \varphi_2 < 360^\circ) \qquad (13.7\text{-}1)$$

Figure 13.7 shows a plot of Eq. (13.7-1).

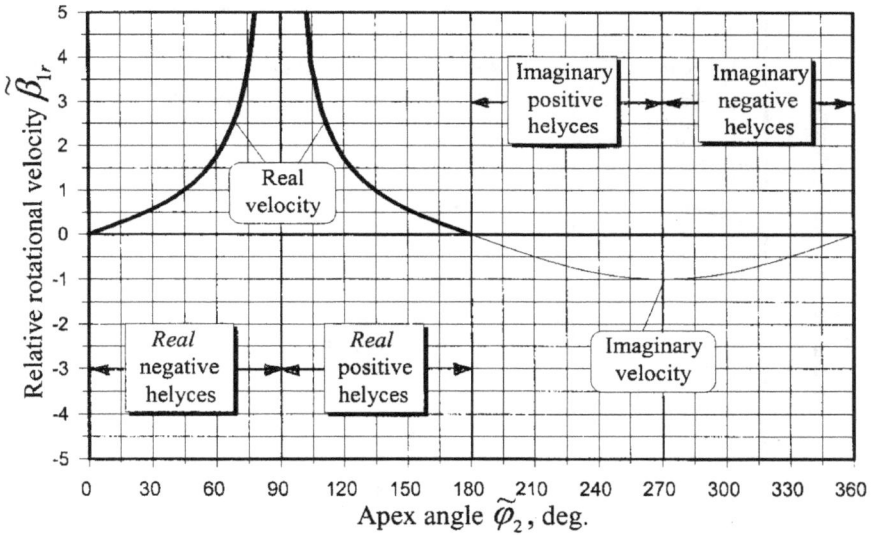

$\widetilde{\varphi}_2$	$360/0^0$	90^0	180^0	270^0
$\widetilde{\beta}_{1r}$	$-0i/+0$	$+\infty$	$+0/-0i$	$-i$

Figure 13.7. Relative rotational velocity of the leading strings $\widetilde{\beta}_{1r}$ as a function of the apex angle of the trailing string $\widetilde{\varphi}_2$.

13.8 Relative Rotational Velocity of Trailing String

The relative rotational velocity of trailing string $\tilde{\beta}_{2r}$ is given by the equation:

$$\tilde{\beta}_{2r} = \sin s\tilde{\varphi}_2 \qquad (0^0 < \varphi_2 < 360^0) \qquad (13.8\text{-}1)$$

Figure 13.8 shows a plot of Eq. (13.8-1).

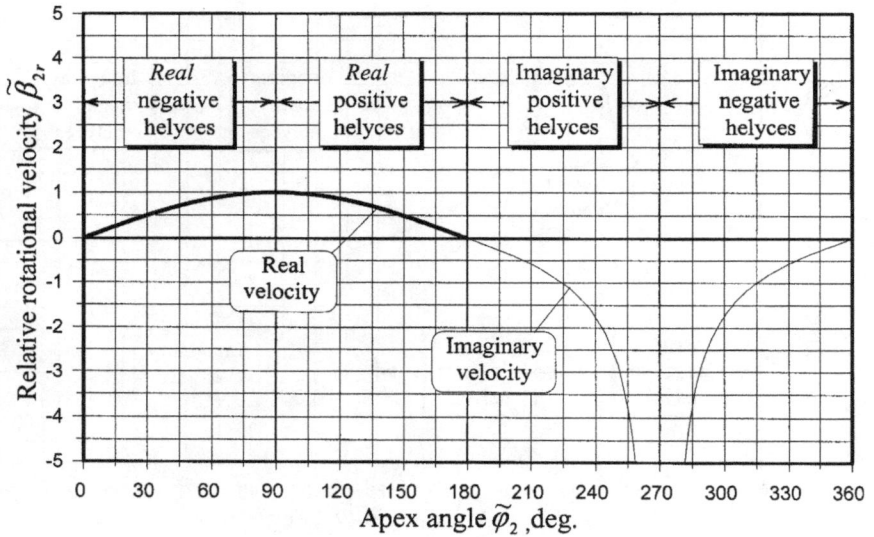

$\tilde{\varphi}_2$	$360/0^0$	90^0	180^0	270^0
$\tilde{\beta}_{2r}$	$-0i/+0$	$+1$	$+0/-0i$	$-\infty i$

Figure 13.8. Relative rotational velocity of the trailing strings $\tilde{\beta}_{2r}$ as a function of the apex angle of the trailing string $\tilde{\varphi}_2$.

13.9 Relative Frequency of Trailing String

The relative frequency of trailing string $\tilde{\delta}_2$ is equal to the relative frequency of the leading string $\tilde{\delta}_1$ and it is given by the equation:

$$\tilde{\delta}_2 = \tilde{\delta}_1 == \frac{(1 - \cos s\tilde{\varphi}_2)^2}{\cos s\tilde{\varphi}_2} \sqrt{\frac{1 + \cos s\tilde{\varphi}_2}{1 - \cos s\tilde{\varphi}_2}} \quad (0^0 < \varphi_2 < 360^0) \quad (13.9\text{-}1)$$

Figure 13.9 shows a plot of Eq. (13.9-1).

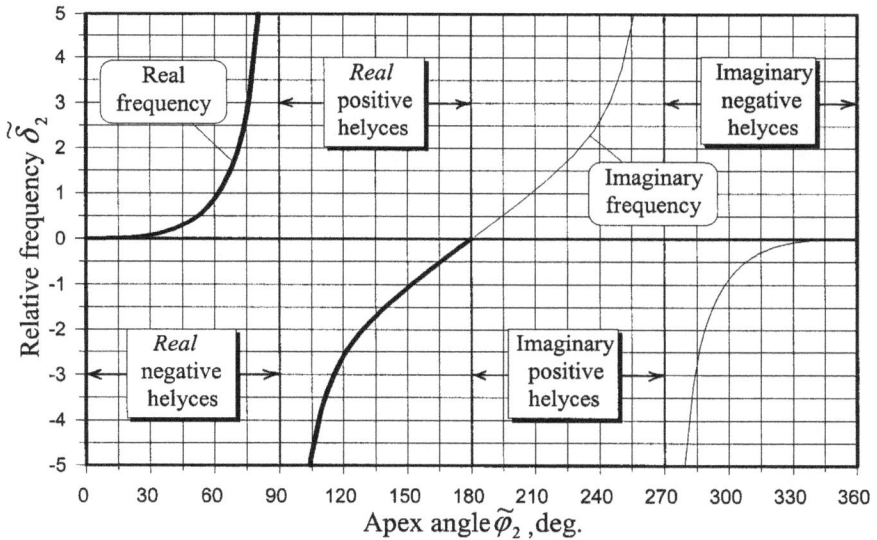

$\tilde{\varphi}_2$	$360/0^0$	90^0	180^0	270^0
$\tilde{\delta}_2$	$-0i/+0$	$+\infty/-\infty$	$-0/+0i$	$+\infty i/-\infty i$

Figure 13.9. Relative frequency of the trailing strings $\tilde{\delta}_2$ as a function of the apex angle of the trailing string $\tilde{\varphi}_2$.

14

INVERSION STATES OF HELYX

Similarly to the toryx, the helyx spacetime topology undergoes through significant transformation as the radius of the helyx leading string changes from positive to negative infinity.

14.1 Metamorphoses of Helyx Topology

Topological transformations of helyces are similar to the transformations of toryces.

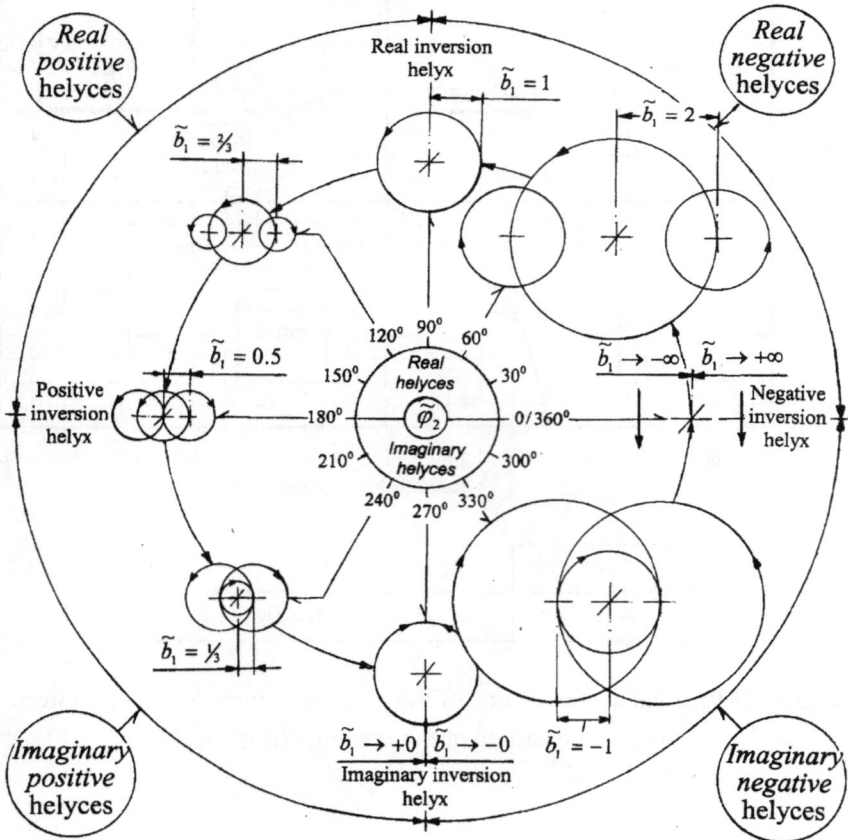

Figure 14.1. Transformations of cross-sections of helyces as a function of the apex angle of the trailing string $\widetilde{\varphi}_2$.

Figure 14.1 shows metamorphoses of cross-sections of the helyx leading and trailing strings as a function of the apex angle of the trailing string $\widetilde{\varphi}_2$. Four kinds of *inversion helyces* are located at the boundaries of the circular diagram between four main groups of helyces.

<u>Negative inversion helyx</u> $(\widetilde{\varphi}_2 \rightarrow +0^{\circ}/360^{\circ})$ - At this point, $\widetilde{b}_1 \rightarrow \pm\infty$, $\widetilde{b}_2 \rightarrow \pm\infty$ and the helyx leading string, trailing string and their wavelengths become inverted. Consequently, as $\widetilde{\varphi}_2$ crosses the borderline at $0^0/360^0$ the helyx vorticity $\widetilde{V}$ remains negative, while the helyx reality $\widetilde{R}$ inverts from imaginary to real. The helyx appears as two parallel lines separated by the distance equal to the diameter real inversion string.

<u>Real inversion helyx</u> $(\widetilde{\varphi}_2 \rightarrow 90^{\circ})$ - At this point, $\widetilde{b}_1 \rightarrow +1$, $\widetilde{b}_2 \rightarrow \pm 0$ and the helyx trailing string becomes inverted. Consequently, as $\widetilde{\varphi}_2$ crosses the borderline at 90^0, the helyx reality $\widetilde{R}$ remains real while its vorticity $\widetilde{V}$ inverts from negative to positive. It appears as a circle with the relative radius $\widetilde{b}_1 \rightarrow +1$.

<u>Positive inversion helyx</u> $(\widetilde{\varphi}_2 \rightarrow 180^{\circ})$ – At this point, $\widetilde{b}_1 \rightarrow +\frac{1}{2}$, $\widetilde{b}_2 \rightarrow -\frac{1}{2}$ and the helyx wavelengths become inverted. Consequently, as $\widetilde{\varphi}_2$ crosses the borderline at 180^0, the helyx vorticity $\widetilde{V}$ remains positive while its reality $\widetilde{R}$ inverts from real to imaginary. At that point, the inner parts of the helyx windings are touching one another.

<u>Imaginary inversion helyx</u> $(\widetilde{\varphi}_2 \rightarrow 270^{\circ})$ – At this point, $\widetilde{b}_1 \rightarrow \pm 0$, $\widetilde{b}_2 \rightarrow -1$ and the helyx leading string becomes inverted. Consequently, as $\widetilde{\varphi}_2$ crosses the borderline at 270^0, the helyx reality $\widetilde{R}$ remains imaginary while its vorticity $\widetilde{V}$ inverts from positive to negative. The helyx appears as a circle with the relative radius approaching -1. The circle is located at the plane perpendicular to the plane of the real inversion string.

Located between the inversion helyces on the circular diagram of Figure 14.1 are the helyces that belong to their four main groups. Shown below are the transformations of cross-sections of helyces within each main group.

14.2 Real Negative Helyces

Figure 14.2 shows cross-sections of real negative helyces. They belong to the top right quadrant of the circular diagram shown in Figure 14.1. The trailing strings of these helyces are wound counter-clockwise outside of real inversion string. As $\tilde{\varphi}_2$ increases, $\tilde{b}_1$ and $\tilde{b}_2$ decrease.

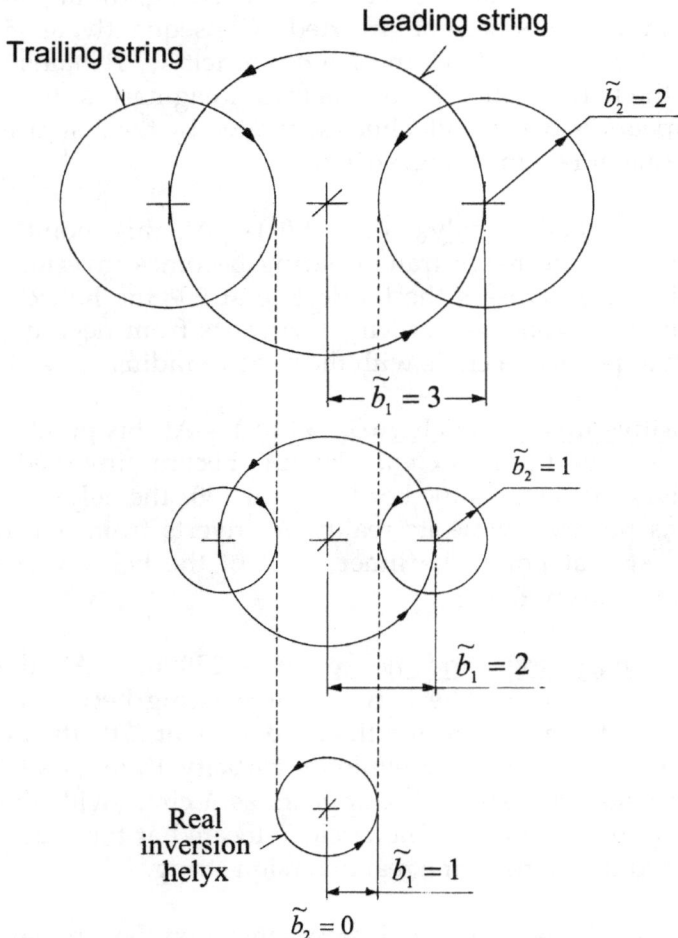

Trailing string

Leading string

$\tilde{b}_2 = 2$

$\tilde{b}_1 = 3$

$\tilde{b}_2 = 1$

$\tilde{b}_1 = 2$

Real
inversion
helyx

$\tilde{b}_1 = 1$

$\tilde{b}_2 = 0$

Figure 14.2. Metamorphoses of cross-sections of real negative helyces.

14.3 Real Positive Helyces

Figure 14.3 shows cross-sections of real positive helyces. They belong to the top left quadrant of the circular diagram shown in Figure 14.1. Within this range, the trailing string is inverted, so that its windings are now wound clockwise inside the real inversion helyx. As $\tilde{\varphi}_2$ increases, $\tilde{b}_1$ decrease, but negative values of $\tilde{b}_2$ increase. Consequently, the helyx appears as an inverted helical spiral.

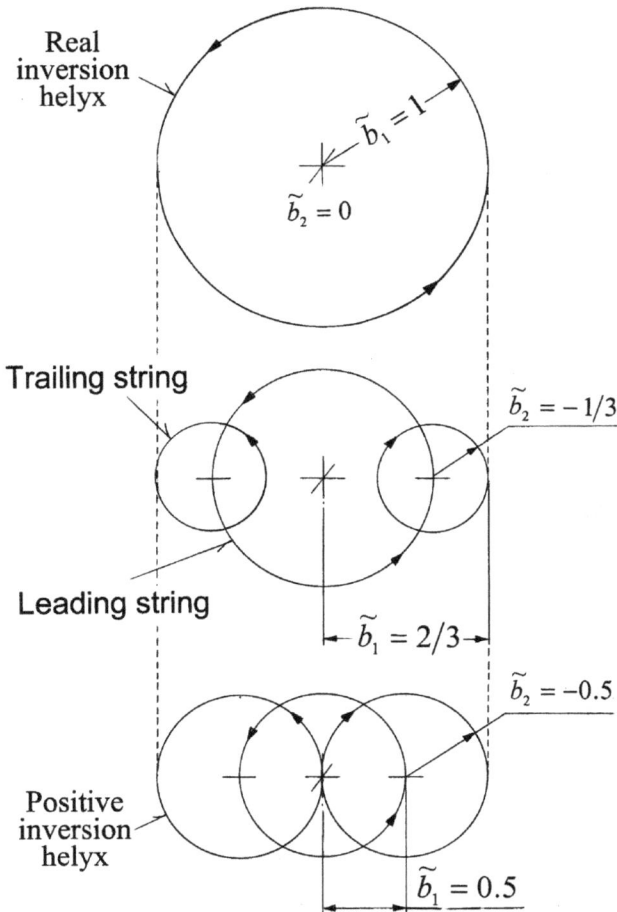

Figure 14.3. Metamorphoses of cross-sections of real positive helyces.

14.4 Imaginary Positive Helyces

Figure 14.4 shows cross-sections of imaginary positive helyces. They belong to the bottom left quadrant of the circular diagram shown in Figure 14.1. As $\widetilde{\varphi}_2$ increases, $\widetilde{b}_1$ decreases, but the negative values of $\widetilde{b}_2$ increase. Within this range, the trailing string is still inverted and its windings are wound inside the imaginary inversion helyx. The opposite parts of windings of the trailing string intersect with one another.

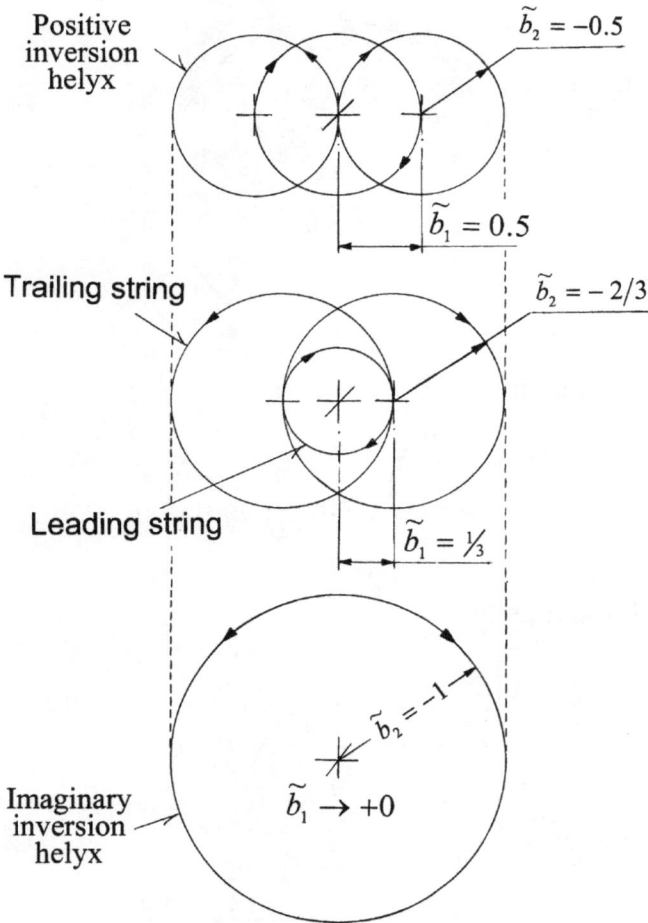

Figure 14.4. Metamorphoses of cross-sections of imaginary positive helyces.

14.5 Imaginary Negative Helyces

Figure 14.5 shows cross-sections of imaginary negative helyces. They belong to the bottom right quadrant of circular diagram shown in Figure 14.1. Here the leading string becomes inverted. As the negative values of $\tilde{b}_1$ increase, the negative values of $\tilde{b}_2$ also increase. Within this range the trailing string propagates outward and its windings are located outside of imaginary inversion helyx.

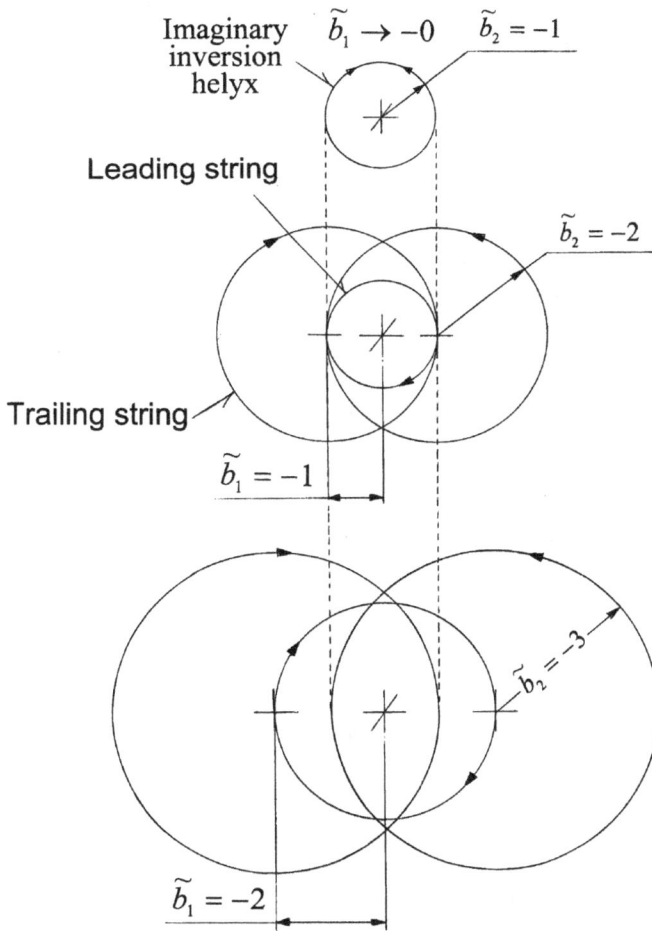

Figure 14.5. Metamorphoses of cross-sections of imaginary negative helyces.

15

FORMATION OF RADIATION PARTICLES

The elementary radiation particles are formed by the unification of polarized helyces emitted by their parental toryces. The elementary radiation particles emitted by the excited trons are called *tons*, and the elementary radiation particles emitted by the oscillated trons are called *tins* as shown in Table 15.

Similarly to the trons, there are two kinds of tons, the *charge-polarized tons* and the *reality-polarized tons*. Both of them are formed when their respective parental trons are transferred from higher to lower quantum states.

Table 15. Types of elementary radiation particles.

Parental matter particles	Elementary radiation particles	
	Excited (tons)	Oscillated (tins)
Electron	Electon	Electin
Positron	Positon	Positin
Ethertron	Etherton	Ethertin
Singulatron	Singulaton	Singulatin

15.1 Emission of Helyces

Figure 15.1 illustrates schematically the emission of helyces when an excited quantum state of a parental toryx is changed from a higher quantum state k to a lower quantum state j. In Bohr's model of electron, the parameters of photons emitted by the atomic electrons are defined based on the law of conservation of energy. According to this law, the energy lost by an atomic electron during emission of a photon must be equal to the photon energy.

In application to the real and imaginary helyces emitted by their respective real and imaginary parental toryces the UST expresses the above balance of energy in spacetime terms by using the following equations:

$$\text{Real toryces:} \quad E_{tk} - E_{tj} = \frac{h\widetilde{f}}{d} \qquad (15.1\text{-}1)$$

$$\text{Imaginary toryces:} \quad E_{tk} - E_{tj} = \frac{ih\widetilde{f}}{d} \qquad (15.1\text{-}2)$$

Considering that

$$m_e c^2 = \frac{\alpha h \widetilde{f}_0}{2}, \qquad (15.1\text{-}3)$$

we obtain from Eqs. (10.4-5), (15.1-1) – (15.1-3), the equations for the relative frequencies of helyces emitted by the real and imaginary toryces:

$$\text{Real toryces:} \quad \widetilde{\delta}_{kj} = -\frac{\alpha}{2d}\left(Q_{qk}\frac{b_{1k}-1}{b_{1k}^2} - Q_{qj}\frac{b_{1j}-1}{b_{1j}^2} \right) \quad (15.1\text{-}4)$$

$$\text{Imagin. toryces:} \quad \widetilde{\widetilde{\delta}}_{kj} = -\frac{\alpha i}{2d}\left(Q_{qk}\frac{\breve{b}_{1k}-1}{\breve{b}_{1k}^2} - Q_{qj}\frac{\breve{b}_{1j}-1}{\breve{b}_{1j}^2} \right) \quad (15.1\text{-}5)$$

where b_{1k} and b_{1j}, Q_{qk} and Q_{qj} are the relative radii of leading string and the oscillation factors of parental toryces corresponding to the quantum states k and j.

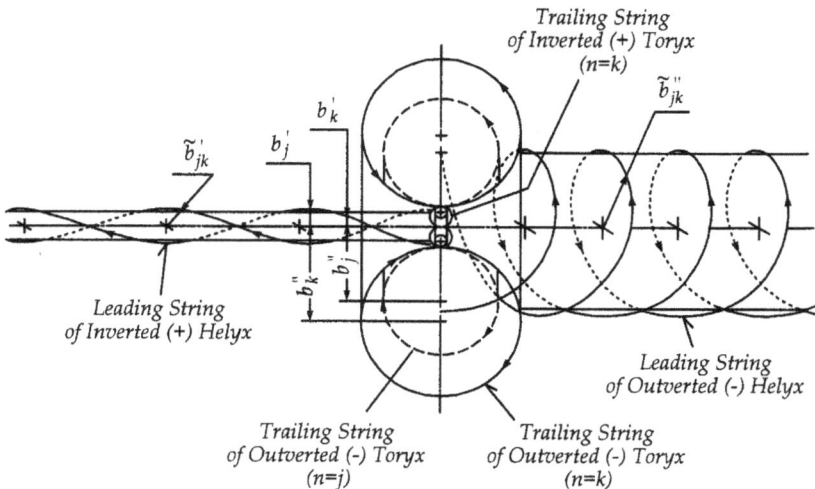

Figure 15.1. Emission of helyces by an excited toryx.

Notably, in Eqs. (15.1-1) - (15.1-5) the constant d depends on the type of parental trons. As it will be shown below, electrons and positrons emit real and imaginary helyces in one direction, and for them $d = 1$. Contrarily, the ethertrons emit real negative and positive helyces in opposite directions. Singulatrons also emit imaginary negative and positive helyces in opposite directions. Consequently, for ethertrons and singulatrons $d = 2$.

15.2 Formation of Elementary Radiation Particles

Figure 15.2 shows schematically the formation of four elementary radiation particles (tons) from their parental matter particles (trons).

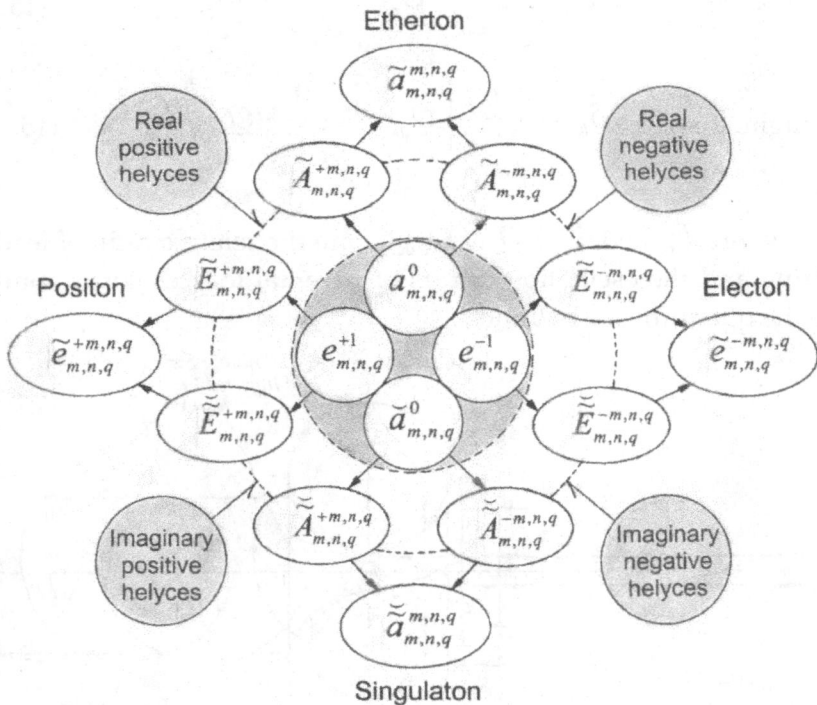

Figure 15.2. Formation of four elementary radiation particles (tons) from their parental particles (trons).

In the symbols of helyces and tons, the right superscripts and subscripts indicate respectively the quantum states m, n and q of their parental toryces and trons prior to and after their emission of helyces and tons. Preceding the right top superscripts are the signs of the helyx and ton vorticities. The structures of the tons are:

Electons $\widetilde{e}_{m,n,q}^{-m,n,q}$ are made up of the real negative helyces $\widetilde{E}_{m,n,q}^{-m,n,q}$ and the imaginary negative helyces $\widetilde{\overline{E}}_{m,n,q}^{-m,n,q}$ emitted by the electrons $e_{m,n,q}^{-1}$.

Positons $\widetilde{e}_{m,n,q}^{+m,n,q}$ are made up of the real positive helyces $\widetilde{E}_{m,n,q}^{+m,n,q}$ and the imaginary positive helyces $\widetilde{\overline{E}}_{m,n,q}^{+m,n,q}$ emitted by the positrons $e_{m,n,q}^{+1}$

Ethertons $\widetilde{a}_{m,n,q}^{m,n,q}$ are made up of the real negative helyces $\widetilde{A}_{m,n,q}^{-m,n,q}$ and the real positive helyces $\widetilde{A}_{m,n,q}^{+m,n,q}$ emitted by the ethertrons.

Singulatons $\widetilde{\overline{a}}_{m,n,q}^{m,n,q}$ are made up of the imaginary negative helyces $\widetilde{\overline{A}}_{m,n,q}^{-m,n,q}$ and the imaginary positive helyces $\widetilde{\overline{A}}_{m,n,q}^{+m,n,q}$ emitted by the singulatrons $\widetilde{\overline{a}}_{m,n,q}^{0}$.

The sections below describe compositions and properties of the tons and tins. The calculations were made by using the following sources:

- The equations for the relative spacetime parameters of the parental toryces of helyces are shown in Table 5.6.1.
- The quantization equations for the radii of leading strings and peripheral circles of the basic harmonic excited parental toryces are shown in Table 9.3.1 and Eq. (9.3-1) as the functions of the exponential excitation quantum states m, the linear excitation quantum states n and the oscillation quantum states q.
- The exponential excitation quantum states m are related to the spacetime levels L as shown in Table 11.1.
- The oscillation factors Q_q of the parental toryces of helyces are defined by Eq. (9.3-20) as a function of the oscillation quantum states q.
- The equations for the helyx spacetime parameters are shown in Table 12.5.1.

15.3 Electons

The parental excited electrons emitting the electons are composed of the negative reality-polarized toryces. As shown in Figure 15.3.1, the real toryx R of the excited electron emits a real helyx in which the translational velocity of trailing string $\tilde{\beta}_{2t}$ is subluminal.

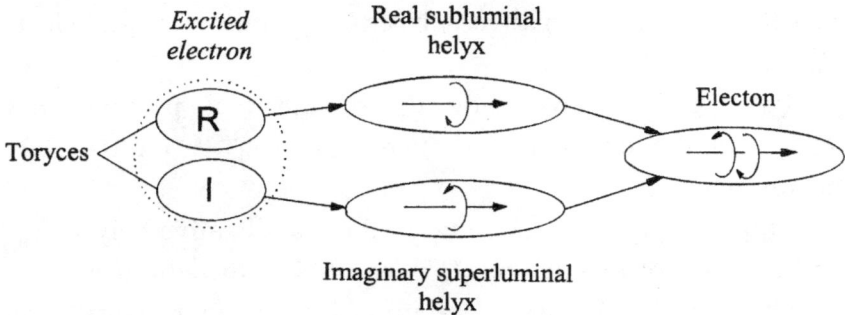

Figure 15.3.1. Formation of an electon.

At the same time, an imaginary toryx I emits in the same direction an imaginary helyx in which $\tilde{\beta}_{2t}$ is superluminal. The real and imaginary helyces have opposite spins, and the reality-polarized electon formed by these helyces propagates with the average velocity equal to the velocity of light. Table 15.3.1 shows the compositions and properties of the excited electrons of the various spacetime levels L.

It follows from Table 15.3.1:

- The relative radii of leading strings $\tilde{b}_1$ of constituent helyces of electons increase with the increase of the spacetime level L.
- In the real helyces of electons, the translational velocities $\tilde{\beta}_{2t}$ are slightly less than the velocity of light, while in the imaginary helyces they are slightly greater than the velocity of light. The differences between these velocities decrease as the spacetime level L increases.
- The frequencies of electons $\tilde{f}_{kj}$ decrease with the increase of the the spacetime level L.
- For the spacetime level $L1$ at which the exponential excitation quantum state $m = 1$, the calculated frequencies of electons are within the frequency range of cosmic X-ray background (CXB) radiations.

- For the spacetime level *L2* at which the exponential excitation quantum state *m* = 2, the calculated frequencies of electons are within the frequency ranges of <u>infrared, visible and ultraviolet radiations</u>.
- For the spacetime level *L3* at which the exponential excitation quantum state *m* = 3, the calculated frequencies of electons are within the frequency ranges of infrared and microwave frequency ranges known in astronomy as <u>cosmic microwave background (CMB) radiation</u>.

Table 15.3.1. Compositions and properties of the excited electons of the various spacetime levels *L* with the quantum states (*m*).

L (*m*)	Electons	Helyces of emitted excited electons				
		Symbols	$\tilde{b}_1$	$\tilde{\beta}_{2t}$	$\tilde{f}_{kj}$ Hz	$\tilde{E}_{kj}$ MeV
L1 (1)	$\tilde{e}_{1,1,0}^{-1,2,0}$	$\tilde{E}_{1,1,0}^{-1,2,0}$	3571.4712	0.99972016	2.24×10^{17}	9.27×10^{-4}
		$\tilde{\tilde{E}}_{1,1,0}^{-1,2,0}$	-3533.5054	1.00028301	-2.28×10^{17}	-9.43×10^{-4}
	$\tilde{e}_{1,2,0}^{-1,3,0}$	$\tilde{E}_{1,2,0}^{-1,3,0}$	7420.4827	0.99986524	7.49×10^{16}	3.10×10^{-4}
		$\tilde{\tilde{E}}_{1,2,0}^{-1,3,0}$	-7374.4504	1.00013560	-7.56×10^{16}	-3.13×10^{-4}
L2 (2)	$\tilde{e}_{2,1,0}^{-2,2,0}$	$\tilde{E}_{2,1,0}^{-2,2,0}$	7.2200×10^4	0.99998615	2.47×10^{15}	1.02×10^{-5}
		$\tilde{\tilde{E}}_{2,1,0}^{-2,2,0}$	-7.2194×10^4	1.00001385	-2.47×10^{15}	-1.02×10^{-5}
	$\tilde{e}_{2,2,0}^{-2,3,0}$	$\tilde{E}_{2,2,0}^{-2,3,0}$	2.2223×10^5	0.99999550	4.57×10^{14}	1.89×10^{-6}
		$\tilde{\tilde{E}}_{2,2,0}^{-2,3,0}$	-2.2222×10^5	1.00000450	-4.57×10^{14}	-1.89×10^{-6}
L3 (3)	$\tilde{e}_{3,1,0}^{-3,2,0}$	$\tilde{E}_{3,1,0}^{-3,2,0}$	1.7317×10^6	0.99999942	2.10×10^{13}	8.69×10^{-8}
		$\tilde{\tilde{E}}_{3,1,0}^{-3,2,0}$	-1.7317×10^6	1.00000058	-2.10×10^{13}	-8.69×10^{-8}
	$\tilde{e}_{3,2,0}^{-3,3,0}$	$\tilde{E}_{3,2,0}^{-3,3,0}$	8.0078×10^6	0.99999988	2.11×10^{12}	8.74×10^{-9}
		$\tilde{\tilde{E}}_{3,2,0}^{-3,3,0}$	-8.0078×10^6	1.00000013	-2.11×10^{12}	-8.74×10^{-9}

It was found that the experimental data for the frequencies $\tilde{f}_{kj}$ of some spectra lines for the hydrogen atom (Figs. 15.3.2 and 15.3.3) are very accurately described by the Rydberg's equation:

$$\tilde{f}_{kj} = R_\infty c \left(\frac{1}{n_j^2} - \frac{1}{n_k^2} \right) \tag{15.3-1}$$

where R_∞ is the Rydberg constant.

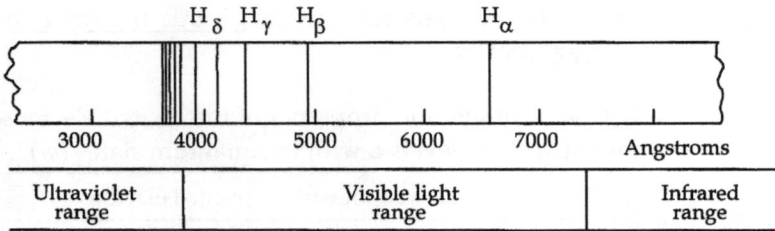

Figure 15.3.2. The line spectrum for atomic hydrogen.
Adapted from G. Gamow (1985).

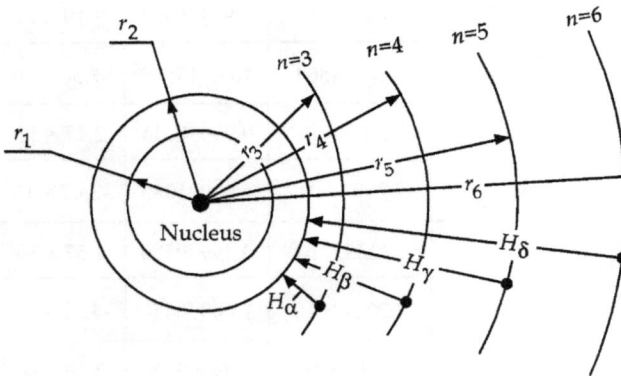

Figure 15.3.3. Electron orbits and spectra lines of a hydrogen atom.
Adapted from H.E. White (1934).

Table 15.3.2 shows a comparison of the frequencies $\tilde{f}_{kj}$ of electrons emitted by the atomic electrons of the hydrogen atom calculated from the UST's Eq. (15.1.4) with those obtained from the Rydberg's Eq. (15.3-1).

Table 15.3.2. Comparison of frequencies of spectra lines for hydrogen atom calculated from the UST's Eq. (15.1-4) with those obtained from the Rydberg's equation (15.3-1).

Quantum states n		Spectra lines of hydrogen	Frequencies $\tilde{f}_{kj}$, Hz		UST/ Rydbers's ratio
k	j	Symbol	Calculated	Rydberg's equation	
3	2	H_α	4.571583×10^{14}	4.569225×10^{14}	1.000516
4	2	H_β	6.171644×10^{14}	6.168454×10^{14}	1.000517
5	2	H_γ	6.912246×10^{14}	6.908668×10^{14}	1.000518
6	2	H_δ	7.314548×10^{14}	7.310760×10^{14}	1.000518

15.4 Positons

The parental excited positrons emitting the positons are composed of the positive reality-polarized toryces. As shown in Figure 15.4, the real toryx R of the excited positron emits a real subluminal helyx in which the translational velocity of trailing string $\tilde{\beta}_{2t}$ is subluminal. At the same time, an imaginary toryx I emits in the same direction an imaginary helyx in which $\tilde{\beta}_{2t}$ is superluminal. The real and imaginary helyces have opposite spins, and the reality-polarized positon formed by these helyces propagates with the average velocity equal to the velocity of light.

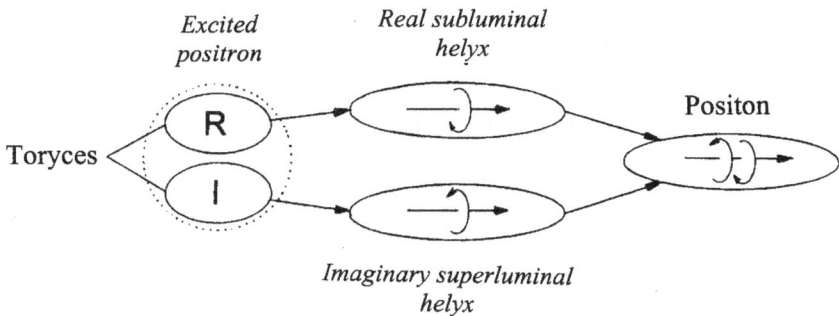

Excited positron   *Real subluminal helyx*   Positon

Toryces   R   I

Imaginary superluminal helyx

Figure 15.4. Formation of a positon.

Table 15.4 shows the compositions and properties of the excited positons of the various spacetime levels L.

Table 15.4. Compositions and properties of excited
positons of the various spacetime levels L with the quantum states (m).

L (m)	Positrons	Helyces of emitted excited positons				
		Symbols	$\widetilde{b}_1$	$\widetilde{\beta}_{2t}$	$\widetilde{f}_{kj}$ Hz	$\widetilde{E}_{kj}$ MeV
L1 (1)	$\widetilde{e}_{1,1,0}^{+1,2,0}$	$\widetilde{E}_{1,1,0}^{+1,2,0}$	0.5000000000124	> -1.0	6.75×10^{17}	2.79×10^{-4}
		$\widetilde{\widetilde{E}}_{1,1,0}^{+1,2,0}$	0.4999999999873	< -1.0	-6.81×10^{17}	-2.82×10^{-4}
	$\widetilde{e}_{1,2,0}^{+1,3,0}$	$\widetilde{E}_{1,2,0}^{+1,3,0}$	0.5000000000014	> -1.0	2.25×10^{17}	9.32×10^{-4}
		$\widetilde{\widetilde{E}}_{1,2,0}^{+1,3,0}$	0.4999999999986	< -1.0	-2.26×10^{17}	-9.36×10^{-4}
L2 (2)	$\widetilde{e}_{2,1,0}^{+2,2,0}$	$\widetilde{E}_{2,1,0}^{+2,2,0}$	> 0.50000000000	> -1.0	7.41×10^{15}	3.06×10^{-5}
		$\widetilde{\widetilde{E}}_{2,1,0}^{+2,2,0}$	< 0.50000000000	< -1.0	-7.41×10^{15}	-3.06×10^{-5}
	$\widetilde{e}_{2,2,0}^{+2,3,0}$	$\widetilde{E}_{2,2,0}^{+2,3,0}$	> 0.50000000000	> -1.0	1.37×10^{15}	5.67×10^{-6}
		$\widetilde{\widetilde{E}}_{2,2,0}^{+2,3,0}$	< 0.50000000000	< -1.0	-1.37×10^{15}	-5.67×10^{-6}
L3 (3)	$\widetilde{e}_{3,1,0}^{+3,2,0}$	$\widetilde{E}_{3,1,0}^{+3,2,0}$	> 0.50000000000	> -1.0	6.31×10^{13}	2.61×10^{-7}
		$\widetilde{\widetilde{E}}_{3,1,0}^{+3,2,0}$	< 0.50000000000	< -1.0	-6.31×10^{13}	-2.61×10^{-7}
	$\widetilde{e}_{3,2,0}^{+3,3,0}$	$\widetilde{E}_{3,2,0}^{+3,3,0}$	> 0.50000000000	> -1.0	6.34×10^{12}	2.62×10^{-8}
		$\widetilde{\widetilde{E}}_{3,2,0}^{+3,3,0}$	< 0.50000000000	< -1.0	-6.34×10^{12}	-2.62×10^{-8}

It follows from Table 15.4:

- The magnitudes of the relative radii of leading strings $\widetilde{b}_1$ of both the real and imaginary constituent helyces of positons approach closer to 0.5 with the increase of the spacetime level L.
- In the real helyces of positons, the translational velocities $\widetilde{\beta}_{2t}$ are slightly less than the velocity of light, while in the imaginary helyces they are slightly greater than the velocity of light. The differences between these velocities decrease as the spacetime level L increases.
- The frequencies of both real and imaginary helyces of positons $\widetilde{f}_{kj}$ decrease with the increase of the spacetime level L and they have the same magnitudes, but opposite signs (spins).

15.5 Ethertons

The parental excited ethertrons emitting the ethertons are composed of the real vorticity-polarized toryces. As shown in Figure 15.5, the real negative toryx (-) of the excited ethertron emits the real negative subluminal helyx in which the translational velocity of trailing string $\tilde{\beta}_{2t}$ is subluminal. At the same time, the real positive toryx (+) of the excited ethertron emits in the opposite direction the real positive helyx in which $\tilde{\beta}_{2t}$ is also subluminal. The real negative and positive helyces have opposite spins, and they serve respectively as the real negative and positive ethertons.

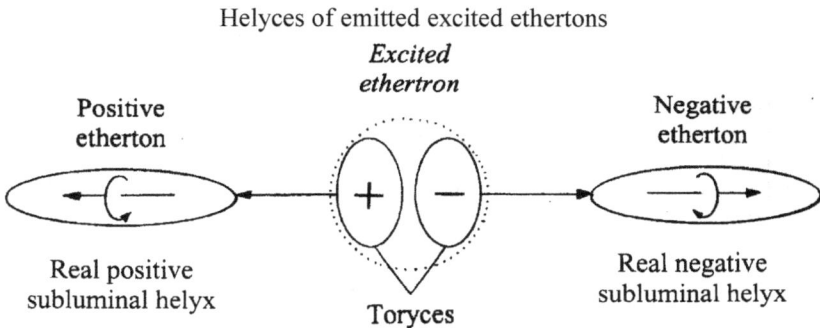

Helyces of emitted excited ethertons

Figure 15.5. Formation of the real positive and negative subluminal ethertons.

Table 15.5 shows the compositions and properties of the excited ethertons of the various spacetime levels L.

It follows from this Table 15.5:

- The magnitudes of relative radii of leading strings $\tilde{b}_1$ of positive ethertons approach closer to 0.5 with the increase of the spacetime level L.
- In both real negative and positive ethertons, the translational velocities of are slightly less than the velocity of light. The differences between these speeds decrease as the spacetime level L increases.
- The frequencies of both negative and positive ethertons $\tilde{f}_{kj}$ decrease with the increase of the spacetime level L and they have the same magnitudes, but opposite signs (spins).

Table 15.5. Compositions and properties of the excited ethertons of the various spacetime levels L.

L (m)	Ether-tons	Helyces of emitted ethertons				
		Symbols	$\tilde{b}_1$	$\tilde{\beta}_{2t}$	$\tilde{f}_{kj}$ Hz	$\tilde{E}_{kj}$ MeV
L2 (1)	$\tilde{a}_{1,1,0}^{1,2,0}$	$\tilde{A}_{1,1,0}^{-1,2,0}$	5672.238287	0.99982370	1.12×10^{17}	4.64×10^{-4}
		$\tilde{A}_{1,1,0}^{+1,2,0}$	> 0.50000000	< -1.0000000	-1.13×10^{17}	-4.69×10^{-4}
	$\tilde{a}_{1,2,0}^{1,3,0}$	$\tilde{A}_{1,2,0}^{-1,3,0}$	11778.98830	0.99991510	3.75×10^{16}	1.55×10^{-4}
		$\tilde{A}_{1,2,0}^{+1,3,0}$	> 0.50000000	< -1.0000000	-3.77×10^{16}	-1.56×10^{-4}
L3 (2)	$\tilde{a}_{2,1,0}^{2,2,0}$	$\tilde{A}_{2,1,0}^{-2,2,0}$	114610.5386	0.99999127	1.23×10^{15}	5.11×10^{-6}
		$\tilde{A}_{2,1,0}^{+2,2,0}$	> 0.50000000	< -1.0000000	-1.23×10^{15}	-5.11×10^{-6}
	$\tilde{a}_{2,2,0}^{2,3,0}$	$\tilde{A}_{2,2,0}^{-2,3,0}$	352759.6677	0.99999717	2.29×10^{14}	9.45×10^{-7}
		$\tilde{A}_{2,2,0}^{+2,3,0}$	> 0.50000000	< -1.0000000	-2.29×10^{14}	-9.45×10^{-7}

15.6 Singulatons

The parental excited singulatons are composed of the imaginary vorticity-polarized toryces.

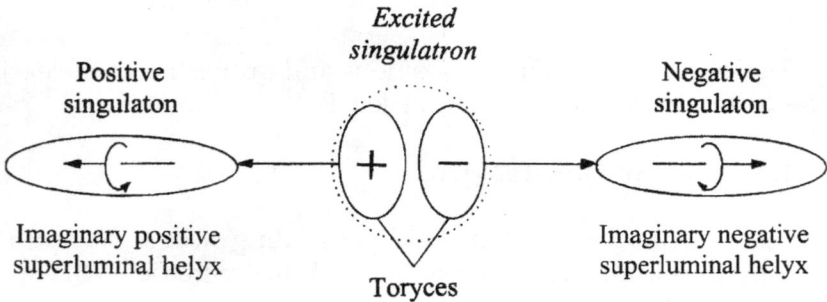

Figure 15.6.1. Formation of the imaginary positive and negative superluminal singulatons.

As shown in Figure 15.6.1, the imaginary negative toryx (-) of the excited singulatron emits the imaginary negative helyx in which the translational velocity of trailing string $\tilde{\beta}_{2t}$ is superluminal. At the same time, the imaginary positive toryx (+) of the excited singulatron emits in the opposite direction the imaginary positive helyx in which $\tilde{\beta}_{2t}$ is also subluminal. The imaginary negative and positive helyces have opposite spins, and they serve respectively as the imaginary negative and positive singulatons.

Table 15.6 shows the compositions and properties of the excited singulatons of the various spacetime levels L.

Table 15.6. Compositions and properties of the excited singulatons of the various spacetime levels L.

L (m)	Singula-tons	Helyces of emitted singulatons				
		Symbols	$\tilde{b}_1$	$\tilde{\beta}_{2t}$	$\tilde{f}_{kj}$ Hz	$\tilde{E}_{kj}$ GeV
$L2$ (1)	$\tilde{\tilde{a}}_{1,1,0}^{1,2,0}$	$\tilde{A}_{1,1,0}^{-1,2,0}$	-0.002436687	411.393376	1.39×10^{25}	57.4756
		$\tilde{A}_{1,1,0}^{+1,2,0}$	0.002436688	-409.393124	1.39×10^{25}	57.6156
	$\tilde{\tilde{a}}_{1,2,0}^{1,3,0}$	$\tilde{A}_{1,2,0}^{-1,3,0}$	-0.001461303	685.320961	2.32×10^{25}	95.8393
		$\tilde{A}_{1,2,0}^{+1,3,0}$	0.001461303	-683.320787	2.32×10^{25}	95.9794
	$\tilde{\tilde{a}}_{1,3,0}^{1,4,0}$	$\tilde{A}_{1,3,0}^{-1,4,0}$	-0.00104357	959.248720	3.25×10^{25}	134.2031
		$\tilde{A}_{1,3,0}^{+1,4,0}$	0.001043571	-957.248444	3.25×10^{25}	134.3431
$L3$ (2)	$\tilde{\tilde{a}}_{2,1,0}^{2,2,0}$	$\tilde{A}_{2,1,0}^{-2,2,0}$	-2.593×10^{-8}	3.856×10^{7}	1.31×10^{30}	5.40×10^{6}
		$\tilde{A}_{2,1,0}^{+2,2,0}$	2.593×10^{-8}	-3.856×10^{7}	1.31×10^{30}	5.40×10^{6}
	$\tilde{\tilde{a}}_{2,2,0}^{2,3,0}$	$\tilde{A}_{2,2,0}^{-2,3,0}$	-5.985×10^{-9}	1.671×10^{8}	5.66×10^{30}	2.34×10^{7}
		$\tilde{A}_{2,2,0}^{+2,3,0}$	5.985×10^{-9}	-1.671×10^{8}	5.66×10^{30}	2.34×10^{7}
	$\tilde{\tilde{a}}_{2,3,0}^{2,4,0}$	$\tilde{A}_{2,3,0}^{-2,4,0}$	-2.223×10^{-9}	4.499×10^{8}	1.52×10^{31}	6.30×10^{7}
		$\tilde{A}_{2,3,0}^{+2,4,0}$	2.223×10^{-9}	-4.499×10^{8}	1.52×10^{31}	6.30×10^{7}

It follows from Table 15.6:

- In the helyces of the singulatons of the spacetime levels $L2$ with $m = 1$ (ordinary matter), the translational velocities of trailing strings $\tilde{\beta}_{2t}$ exceed the velocity of light by the factors 411, 685 and 951.

- In the helyces of the singulatons of the spacetime level *L3*, the translational velocities of trailing strings $\tilde{\beta}_{2t}$ exceed the velocity of light by the factors 3.86, 167 and 450 millions.
- The frequencies of singulatons $\tilde{f}_{kj}$ increases with the increase of the spacetime level *L*.
- When a parental singulatron of the spacetime level *L2* (ordinary matter) is transferred from the excitation quantum state $n = 4$ to $n = 3$, the calculated energy of the emitted singulaton is approximately 6.8% greater than the measured energy of a new particle discovered during the proton-proton collisions experiments at CERN in 2012 that has some properties expected to be in the Higgs boson.

15.7 Oscillated Excited Tins

The oscillated excited tins are emitted by the oscillated excited electrons and positrons. They represent neutrinos of the Standard Model. The compositions and properties of negative excited tins are shown in Table 15.7.

Table 15.7. Parameters of the emitted oscillated electins when the linear excitation quantum states of their parental electrons are reduced from the higher to the lower oscillation quantum states *q*.

Electins		Helyces of emitted excited electins			
Names	Symbol	$\tilde{b}_1$	β_{2t}	$\tilde{f}_{kj}$ Hz	$\tilde{E}_{kj}$ MeV
(e-neutrino) $\tilde{e}_{2,1,0}^{-2,1,1}$	$\tilde{E}_{2,1,0}^{-2,1,1}$	37545.74234	0.99997337	-6.583×10^{15}	0.00003
	$\tilde{E}_{2,1,0}^{-2,1,1}$	-37542.75000	1.00002664	6.584×10^{15}	0.00003
(μ-neutrino) $\tilde{e}_{2,1,1}^{-2,1,2}$	$\tilde{E}_{2,1,1}^{-2,1,2}$	1728.822293	0.99942157	-6.665×10^{17}	0.00276
	$\tilde{E}_{2,1,1}^{-2,1,2}$	-1727.730500	1.00057879	6.666×10^{17}	0.00276
(τ-neutrino) $\tilde{e}_{2,1,2}^{-2,1,3}$	$\tilde{E}_{2,1,2}^{-2,1,3}$	268.64410	0.99627760	1.091×10^{19}	0.04511
	$\tilde{E}_{2,1,2}^{-2,1,3}$	-267.629770	1.00373651	1.091×10^{19}	0.04511
(x-neutrino) $\tilde{e}_{2,1,3}^{-2,1,4}$	$\tilde{E}_{2,1,3}^{-2,1,4}$	59.395040	0.98316358	1.060×10^{20}	0.43824
	$\tilde{E}_{2,1,3}^{-2,1,4}$	-58.391910	1.01712566	1.060×10^{20}	0.43828

It follows from this Table 15.7:

- The magnitudes of the relative radii of leading strings $\widetilde{b}_1$ of the constituent helyces of oscillated excited electons decrease with the increase of the oscillation quantum state q increases.
- In the real helyces of electons, the translational velocities are slightly less than the velocity of light, while in the imaginary helyces they are slightly greater than the velocity of light. The differences between these speeds increase as the oscillation quantum state q increases.
- The frequencies and energy of electons $\widetilde{f}_{kj}$ increase with the increase of the oscillation quantum state q.

Your Notes

PART 4

Properties
of the Macro-World

16

MACRO-TORYCES, TRONS & GRAVITONS

In previous chapters we described a role of the micro-toryces in the formation of matter particles and atoms of the micro-world. In the macro-world, the assemblies of atoms contained in each celestial body form the macro-spacetimes called the *macro-toryces* intimately associated with this body.

16.1 Macro-Toryces

Figure 16.1 shows a macro-toryx associated with the body A encompassing the satellite body B. Structurally, the macro-toryces are the same as the micro-toryces.

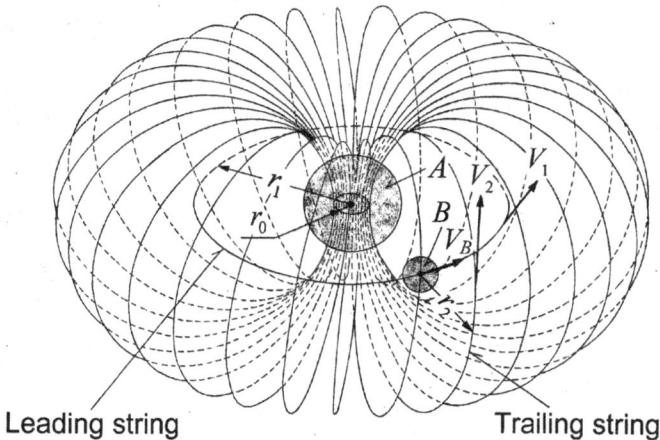

Leading string   Trailing string

Figure 16.1. Central body A with a macro-toryx encompassing the satellite body B.

Similarly to the micro-toryx (Figs. 5.1.1 – 5.1.3), the macro-toryx contains two strings, a *leading string* and a *trailing string*. Leading string is double-circular; it is moving at the velocity V_1 along a circle with the radius r_1. Trailing string is double-toroidal with the radius r_2. Its each toroidal branch propagates around one of the circular branches of the double-circular leading string.

Both branches of the toryx trailing string propagate along their toroidal spiral paths at the same spiral velocity V_2.

Spacetime postulates of the macro-toryx are the same as the spacetime postulates of the micro-toryx described by Eqs. (5.3-1) – (5.3-3).

Macro-Toryx Spacetime Postulates
(in absolute values of the toryx spacetime parameters)

- The length of one winding of trailing string L_2 is equal to the length of one winding of leading string L_1:

$$L_2 = L_1 = 2\pi r_1 \quad (+\infty > r_1 > -\infty) \qquad (16.1\text{-}1)$$

- The toryx eye radius r_0 is equal to a real positive constant:

$$r_0 = r_1 - r_2 = const. \quad (-\infty < r_1 < +\infty) \qquad (16.1\text{-}2)$$

- The spiral velocity of the trailing string V_2 is constant and equals to the velocity of light in vacuum c at each point of its spiral path. Its components, the translational velocity V_{2t} and the rotational velocity V_{2r}, relate to the spiral velocity V_2 by the Pythagorean Theorem:

$$V_2 = \sqrt{V_{2t}^2 + V_{2r}^2} = c = const. \quad (-\infty < r_1 < +\infty) \quad (16.1\text{-}3)$$

- The cosine of apex angle of trailing string $\cos s\widetilde{\varphi}_2$ is equal to the ratio of the radii of trailing and leading strings:

$$\cos s\widetilde{\varphi}_2 = \frac{\widetilde{r}_2}{\widetilde{r}_1} \quad (-\infty < \widetilde{r}_1 < +\infty) \qquad (16.1\text{-}4)$$

Eqs. (16.1-1) – (16.1-3) yield the same spacetime equations as the ones applicable to the micro-toryces – see Table 5.6-1, including Eq. (5.6-8a) describing the Spacetime Law of Planetary Motion:

$$\beta_1 = \frac{V_1}{c} = \frac{\sqrt{2b_1 - 1}}{b_1} \qquad (16.1\text{-}5)$$

where b_1 is the relative radius of the macro-toryx leading string that is equal to:

$$b_1 = \frac{r_1}{r_0} \qquad (16.1\text{-}6)$$

Only one spacetime equation is different for the macro- and macro-toryces. It is related to the toryx eye radius r_0. According to Eq. (10.1-3), the eye radius of the micro-toryx r_0 is equal to the radius of the real inversion micro-toryx r_i. When the atomic number $Z = 1$, it is equal to:

$$r_i = r_0 = \frac{e^2}{8\pi\varepsilon_0 m_e c^2} = 1.408970163 \times 10^{-15} m \qquad (16.1\text{-}7)$$

16.2 The Macro-Toryx Eye Radius

The macro-toryx eye radius r_0 is one of major parameters allowing us to establish a correlation between physical and spacetime properties of macro-toryces. It is possible to express the radius r_0 in physical terms by comparing the law of planetary motion based on classical mechanics and with the Spacetime Law of Planetary motion.

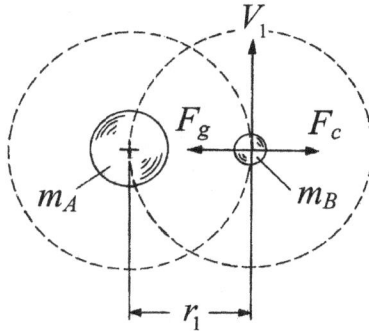

Figure 16.2. Forces applied to a satellite body
in a two-body planetary system according to classical mechanics.

The law of planetary motion based on classical mechanics can be derived by equating two forces (Fig. 16.2):

- The attraction gravitational force F_g between the bodies A and B with the respective masses m_A and m_B separated by the distance r_1

- The centrifugal forces F_c applied to the body B orbiting around the body A with the orbital velocity V_1.

The gravitational force F_g can be calculated by using Newton's law of universal gravitation. According to this law, the gravitational force F_g between a body with the mass m_A and a body with the mass m_B separated by the distance r_1 is equal to:

$$F_g = \frac{m_A m_B G}{r_1^2}$$
(16.2-1)

The Dutch physicist and astronomer Christiaan Huygens discovered that the centrifugal force F_c applied to the body with the mass m_B moving with the orbital velocity V_1 is equal to:

$$F_c = -\frac{m_B V_1^2}{r_1}$$
(16.2-2)

A body with the mass m_A will be in equilibrium state with the body m_B when the sum of the gravitational and centrifugal forces is equal to zero. Thus, we obtain from Eqs. (16.2-1) and (16.2-2) that in the equilibrium state, the orbital velocity V_1 is inversely proportional to the square root of the orbital radius r_1:

$$V_1 = \sqrt{\frac{m_A G}{r_1}}$$
(16.2-3)

Eq. (16.2-3) expresses the law of planetary motion based on classical mechanics. It is possible to simplify this equation by presenting both the orbital velocity V_1 and the orbital radius r_1 in relative terms. In respect to the velocity of light c the relative orbital velocity β_1 is equal to:

$$\beta_1 = \frac{V_1}{c}$$
(16.2-4)

Consequently, we obtain from Eqs. (16.1-6), (16.2-3) and (16.2-4) an equation expressing parameters of the law of planetary motion based on classical mechanics in relative terms:

$$\beta_1 = \sqrt{\frac{m_A G}{2c^2} \frac{2}{b_1 r_0}}$$ (16.2-5)

For the case when the radius of the real inversion macro-toryx r_j is equal to the macro-toryx eye radius r_0, it is described by the equation:

$$r_j = r_0 = \frac{m_A G}{2c^2},$$ (16.2-6)

When $b_1 >> 1$, Eq. (16.2-6) reduces to the equation expressing the law of planetary motion based on classical mechanics – see Eq. (5.8-2).

16.3 Macro-Trons

Topological transformations of the macro-toryces making up the macro-trons are the same as the topological transformations of the micro-toryces making up the micro-trons shown in Figs. 8.4, 8.5.1 – 8.5.4 and 8.6.

Similarly to the formation of the micro-trons (Fig. 9.1.1), the macro-trons are also made up by the unification of four kinds of topologically-polarized macro-toryces shown in Fig. 16.3.

- Real negative macro-toryces
- Real positive macro-toryces
- Imaginary positive macro-toryces
- Imaginary negative macro-toryces.

Similarly to the classification of the micro-trons, the macro-trons are also divided into four groups:

- Macro-electrons
- Macro-positrons
- Macro-ethertrons
- Macro-singulatrons.

Macro-ethertron

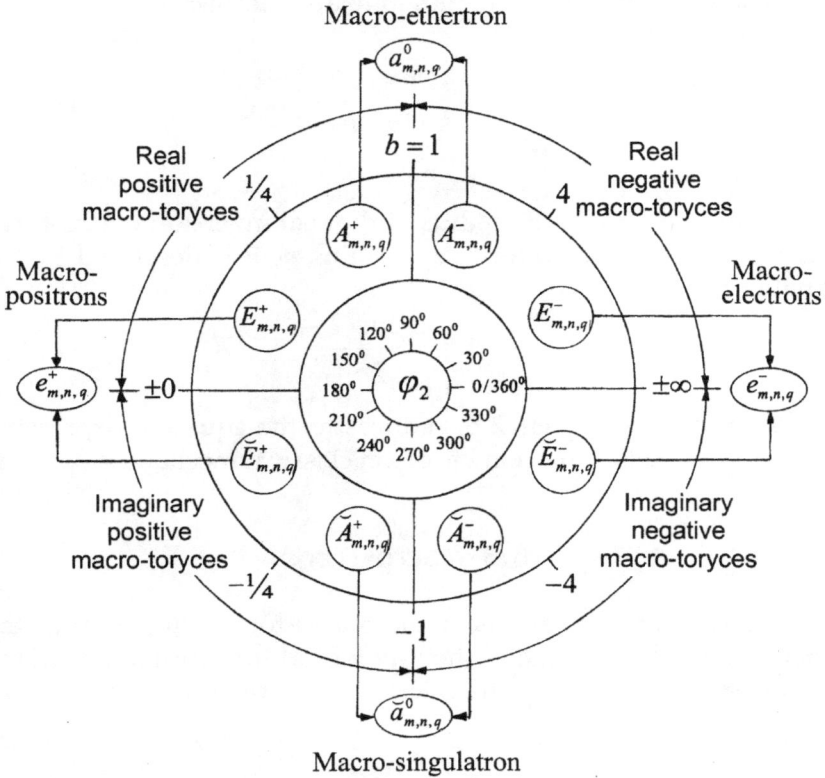

Figure 16.3. General presentation of formation of four macro-trons from polarized macro-toryces.

16.4 Outverted and Inverted Stars

Depending on the relationship between the outer radius r_b of a star body and the eye radius r_{0m} of a macro-toryx associated with a star, the stars can be divided into three kinds: *outverted stars, inverted stars* and *imaginary stars* as shown in Figure 16.4.

Outverted stars $(r_b \gg r_{0m})$ – In the outverted stars, the outer radius r_b of the star body is much greater than the eye radius r_{0m} of a macro-toryx associated with this star.

Marginally-outverted stars $(r_b > r_{0m})$ – In the marginally-outverted stars, the outer radius r_b of the star body is slightly greater than the eye radius r_{0m} of a macro-toryx associated with this star.

Inverted stars $(0.5r_b < r_b < r_{0m})$ – In the inverted stars, the outer radius r_b of a star body is greater than one half of the eye radius r_{0m}, but less than the eye radius r_{0m} of macro-toryx associated with this star.

Imaginary stars $(r_b < 0.5r_b)$ – In the imaginary stars, the outer radius r_b of a star body is less than one half of the eye radius r_{0m} of a macro-toryx associated with this star.

Real star   Inverted star   Imaginary star
$(r_b > r_{0m})$ $(0.5r_{0m} < r_b < r_{0m})$ $(r_b < 0.5r_{0m})$

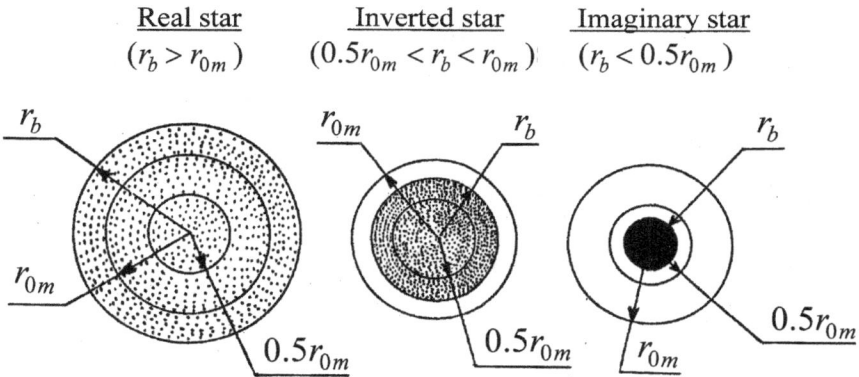

Figure 16.4. Three kinds of stars.

The proposed theory considers the following examples of the stars:

- Our Sun is a typical example of an outverted star
- The **neutron stars** and **pulsars** are typical examples of the marginally-outverted stars

The inverted and imaginary stars represent two phases of **black holes**.

16.5 Quantization Equations of Macro-Toryces

As we described in Chapter 9, the quantum states of the polarized micro-toryces are described by Eqs. (9.3-1) - (9.3-17) as a function of the micro-toryx quantization parameter z. The macro-trons are assumed to be made up of the polarized macro-toryces which quantum states are described as a function of the *macro-toryx quantization parameter* Z:

$$Z = 2(Kn\Lambda)^m \qquad (16.5-1)$$

where
K = macro-toryx quantization factor
m = macro-toryx exponential excitation quantum state
n = macro-toryx linear excitation quantum state.

Table 16.5.1 shows the values of K, the masses m_g, the equatorial body radius r_b and the radii of the real inversion macro-toryces r_j for the largest bodies of our solar system.

Table 16.5.1. Parameters of largest celestial bodies of our solar system.

Celestial body	K	m_g, kg	r_b, m	r_j, m
Sun	15	1.98910×10^{30}	6.9550×10^{08}	738.566556
Jupiter	32	1.89973×10^{27}	7.1492×10^{07}	0.705383
Saturn	55	5.68598×10^{26}	6.0268×10^{07}	0.211124
Uranus	100	8.68900×10^{25}	2.5559×10^{07}	0.032263
Neptune	100	1.02966×10^{26}	2.4764×10^{07}	0.038232

Table 16.5.2 shows quantization equations for the relative radii of leading strings and peripheral circles of the basic excited macro-toryces forming the basic excited macro-trons. These equations are similar to Eqs. (9.3-2) – (9.3-17) applied to the basic excited micro-trons. The brackets $\langle\,\rangle$ in the symbols of the macro-trons and macro-toryces differentiate them from the micro-toryces and micro-trons.

Table 16.5.2. Quantization equations for relative radii of peripheral circles and leading strings of basic excited macro-toryces forming the basic macro-trons.

Macro-Tron	Macro-toryx	Relative radius of leading string	Relative radius of peripheral circle
Macro-electron $\langle e \rangle_{m,n,q}^{-1}$	$\langle E \rangle_{m,n,q}^{-}$	$b_1^- = Z$ $\quad$ (16.5-2)	$b^- = 2Z - 1$ $\quad$ (16.5-10)
	$\langle \breve{E} \rangle_{m,n,q}^{-}$	$\breve{b}_1^- = 1 - Z$ $\quad$ (16.5-3)	$\breve{b}^- = 1 - 2Z$ $\quad$ (16.5-11)
Macro-positron $\langle e \rangle_{m,n,q}^{+1}$	$\langle E \rangle_{m,n,q}^{+}$	$b_1^+ = \dfrac{Z}{2Z - 1}$ $\quad$ (16.5-4)	$b^+ = \dfrac{1}{2Z - 1}$ $\quad$ (16.5-12)
	$\langle \breve{E} \rangle_{m,n,q}^{+}$	$\breve{b}_1^- = \dfrac{1-Z}{1-2Z}$ $\quad$ (16.5-5)	$\breve{b}^+ = \dfrac{1}{1-2Z}$ $\quad$ (16.5-13)
Macro-ethertron $\langle a \rangle_{m,n,q}^{0}$	$\langle A \rangle_{m,n,q}^{-}$	$b_1^- = \dfrac{Z}{Z-1}$ $\quad$ (16.5-6)	$b^- = \dfrac{Z+1}{Z-1}$ $\quad$ (16.5-14)
	$\langle A \rangle_{m,n,q}^{+}$	$b_1^+ = \dfrac{Z}{Z+1}$ $\quad$ (16.5-7)	$b^+ = \dfrac{Z-1}{Z+1}$ $\quad$ (16.5-15)
Macro-singulatron $\langle \breve{a} \rangle_{m,n,q}^{0}$	$\langle \breve{A} \rangle_{m,n,q}^{-}$	$\breve{b}_1^- = \dfrac{1}{1-Z}$ $\quad$ (16.5-8)	$\breve{b}^- = \dfrac{1+Z}{1-Z}$ $\quad$ (16.5-16)
	$\langle \breve{A} \rangle_{m,n,q}^{+}$	$\breve{b}_1^+ = \dfrac{1}{1+Z}$ $\quad$ (16.5-9)	$\breve{b}^+ = \dfrac{1-Z}{1+Z}$ $\quad$ (16.5-17)

According to the UST, the quantum states of the orbital radii of the planets in our solar system correspond to the quantum states of the macro-toryces making up the macro-electron described by Eqs. (16.5-1) - (16.5-3) when $m = 2$.

16.6 Quantum States of Planets & Moons

Table 16.6.1 compares calculated and mean relative distances between the Sun and planets in our solar system. Table 16.6.1 includes data for 29 planets with their status shown in Table 16.6.2. Tables 16.6.3 – 16.6.6 compare calculated and mean relative distances between the planets the centers of Jupiter, Saturn, Uranus and Neptune and their respective moons.

Macro-Toryces, Trons & Gravitons

Table 16.6.1. Parameters of planets in our solar system.
($K = 15$, $m_g = 1.98910 \times 10^{30}$ kg, $r_b = 6.9550 \times 10^{08}$ m, $r_j = 738.566556$ m).

N	Planets	Mean distance to Sun, r_{1m}, m	Relative mean dist. to Sun b_{1m}	Relative calculated dist. to Sun b_{1c}	Mean/ calc. ratio b_{1m}/b_{1c}
1	Sun1	6.2378×10^{09}	8.4461×10^{06}	8.4461×10^{06}	1.0000
2	Sun 2	2.4951×10^{10}	3.3784×10^{08}	3.3784×10^{08}	1.0000
3	**Mercury**	5.7909×10^{10}	7.8408×10^{07}	7.6014×10^{07}	**1.0315**
4	**Venus**	1.0821×10^{11}	1.4651×10^{08}	1.3514×10^{08}	**1.0842**
5	**Earth**	1.4960×10^{11}	2.0255×10^{08}	2.1115×10^{08}	**0.9593**
6	**Mars**	2.2794×10^{11}	3.0863×10^{08}	3.0406×10^{08}	**1.0150**
7	Hungaria	3.0567×10^{11}	4.1386×10^{08}	4.1386×10^{08}	1.0000
8	Phocaea	3.9923×10^{11}	5.4055×10^{08}	5.4055×10^{08}	1.0000
9	Cybele	5.0527×10^{11}	6.8413×10^{08}	6.8413×10^{08}	1.0000
10	Hilda	6.2382×10^{11}	8.4461×10^{08}	8.4461×10^{08}	1.0000
11	**Jupiter**	7.7833×10^{11}	1.0538×10^{09}	1.0220×10^{09}	**1.0312**
12	Io	8.9826×10^{11}	1.2162×10^{09}	1.2162×10^{09}	1.0000
13	Europe	1.0542×10^{12}	1.4274×10^{09}	1.4274×10^{09}	1.0000
14	Ganymede	1.2227×10^{12}	1.6554×10^{09}	1.6554×10^{09}	1.0000
15	**Saturn**	1.4270×10^{12}	1.9321×10^{09}	1.9004×10^{09}	**1.0167**
16	Tethys	1.5970×10^{12}	2.1622×10^{09}	2.1622×10^{09}	1.0000
17	Dione	1.1029×10^{12}	2.4410×10^{09}	2.4410×10^{09}	1.0000
18	Rhea	2.0210×10^{12}	2.7365×10^{09}	2.7365×10^{09}	1.0000
19	Titan	2.2518×10^{12}	3.0490×10^{09}	3.0490×10^{09}	1.0000
20	Iapetus	2.4951×10^{12}	3.3784×10^{09}	3.3784×10^{09}	1.0000
21	**Uranus**	2.8696×10^{12}	3.8853×10^{09}	3.7247×10^{09}	**1.0431**
22	Miranda	3.0190×10^{12}	4.0879×10^{09}	4.0879×10^{09}	1.0000
23	Ariel	3.2998×10^{12}	4.4680×10^{09}	4.4680×10^{09}	1.0000
24	Umbriel	3.5929×10^{12}	4.8649×10^{09}	4.8649×10^{09}	1.0000
25	Titania	3.8988×10^{12}	5.2788×10^{09}	5.2788×10^{09}	1.0000
26	Oberon	4.2170×10^{12}	5.7095×10^{09}	5.7095×10^{09}	1.0000
27	**Neptune**	4.4966×10^{11}	6.0883×10^{08}	6.0883×10^{08}	**0.9888**
28	Proteus	4.8904×10^{12}	5.6217×10^{09}	5.6217×10^{09}	1.0000
29	Triton	5.2460×10^{12}	7.1029×10^{09}	7.1029×10^{09}	1.0000

Table 16.6.2. Status of planets of our solar system.

N	Planets	Planet status
1	Sun1	*Captured and swallowed by the Sun*
2	Sun2	*Captured and swallowed by the Sun*
3	**Mercury**	**Currently existing**
4	**Venus**	**Currently existing**
5	**Earth**	**Currently existing**
6	**Mars**	**Currently existing**
7	Hungaria	*Could be formed from Hungaria family of asteroid belt*
8	Phocaea	*Could be formed from Phocaea family of asteroid belt*
9	Cybele	*Could be formed from Cybele family of asteroid belt*
10	Hilda	*Could be formed from Hilda family of asteroid belt*
11	**Jupiter**	**Currently existing**
12	Io	*Captured by the Jupiter and became its moon*
13	Europe	*Captured by the Jupiter and became its moon*
14	Ganymede	*Captured by the Jupiter and became its moon*
15	**Saturn**	**Currently existing**
16	Tethys	*Captured by the Saturn and became its moon*
17	Dione	*Captured by the Saturn and became its moon*
18	Rhea	*Captured by the Saturn and became its moon*
19	Titan	*Captured by the Saturn and became its moon*
20	Iapetus	*Captured by the Saturn and became its moon*
21	**Uranus**	**Currently existing**
22	Miranda	*Captured by the Uranus and became its moon*
23	Ariel	*Captured by the Uranus and became its moon*
24	Umbriel	*Captured by the Uranus and became its moon*
25	Titania	*Captured by the Uranus and became its moon*
26	Oberon	*Captured by the Uranus and became its moon*
27	**Neptune**	**Currently existing**
28	Proteus	*Captured by the Neptune and became its moon*
29	Triton	*Captured by the Neptune and became its moon*

Table 16.6.3. Comparisons between calculated and mean relative distances between the Jupiter and its moons. ($K = 32$, $m_g = 1.89973 \times 10^{27}$ kg, $r_b = 7.1492 \times 10^{07}$ m, $r_j = 0.705383$ m).

N	Jupiter's moons	Mean distance to Jupiter, m	Relative mean distance to Jupiter	Relative calculated distance to Jupiter	Mean/ calc. ratio
		r_{1m}, m	b_{1m}	b_{1c}	b_{1m}/b_{1c}
1	Jupiter 1*	2.7115×10^{07}	3.8439×10^{07}	3.8439×10^{07}	1.0000
2	Jupiter 2*	1.0846×10^{08}	1.5376×10^{08}	1.5376×10^{08}	1.0000
3	Io	2.6200×10^{08}	3.7143×10^{08}	3.4595×10^{08}	1.0737
4	Europa	4.1690×10^{08}	5.9103×10^{08}	6.1502×10^{08}	0.9610
5	Ganymede	6.6490×10^{08}	9.4261×10^{08}	9.6097×10^{08}	0.9809
7	Callisto**	1.1701×10^{09}	1.6588×10^{09}	1.8835×10^{09}	0.8807

*) Jupiter 1 and Jupiter 2 are assumed to be the moons swallowed by the Jupiter.

**) Callisto is a large asteroid that is assumed to come from outside of our solar system and captured by the Jupiter to become its moon.

Table 16.6.4. Comparisons between calculated and mean relative distances between the Saturn and its moons. ($K = 55$, $m_g = 5.68598 \times 10^{26}$ kg, $r_b = 6.0268 \times 10^{07}$ m, $r_j = 0.211124$ m).

n	Saturn's moons	Mean distance to Saturn, m	Relative mean distance to Saturn	Relative calculated distance to Saturn	Mean/ calc. ratio
		r_{1m}, m	b_{1m}	b_{1c}	b_{1m}/b_{1c}
1	Saturn 1*	2.3974×10^{07}	1.1355×10^{08}	1.1355×10^{08}	1.0000
2	Saturn 2*	9.5890×10^{07}	4.5421×10^{08}	4.5421×10^{08}	1.0000
3	Tethys	2.9462×10^{08}	1.3955×10^{09}	1.0220×10^{09}	1.3655
4	Dione	3.7740×10^{08}	1.7876×10^{09}	1.8168×10^{09}	0.9839
5	Rhea	5.2711×10^{08}	2.4967×10^{09}	2.8388×10^{09}	0.8795
7	Titan	1.2219×10^{09}	5.7877×10^{09}	5.5641×10^{09}	1.0402
12	Iapetus	3.5608×10^{09}	1.6866×10^{10}	1.6352×10^{10}	1.0315

*) Saturn 1 and Saturn 2 are assumed to be the moons swallowed by the Saturn.

Table 16.6.5. Comparisons between calculated and mean relative distances between the Uranus and its moons. ($K = 100$, $m_g = 8.6890 \times 10^{25}$ kg, $r_b = 2.5559 \times 10^{07}$ m, $r_j = 0.032263$ m).

N	Uranus' Moons	Mean distance to Uranus, m	Relative mean distance to Uranus	Relative calculated distance to Uranus	Mean/ calc. ratio
		r_{1m}, m	b_{1m}	b_{1c}	b_{1m}/b_{1c}
1	Uranus 1*	1.2111×10^{07}	3.7538×10^{08}	3.7538×10^{08}	1.0000
2	Uranus 2*	4.8446×10^{07}	1.5015×10^{09}	1.5015×10^{09}	1.0000
3	Miranda	1.2978×10^{08}	4.0225×10^{09}	3.3784×10^{09}	1.1907
4	Ariel	1.9102×10^{08}	5.9207×10^{09}	6.0061×10^{09}	0.9858
5	Umbriel	2.6630×10^{08}	8.2540×10^{09}	9.3845×10^{09}	0.8795
6	Titania	4.3591×10^{08}	1.3511×10^{10}	1.3514×10^{10}	0.9998
7	Oberon	5.8352×10^{08}	1.8086×10^{10}	1.8394×10^{10}	0.9833

*) Uranus 1 and Uranus 2 are assumed to be the moons swallowed by the Uranus.

Table 16.6.6. Comparisons between calculated and mean relative distances between the Neptune and its moons. ($K = 100$, $m_g = 1.02966 \times 10^{26}$ kg, $r_b = 2.4764 \times 10^{07}$ m, $r_j = 0.038232$ m).

n	Neptune moons	Mean distance to Neptune, m	Relative mean distance to Neptune	Relative calculated distance to Neptune	Mean/ calc. ratio
		r_{1m}, m	b_{1m}	b_{1c}	b_{1m}/b_{1c}
1	Neptune 1*	1.4352×10^{07}	3.7538×10^{08}	3.7538×10^{08}	1.0000
2	Neptune 2*	4.8446×10^{07}	1.5015×10^{09}	1.5015×10^{09}	1.0000
3	Proteus	1.1765×10^{08}	3.0772×10^{09}	3.3784×10^{09}	0.9108
5	Triton	3.5476×10^{08}	9.2791×10^{09}	9.3845×10^{09}	0.9888

*) Neptune 1 and Neptune 2 are assumed to be the moons swallowed by the Neptune.

16.7 Gravitons

The gravitons are the radiation particles of the macro-world. They are composed of the reality-polarized and charge-polarized matched macro-helyces. The gravitons are emitted after the orbit of either a planet or a satellite collapses to a lower quantum energy state. The names of gravitons are similar to the names of their parental macro-trons responsible for the creation of the macro-helyces.

The gravitons emitted by the excited macro-trons are called *macro-tons*, while the gravitons emitted by the oscillated macro-trons are called *macro-tins* as shown in Table 16.7. Similarly to the micro-trons, the macro-electrons, macro-positrons and macro-ethertrons emit radiation particles propagating with luminal velocity, while the macro-singulatrons emit radiation particles propagating with superluminal velocity.

Table 16.7. Types of gravitons.

Parental macro-trons	Gravitons	
	Macro-tons (emitted by excited macro-trons)	**Macro-tins** (emitted by oscillated macro-trons)
Macro-electron	Macro-electon	Macro-electin
Macro-positron	Macro-positon	Macro-positin
Macro-ethertron	Macro-etherton	Macro-ethertin
Macro-singulatron	Macro-singulaton	Macro-singulatin

17

INTERACTIONS IN THE MACRO-WORLD

We will provide in this Chapter a comparison of motions of celestial bodies and interactions between them described by the UST and classical mechanics.

17.1 Kepler's Third Law of Planetary Motion

One of the most interesting surprises of the UST is that it predicts Kepler's Third Law of planetary motion discovered by the German astronomer and physicist Johannes Kepler. This law states that the cubes of the mean distances of planets from the Sun r_1 are proportional to the squares of their periods of revolution T_1, thus:

$$r_1^3 = kT_1^2 \tag{17.1-1}$$

where k is a constant.

Let us show that Kepler's Third Law of planetary motion is merely a particular case of the Spacetime Law of Planetary Motion described by Eqs. (16.1-4) and (16.1-5). For the case when $b_1 \gg 1$, we obtain:

$$V_1 = c\sqrt{\frac{2r_j}{r_1}} \tag{17.1-2}$$

Consider a case (Fig. 16.1) in which a planet B orbit around the Sun A with the mass m_A. Since the orbital velocity of the planet B around the Sun A is equal to:

$$V_1 = \frac{2\pi r_1}{T_1} , \tag{17.1-3}$$

We can express Kepler's Third Law from Eqs. (17.1-2) and (17.1-3) in the form:

$$r_1^3 = \frac{r_j c^2}{2\pi^2} T_1^2 \tag{17.1-4}$$

After comparing Eq. (17.1-4) with Eq. (17.1-1) and taking into account Eq. (16.2-6), we find that they are identical when the constant k is equal to:

$$k = \frac{r_j c^2}{2\pi^2} = \frac{m_A G}{4\pi^2} \tag{17.1-5}$$

Thus, Kepler's Third Law of planetary motion is a particular case of the Spacetime Law of Planetary Motion for the case when $b_1 \gg 1$.

17.2 Acceleration of a Satellite Body

Consider a case when the satellite body B moves around the body A along a circular path with the radius r_1 at the orbital velocity V_B as shown in Figure 16.1. The body B is embraced by a macro-toryx associated with the central body A in such a way that both the satellite body B and a leading string of the macro-toryx follow the same circular path.

The behavior of the satellite body B depends on a relationship between its orbital velocity V_B and the spiral velocity of leading string V_1 of the macro-toryx associated with the central body A. When these velocities are not exactly the same, it means that motion of the satellite body B does not follow the Spacetime Law of Planetary Motion described by Eq. (16.1-4). Consequently, the body B will move in a radial direction either towards to or away from the body A with the acceleration a_b equal to:

$$a_b = \frac{V_1^2 - V_B^2}{r_1} = \frac{V_1^2(1 - \gamma_V^2)}{r_1} \tag{17.2-1}$$

where γ_V is the velocity ratio that is equal to:

$$\gamma_V = \frac{V_B}{V_1} \tag{17.2-2}$$

Table 17.2 shows that the mean orbital velocities V_B of planets of our solar system are very close to the velocity of propagation of the leading string V_1 of the macro-toryx associated with the Sun.

Table 17.2. Mean orbital velocities V_B of the planets of the solar system and the velocities of propagation of the leading string V_1 of the macro-toryx associated with the Sun.

Planets	V_1	V_B	$\gamma_V = V_B / V_1$
Mercury	47880.0	47890.0	1.0002
Venus	35026.5	35030.0	1.0001
Earth	29789.6	29783.0	0.9998
Mars	24133.3	24130.0	0.9999
Jupiter	13060.1	13060.0	1.0000
Saturn	9645.3	9640.0	1.0004
Uranus	6801.7	6810.0	1.0012

It follows from Eq. (17.2-1) that when $V_B > V_1$ the body B will move away from the central body A with negative acceleration. Conversely, when $V_B < V_1$ the body B will move towards the central body A with positive acceleration. Importantly, the acceleration a_b calculated from Eq. (17.2-1) is applied to the center of gravity of the body B.

From Eqs. (16.1-4), (16.1-5), (16.2-4), (16.2-6) and (17.2-1), the acceleration a_b is equal to:

$$a_b = \frac{m_A G}{r_1^2} \frac{2b_1 - 1}{2b_1} (1 - \gamma_V^2) \qquad (17.2\text{-}3)$$

The acceleration a_b is called the *spacetime acceleration*. For a particular case when $b_1 >> 1$ and $\gamma_V = 0$, Eq. (17.2-3) reduces to the form representing the classical acceleration:

$$a_b = \frac{m_A G}{r_1^2} \qquad (17.2\text{-}4)$$

17.3   The Spacetime Law of Gravitation

Consider a case when the satellite body B with the mass m_B (Fig. 16.2) does not comply with the Spacetime Law of Planetary Motion and moves freely either towards to or away from the body A with the acceleration a_b. Let the dimensions of the body B to be negligibly small in comparison with its distance from the body A. In that case, the body B will not experience any force applied to it until after its free motion is affected by either internal or external causes. In that moment a holding force will be applied to the body B. This force is called the *spacetime gravitational force* F_{gs}, and it is described by the equation:

$$F_{gs} = m_B a_b \tag{17.3-1}$$

Similarly to Eq. (10.2-3), when $Q_q = 1$, the inertial mass of the body B is equal to:

$$m_{Bi} = m_B \frac{2(b_1 - 1)}{2b - 1_1} \tag{17.3-2}$$

Consequently, from Eqs. (17.2-3), (17.3-1) and (17.3-2) the spacetime gravitational force F_{gs} is equal to:

$$F_{gs} = \frac{m_A m_B G}{r_1^2} \frac{b_1 - 1}{b_1} (1 - \gamma_V^2) \tag{17.3-3}$$

Eq. (17.3-3) is called the *Spacetime Law of Gravitation*. For a particular case when $b_1 \gg 1$ and $\gamma_V = 0$, Eq. (17.3-3) reduces to the form representing the Newton's universal law of gravitation:

$$F_{gN} = \frac{m_A m_B G}{r_1^2} \tag{17.3-4}$$

Based on Eqs. (17.2-1) - (17.2-3), (17.3-3) and Fig. 17.3, we may conclude:

- When the velocity V_B of the satellite body B around the central body A is the same as the velocity of propagation of the leading string V_1 of the macro-toryx associated with the central body A ($\gamma_V = 1$), the body B will orbit the body A according to the Spacetime Law of Planetary motion and, consequently, the velocity V_B and its distance to the satellite body B will remain unchanged.
- When the velocity V_B of the satellite body B around the central body A is different than the velocity of propagation of the leading string V_1 of the macro-toryx associated with the central body A, the satellite body B accelerates either towards to or away from the body A.

- If the radius of the satellite body B is negligibly small than its distance to the central body A then the "gravitational force" will be "felt" by the satellite body B only after its acceleration towards or away from the central body A will be obstructed by either internal or external means.

Figure 17.3 shows the ratio of the spacetime to Newton's gravitational force F_{gs} / F_{gN} defined by Eqs. (17.3-3) and (17.3-4) as a function of the relative distance b_1 between the bodies A and B when velocity of the body B is equal to zero ($\gamma_V = 0$).

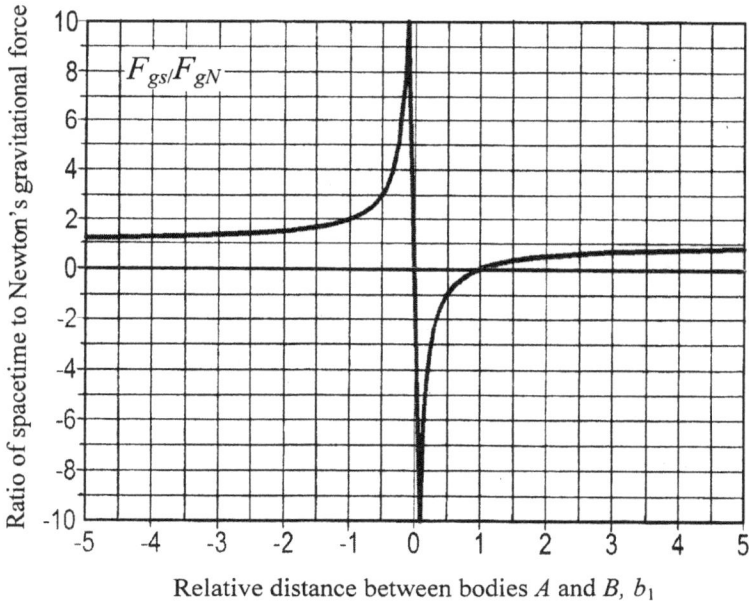

Figure 17.3. Ratio of the spacetime to Newton's gravitational force F_{gs} / F_{gN} as a function of the relative distance b_1 between bodies A and B.

Based on Fig. 17.3, we may conclude:

- When the relative distance b_1 between the bodies A and B is much greater than 5.0, the ratio between the spacetime and Newton's gravitation forces F_{gs} / F_{gN} approaches to 1.0, ie. the values of these forces become close to one another.
- As b_1 decreases the ratio F_{gs} / F_{gN} decreases too and reduces to zero at $b_1 = 1.0$, this ratio then changes its sign and approaches negative infinity when b_1 approaches positive infinity (+0).

- When b_1 approaches negative infinility (-0), the ratio F_{gs} / F_{gN} approaches positive infinity; this ratio then decreases and approaches to 1.0 when b_1 approaches negative infinity.

It appears that Eq. (17.3-3) for the spacetime gravitational force F_{gs} might explain better behavior of a body entering a black hole than Eq. (17.3-4) for Newton's gravitational force F_{gN}.

PHYSICAL & SPACETIME CONSTANTS

Physical Constants

Names		Values
Elementary charge	e	$1.602\ 176\ 565 \times 10^{-19}$ C
Electric constant	ε_0	$8.854\ 187\ 817 \times 10^{-12}$ C^2/N/m^2
Electron mass	m_e	$9.109\ 382\ 91 \times 10^{-31}$ kg
Proton relative mass	m_p/m_e	$1836.152\ 671\ 95$
Neutron relative mass	m_n/m_e	$1838.683\ 660\ 08$
Muon relative mass	m_μ/m_e	$206.768\ 284\ 26$
Tau relative mass	m_τ/m_e	$3477.151\ 011\ 54$
Newtonian constant of gravitation	G	$6.674\ 280 \times 10^{-11}$ m^3/kg/s^2
Planck constant	h	$6.626\ 069\ 605 \times 10^{-34}$ J s
Bohr magneton	μ_B	$9.274\ 009\ 68 \times 10^{-24}$ J/T
Muon magneton	μ_μ	$4.485\ 218\ 58 \times 10^{-26}$ J/T
Tau magneton	μ_τ	$2.666\ 876\ 44 \times 10^{-24}$ J/T
Electron magnetic moment to Bohr magneton ratio	μ_e/μ_B	$-1.001\ 159\ 652\ 180\ 76$
Muon magnetic moment to muon magneton ratio	μ_μ/μ_μ	$-1.001\ 165\ 923$
Tau magnetic moment to tau magneton ratio	μ_τ/μ_τ	$-0.987\ 629\ 486$
Nuclear magneton	μ_N	$5.050\ 783\ 54 \times 10^{-27}$ J/T
Proton magnetic moment to nuclear magneton ratio	μ_p/μ_N	$2.792\ 847\ 356$
Neutron magnetic moment to nuclear magneton ratio	μ_n/μ_N	$-1.913\ 042\ 72$

Spacetime Constants

Names		Values
Speed of light in vacuum	c	$2.997\ 924\ 58 \times 10^{08}$ m/s
Classical electron radius	r_e	$2.817\ 940\ 327 \times 10^{-15}$ m
Micro-toryx eye radius	r_0	$1.408\ 970\ 164 \times 10^{-15}$ m
Toryx basic frequency	f_0	$3.386\ 406\ 102 \times 10^{22}$ s^{-1}
Inverse fine structure constant	α^{-1}	$137.035\ 999\ 11$
Rydberg constant	R_x	$1.097\ 373\ 157 \times 10^{07}$ m^{-1}

PUBLICATIONS

LIST OF PUBLICATIONS CONSULTED

A

Aczel, A.D., *The Mystery of the Aleph*, Washington Square Press, Published by Pocket Books, New York, London, Toronto, Sydney, 2001.

Aczel, A.D., *Entanglement*, A Plume Book, Penguin Group, New York, 2003.

Adair, R.K., *The Great Design – Particles, Fields and Creation*, Oxford University Press, New York, Oxford, 1987.

Agassi, J., *Faraday as a Natural Philosopher*, The University of Chicago Press, Chicago, ILL, 1971.

Aiton, E.J., *The Vortex Theory of Planetary Motions*, American Elsevier, Inc., New York, 1972.

Aiton, E.J., *Leibniz - A Biography*, Adam Hilger Ltd, Bristol and Boston, 1985.

Akimov, A.E. and Shipov, G.I., "Torsion Fields and Their Experimental Manifestations," *Proc. of the Int'l Conference on New Ideas in Natural Sciences*, St. Petersburg, Russia, June 1996.

Akimov, A.E. and Tarasenko, V.Y., "Models of Polarized States of the Physical Vacuum and Torsion Fields," *Fizika*, No. 3, March 1992.

Albert, D.Z., "Bohm's Alternative to Quantum Mechanics," *Scientific American*, May 1994.

Albert, D.Z., *Quantum Mechanics and Experience*, Harvard University Press, Cambridge, Massachusetts, London, England, 1992,

Alexandersson, O., *Living Water - Victor Schauberger and the Secrets of Natural Energy*, Gateway Books, Bath, UK, 1996.

Alfven, H., *Worlds - Antiworlds: Antimatter in Cosmology*, W.H. Freeman and Company, San Francisco, 1966.

Allen, H.S., "The Case for a Ring Electron, *Proc. Phys. Soc. London*, vol. 31, pp 49-68 (1919).

Allen, R.E., *Greek Philosophy: Thales to Aristotle*, The Free Press, New York, 1966.

Allexander, A., *Infinitesimal*, Scientific American, Farrar, Straus and Giroux, New York, 2014.

Andrade, E.N. da C., *Rutherford and the Nature of the Atom*, Doubleday & Company, Inc., New York, 1964.

Andrulis, E.D., "Theory of the Origin, Evolution, and Nature of Life," (2012), *Life* **2012**, *2*, 1-30.

Arp, H., *Seeing Red - Redshifts, Cosmology and Academic Science*, Apeiron, Montreal, Canada, 1998.

Ash, D. and Hewitt, P., *The Vortex - Key to Future Science*, Gateway Books, Bath, England, 1991.

B

Babbitt, E.D., *The Principles of Light and Color*, Babbitt & Co., Kessinger Legacy Reprints, 1878.

Baggott, J., *The Meaning of Quantum Theory*, Oxford University Press, Oxford, UK, 1992.

Baker, J., *50 Physics Ideas You Really Need to Know,* Quercus, London, 2007.

Barrow, J.D. and Silk, J., *The Left Hand of Creation*, Oxford University Press, New York, 1983.

Barrow, J.D., *The Constants of Nature – The Numbers that Encode the Deepest Secrets of the Universe,* Vintage Books A Division of Random House, Inc., New York, 2002.

Barrow, J.D., *The Infinite Book – A Short Guide to the Boundless, Timeless and Endless*, Pantheon Books, New York, 2005.

Barrow, J.D., *New Theories of Everything – The Quest for Ultimate Explanation*, Oxford University Press Inc., New York, 2007.

Bartusiak, M., "Loops of Space," *Discover*, April 1993.

Bartusiak, M., "Gravity Wave Sky," *Discover*, July 1993.

Bastrukov, S.I. at al, "Spiral Magneto-Electron Waves in Interstellar Gas," Journal of Experimental and Theoretical Physics, Volume 93, pp 671-676, October 2001.

Beckmann, P., *A History of PI*, Dorset Press, New York, 1989.

Beiser, G., *The Story of Gravity - An Historical Approach to the Study of the Force That Holds the Universe Together*, E.P. Dutton & Co., Inc., New York, 1968.

Bekenstein, J.D., "Information in the Holographic Universe," *Scientific American*, pp. 59-65, August 2003.

Bentov, I., *Stalking the Wild Pendulum – On the Mechanics of Consciousness*, Density Book, Rochester, New York, 1977.

Bergman, D.L. and Wesley, J.P., "Spinning Charge Ring Model of Electron Yielding Anomalous Magnetic Moment," *Galilean Electrodynamics*, Vol. 1, No. 5, Sept./Oct. 1990.

Berke, J.P., Author, Editor, *Nanotubes and Nanowires (Selected Topics in Electronics and Systems*, World Scientific Publishing Company, Singapore, 2007.

Bernauer, J.C. and Paul, R., "The Proton Radius Problem," *Scientific American*, February 2014.

Biedermannn, H., *Dictionary of Symbolism –Cultural Icons and the Meaning behind Them*, A Meridian Book, New York, 1994.

Bhadkamkar, A. and Fox, H., "Electron Charge Cluster Sparking in Aqueous Solutions," *Journal of New Energy*, Vol. 1, No. 4, 1996.

Blackwood, O.H., et al, *An Outline of Atomic Physics*, John Wiley & Sons, Inc., New York, 1955.

Bloyd, J.G., *Broken Arrow of Time - Rethinking the Revolution in Modern Physics*, Writers Club Press, San Jose, CA, 2001.

Born, M., *Atomic Physics*, Dover Publications, Inc., Mineola, NY, 1989.

Boscovich, R.J., *A Theory of Natural Philosophy*, The M.I.T. Press, Cambridge, MA, 1966.

Boslough, J., *Stephen Hawking's Universe - An Introduction to the Most Remarkable Scientist of Our Time*, Quill/William Morrow, New York, 1985.

Boslough, J., *Masters of Time - Cosmology at the End of Innocence*, Addison-

Wesley Publishing Company, Reading, MA, 1992.

Bostick, W., "Mass, Charge, and Current: The Essence of Morphology," *Physics Essays*, Vol. 4, No. 1, pp. 45-59, March 1991.

Bowers, B., *Michael Faraday and Electricity*, Priory Press Ltd., London, 1974.

Brennan, R.P., *Heisenberg Probably Slept Here – The Lives, Times and Ideas of the Great Physicists of the 20th Century*, John Wiley & Sons, Inc., New York, 1997.

Broglie, L., de, *The Revolution in Physics*, The Noonday Press, New York, 1953.

Broglie, L., de, *New Perspectives in Physics*, Basic Books, Inc. Publishers, New York, 1962.

Burger, T.J., *Nature* **271**, 402, 1978.

C

Calladine, C.R. and Drew, H.R., *Understanding DNA – The Molecule and How It Works*, Second Edition, Academic Press, New York, 2002.

Cambier, J-L., at al, "Theoretical Analysis of the Electron Spiral Toroidal Concept," NASA/CR-2000-210654, Dec. 2000.

Capra, F., *The Web of Life*, Anchor Books/Random House, Inc., 1997.

Capra, F., *The Tao of Physics*, Shambhala Publications, Inc., 1999.

Carter, J., *The Other Theory of Physics - A Non-Field Unified Theory of Matter and Motion*, Absolute Motion Press, 2000.

Carrigan, Jr., R.A. and Trower, W.P.. *Particles and Forces at the Heart of the Matter,* W.H. Freeman and Company, New York, 1990.

Carroll, R.L., *The Energy of Physical Creation,* The Carroll Research Institute, P.O. Box 3425, Columbia, S.C., 29230, 1985.

Carter, J., *The Other Theory of Physics – A Non-Field Unified Theory of Matter and Motion,* Absolute Motion Press, Enumclaw, Washington, 2000.

Cartledge, P., *Democritus*, Routledge, New York, 1999.

Caspar, M., *Kepler*, Abelard-Schuman, London and New York, 1992.

Cecil, T.E. and Chern, S., *Tight and Taut Submanifolds*, Cambridge University Press, New York, 1997.

Chalidze, V., *Mass and Electric Charge in the Vortex Theory of Matter*, Universal Publishers, 2001.

Chen, C., at al, "Equilibrium and Stability Properties of Self-Organized Electron Spiral Toroid," Physics of Plasma, Volume 8, Number 10, October 2001.

Chen, Y., Editor, *Nanotubes and Nanosheets: Functionalization and Applications of Boron Nitride and Other Nanomaterials,* CRC Press, London, 2015.

Clark, G., *The Man Who Tapped the Secrets of the Universe*, The University of Science and Philosophy, Swannanoa, Waynesboro, 2000.

Clawson, C.C., *Mathematical Sorcery – Revealing the Secrets of Numbers*, Perseus Books, Cambridge, MA, 2001.

Clawson, C.C., *Mathematical Mysteries – The Beauty and Magic of Numbers*, Perseus Publishing, Cambridge, MA, 1999.

Close, F., *Neutrino,* Oxford University Press, 2010.

Close, F., Marten, M., & Sutton, C., *The Particle Explosion*, Oxford University

Press, New York, 1994.

Coats, C., *Living Energies*, Gateway Books, Bath, UK, 1996.

CODATA Recommended Values, *The NIST Reference on Constants, Units and Uncertainties*, 2011.

Consa, O., "Helical Model of the Electron," The General Science Journal, June 2014.

Cook, N., *The Hunt for Zero Point - Inside the Classified World of Antigravity Technology*, Broadway Books, New York, 2001.

Cook, T.A., *The Curves of Life*, Dover Publications, Inc., New York, 1979.

Collins, H. and Pinch T., *The Golem – What Everyone Should Know about Science,* Cambridge University Press, 1993.

Compton, A., "The Size and Shape of the Electron," *Phys. Rev. Second Series*, vol. 14, no.3, pp 247-259, (1919).

Coxeter, H.S.M., *Introduction to Geometry*, John Wiley & Sons, Inc., New York, 1961.

Coxeter, H.S.M., *The Beauty of Geometry*, Dover Publications, Inc., New York, 1968.

Crandall, B.C., *Nanotechnology – Molecular Speculations on Global Abundance,* The MIT Press, Cambridge, Massachusetts, London, England, 1996.

Crew, H., *The Wave Theory of Light - Memoirs by Huygens, Young and Fresnel*, American Book Company, New York, 1900.

Cushing, J.T., *Philosophical Concepts in Physics – The History Relations between Philosophy and Scientific Theories,* Cambridge University Press, 1998.

D

Dalton, J., et al, *Foundations of the Atomic Theory: Comprising Papers and Extracts*, Alembic Club, Edinburgh, UK, 1968.

Davies, P.C.W. and Brown, J., *Superstrings - A Theory of Everything?*, Cambridge University Press, Cambridge, UK, 1988.

Davies, P. and Gribbin, J., *The Matter Myth - Dramatic Discoveries That Challenge Our Understanding of Physical Reality*, Touchstone Book/Simon & Schuster, New York, 1992.

Davies, P., *About Time - Einstein's Unfinished Revolution*, Simon & Schuster, New York, 1995.

Day, W., *Bridge from Nowhere - The Photonic Origin of Matter*, Rhombics, Cambridge, MA, 1996.

Day, W., *A New Physics - Foundation for New Directions*, Cambridge, MA, 2000.

Derbyshire, J., *Unkn( )wn Quantity - A Real and Imaginary History of Algebra*, A Plume Book, Published by Penguin Group, New York, 2007.

Di Mario, D., "Electrogravity: A Basic Link Between Electricity and Gravity," *Speculations in Science and Technology*, Vol. 20, No. 4, Dec. 1997.

Dibner, B., *Oersted - And the Discovery of Electromagnetism*, Blaisdell Publishing Company, New York, 1962.

Dijksterhuis, E.J., *Archimedes*, Princeton University Press, Princeton, N.J., 1987.

Dirac, PAM, "Quantized Singularities in the Electromagnetic Fields," Scribd.com. 1931-05-29.

Dixon, R., *Mathographics*, Dover Publications, Inc, New York, 1991.

Dmitriyev, V.P., "Mechanical Analogy for the Wave - Particle: Helix on Vortex Filament," *Apeiron*, Vol. 8, No. 2, April 2001.

Domb, C., *Clerk Maxwell and Modern Science - Six Commemorative Lectures*, The Athlone Press, University of London, UK, 1963.

Drake, S., *Galileo at Work, His Scientific Biography*, The University of Chicago Press, Chicago, 1978.

Dresselhaus, M.S. and Eklund, P.C., *Science of Fullerences and Carbon Nanotubes: Their Properties and Applications*, Academic Press, 1996.

Drew, H.R., "The Electron as a Four-Dimensional Helix of Spin-1/2 Symmetry," *Physics Essays*, Vol. 12, No. 4, 1999.

Driscoll, R.B., *United Theory of Ether, Field and Matter*, Published by Author, Portland, OR, 1964.

Driscoll, R.B., *United Theory of Ether, Field and Matter (supplement)*, Published by Author, Oakland, CA, 1965.

Duncan, J.C., *Astronomy – A Textbook*, Fifth Edition, Harper & Brothers Publishers, New York, 1926.

E

Eckhart, L., *Four-Dimensional Space*, Indiana University Press, Bloomington, 1968.

Edwards, E.B., *Pattern and Design with Dynamic Symmetry*, Dover Publications, Inc., New York, 1932.

Edwards, L., *The Vortex of Life – Nature's Patterns in Space and Time*, Floris Books, Edinburgh, UK, 2006.

Ehrlich, R., *Crazy Ideas in Science - If You Might Even be True,* Princeton University Press, Prinston and Oxford, 2002.

Einstein, A. and Hopf, L., Ann. Phys., 33, 1096 (1910a): Ann. Phys., 33, 1105, 1910b.

Einstein, A., *Out of My Later Years*, A Citadel Press Book-Carol Publishing Group, New York, NY, 1991.

Einstein, A., *Relativity - The Special and the General Theory*, Crown Publishers, Inc., New York, 1961.

Einstein, A., Infeld, L., *The Evolution of Physics - From Early Concepts to Relativity and Quanta*, A Touchstone Book/Simon & Schuster, New York, 1966.

Einstein, A., "Aether and the Theory of Relativity," (Address on May 5, 1920, at the University of Leyden), *Journal of New Energy*, Vol. 7, No. 1, 2003.

Elgin, D., *The Living Universe – Where are We? Who are We? Where are We Going?*, Berrett-Koehler Publishers, Inc., San Francisco, CA, 2009.

Epstein, L.C., *Thinking Physics Is Gedanken Physics*, Insight Press, San Francisco, CA, 1983.

Epstein, L.C., *Relativity Visualized*, Insight Press, San Francisco, CA, 1992.

F

Farndon, J,. *The Great Scientists – From Euclid to Stephen Hawking,* Metro Books, New York, 2007.

Farrington, B., *Greek Science - Its Meaning for Us*, Penguin Books, Baltimore, MD, 1971.

Feber, A., "Supertwistors and Conformal Supersymmetry," *Nuclear Physics B* **132**: 55-64, 1978.

Ferguson, K., *Stephen Hawking - Quest for a Theory of Everything*, Bantam Books, New York, 1992.

Feynman, R.P., *Six Easy Pieces and Six Not-So-Easy Pieces,* Perseus Publishing, Cambridge, Massachusetts, 1995.

Flander, T.V., *Dark Matter, Missing Planets & New Comets – Paradoxes Resolved, Origins Illuminated,* Revised Edition, North Atlantic Books, Berkley, California, 1993.

Flood, R. and Lockwood, M., *The Nature of Time*, Basil Blackwell, Inc., Cambridge, MA, 1990.

Folger, T., "Tangled Up In Strings – Two Books Say That Today's Theoretical Physicists Are Way Off Course," *Discover*, Sept. 2006.

Ford, K.W., *101 Quantum Questions*, Harvard University Press, Cambridge, Massachusetts, 2011.

Fowler, P.W. and Manolopoulos, D.E., *An Atlas of Fullerenes*, Dover Publications, 2007.

Frank, P., *Einstein - His Life and Times*, Da Capo Press, Inc., New York, 1947.

Fraser, et al, *The Search for Infinity*, Facts on File, Inc., New York, 1995.

Freedman, D.H., "The Mysterious Middle of the Milky Way," *Discover*, November 1998.

Freeman, K. and McNamara, G., *In Search of Dark Matter*, Springer Praxis Publishing, Chichester, UK, 2006.

Friedman, N., *Bridging Science and Spirit - Common Elements in David Bohm's Rhysics, The Perennial Philosophy and Seth*, Living Lake Books, St. Louis, MO, 1994.

Fritzsch, H., *Quarks - The Stuff of Matter*, Basic Books, Inc., New York, 1983.

Fritzsch, H., *The Creation of Matter - The Universe From Beginning to End*, Basic Books, Inc., New York, 1984.

Funk & Wagnalls New Encyclopedia, Funk & Wagnalls, Inc., USA, 1966.

G

Gamow, G., *The Great Physicists from Galileo to Einstein,* Dover Publications, Inc., New York, 1988.

Gamow, G., *Thirty Years That Shook Physics - The Story of Quantum Theory*, Dover Publications, Inc., New York, 1985.

Gamow, G., *One, Two, Three ... Infinity - Facts and Speculations of Science*, Dover Publications, Inc., New York, 1988.

Gardner, M., *New Mathematical Diversions from Scientific American*, Simon and Schuster, New York, 1966.

Gardner, M., *Knotted Doughnuts and Other Mathematical Entertainments*, W.H. Freeman and Company, New York, 1986.

Gasperini, M., *The Universe Before the Big Bang,* Springer, Berlin, 2010.

Gauthier, R., "Faster-than-light quantum models of the photon and the electron", in M. S. El-Genk, *(*ed.*)* *"Space Technology and Applications International Forum – STAIF 2007"*, American Institute of Physics 978-0-7354-0386-4/07, p1099-1108, 2007.

Gauthier, R., "Transluminal energy quantum models of the photon and the electron", in R.L. Amoroso, P. Rowlands & L.H. Kauffman (eds.) *The Physics of Reality: Space, Time, Matter, Cosmos, 8th Symposium in Honor of Mathematical Physicist Jean-Pierre Vigier*, Hackensack: World Scientific, 2013.

Gauthier, R., "A transluminal energy quantum model of the cosmic quantum", in R.L. Amoroso, P. Rowlands & L.H. Kauffman (eds.) *The Physics of Reality: Space, Time, Matter, Cosmos, 8th Symposium in Honor of Mathematical Physicist Jean-Pierre Vigier*, Hackensack: World Scientific, 2013.

Gautreau, R. and Savin, W., *Schaum's Outline of Theory and Problems of Modern Physics,* McGeaw-Hill, New York, 1978.

Gazale, M.J., *Gnomon*, Princeton University Press, Princeton, N.J., 1999.

Geerlings, G.K., *Wrought Iron In Architecture – An Illustrated Survey,* Dover Publications, Inc, New York, 1983.

Gell-Mann, M., *Quark and the Jaguar – Adventures in the Simple and the Complex*, A W, H. Freeman/Owl Book, Henry Holt and Company, LLC, New York, 1994.

Gell-Mann, M., *Complexity*, Vol. 1, no 5, John Wiley and Sons, Inc., New York, 1995/96.

Genz, H., *Nothingness - The Science of Empty Space*, Perseus Books, Reading, MA, 1999.

Geymonat, L., *Galileo Galilei: A Biography and Inquiry Into His Philosophy of Science*, McGraw-Hill Book Company, 1965.

Ghosh, A., *Origin of Inertia*, Apeiron, Montreal, Canada, 2000.

Ghyka, M., *The Geometry of Art and Life*, Dover Publications, Inc, New York, 1977.

Gillispie, C.C., *Dictionary of Scientific Biography*, Vol. IV, Charles Scribner's Sons, New York, 1971.

Ginzburg, V.L., *Theoretical Physics and Astrophysics*, Pergamon Press, 1979.

Ginzburg, V.L., *Physics and Astrophysics. A Selection of Key Problems*, Pergamon Press, 1985.

Gleick, J., *Chaos - Making a New Science*, Penguin Books, New York, 1988.

Gleick, J., *Genius - The Life and Science of Richard Feynman*, Pantheon Books, New York, 1992.

Gorini, C.A., *Geometry*, Facts On File, Inc., New York, 2003.

Goswami, A., *The Self-Aware Universe - How Consciousness Creates the Material World*, Penquin Putnam Inc., 1993.

Graver, J.E., "The Structure of Fullerene Signatures," *DIMACS Series in Discrete Mathematics and Theoretical Computer Science*, American Mathematical Society, 2005.

Gray, A., *Lord Kelvin - An Account of His Scientific Life and Work*, E.P. Dutton & Co., 1908.

Gray, A., *Modern Differential Geometry of Curves and Surfaces*, CRC Press, Boca Raton, 1993.

Greene, B., *The Elegant Universe - Superstrings, Hidden Dimensions, and the Quest for the Ultimate Theory*, W.W. Norton & Company, New York, 1999.

Greene, B., *The Fabric of the Cosmos*, Alfred A. Knopf, New York, 2004.

Gribbin, J., *Q Is For Quantum, An Encyclopedia of Particle Physics*, A Touchstone Book/ Simon & Schuster, New York, 2000.

Gribbin, J., *Schrödinger's Kittens and the Search for Reality*, Little, Brown & Company (Canada) Limited, 1995.

Guillen, M., *Five Equations that Changed the World*, Hyperion, New York, 1995.

H

Haisch, B., *The Purpose-Guided Universe – Believing in Einstein, Darwin, and God*, The Career Press, Inc., Franklin Lakakes, NJ, 2010.

Haisch, B., *The God Theory - Universes, Zero-Point Fields, and What's Behind It All*, Weser Books, San Francisco, CA, 2006..

Hall, A.R., *Isaac Newton - Adventurer in Thought*, Blackwell Publishers, Oxford, UK, 1994.

Hambidge, J., *Practical Applications of Dynamic Symmetry*, The Devin-Adair Company, New York, 1967.

Hargittai, I., Pickover, C.A., Editors, *Spiral Symmetry*, World Scientific Publishing Co., Singapore, 1992.

Harrington, P.S, *Star Watch*, John, Wiley & Sons, Inc., Hoboken New Jersey, 2003.

Harris, P.J.F., *Carbon Nanotube Science: Synthesis, Properties and Applications*, Cambridge University Press, Cambridge, UK, 2011.

Harrison, L.P., *Meteorology*, National Aeronautics Council, Inc., New York, 1942.

Hatch, E., *Modern Physics from a Classical Scale Perspective- Part 1 Concept Confirmed*, RWWAA Publication, Auburn, California, 2004.

Hawking, S., *Black Holes and Baby Universes and Other Essays*, Bantam Books, New York, 1993.

Hawking, S., *A Brief History of Time - From the Big Bang to Black Holes*, Bantam Books, New York, 1990.

Hawking, S., *On the Shoulders of Giants*, Running Press, Philadelphia, London, 2004.

Hawking, S. and Penrose, R., *The Nature of Space and Time*, Princeton University Press, Princeton, NJ, 2000.

Heath, J.L., *The Works of Archimedes*, Dover Publications, Inc., New York, 1953.

Heilbron, J.L., *The Dilemmas of an Upright Man - Max Planck as Spokesman for German Science*, University of California Press, Berkeley, 1986.

Heisenberg, E., *Inner Exile - Recollections of a Life with Werner Heisenberg*, Birkhauser Boston, MA, 1980.

Helmholtz, H., *On the Sensation of Tone – As a Physiological Basis for the Theory of Music*, Dover Publications, New York, 1954.

Henderson, L.D., *The Fourth Dimension and Non-Euclidean Geometry in Modern Art*, Princeton, 1983.

Herzberg, G., *Atomic Spectra and Atomic Structure*, Dover Publications, Inc., New York, 1944.

Hey, N., *Solar System*, Weidenfeld & Nicolson, The Orion Publishing Group, Wellington House, London, UK.

Hey, T. and Walters, P., *The New Quantum Universe*, Cambridge University Press, UK, 2003.

Hippel von, F., *Citizen Scientist*, A Touchstone Book/Simon & Schuster, New York, 1991.

Hoffmann, B., *Albert Einstein - Creator and Rebel*, New American Library, New York, 1972.

Hoffman, R.N., "Controlling Hurricanes – Can Hurricanes and Other Severe Tropical Storms be Moderated or Deflected?" *Scientific American*, Oct. 2004.

Hooft, G., *In Search of the Ultimate Building Blocks*, Cambridge University Press, UK, 1997.

Horgan, J., "Gravity Quantized? - A Radical Theory of Gravity Weaves Space From Tiny Loops," *Scientific American*, September 1992.

Hotson, D.L., "Dirac's Equation and the Sea of Negative Energy, Part 1," *Infinite Energy*, Vol. 8, Issue 43, 2002.

Hotson, D.L., "Dirac's Equation and the Sea of Negative Energy, Part 2," *Infinite Energy*, Vol. 8, Issue 44, 2002.

Hurley, W.M., *Prehistoric Cordage – Aldine Manuals on Archeology 3*, Taraxacum, Washington, 1979.

I

Icke, V., "From Expansion to Intelligence in the Universe," *Speculations in Science and Technology*, Vol. 14, No. 4, 1991.

Ipsen, D.C., *Archimedes: Greatest Scientist of the Ancient World*, Enslow Publishers, Inc., Hillside, N.J., 1988.

J

Jammer, M., *Concepts of Mass in Classical and Modern Physics*, Dover Publications, Inc., Mineola, New York, 1997.

Jammer, M., *Concepts of Force*, Dover Publications, Inc., Mineola, New York, 1999.

Jean, Sir, J, *Science & Music,* Dover Publications, Inc., New York, 1968

Johnson, G., "The Inelegant Universe – Two New Books Argue That It Is Time For String Theory To Give Way," *Scientific American*, September, 2006.

Jefimenko, O.D., *Gravitation and Cogravitation*, Electret Scientific Company, Star City, West Virginia, 2006.

Jones, B.Z., *The Golden Age of Science,* Simon and Schuster, New York, 1966.

Jonsson, I., *Emanuel Swedenborg*, Twayne Publishers Inc., New York, 1971.

K

Kafatos, M. and Nadeau, R., *The Conscious Universe - Part and Whole in*

Modern Physical Theory, Springer-Verlag New York, Inc., New York, 1990.

Kaku, M., *Physics of the Impossible,* Anchor Book, A Division of Random House, Inc., 2008.

Kaku, M., *Visions – How Science Will Revolutionize the 21st Century,* Anchor Books, Doubleday, New York, London, 1997.

Kaku, M., *Beyond Einstein - The Cosmic Quest for Theory of the Universe*, Anchor Books/Doubleday, New York, 1995.

Kaku, M., *Hyperspace*, Anchor Books/Doubleday, New York, 1994.

Kaku, M., *Introduction to Superstrings*, Springer-Verlag, New York, 1988.

Kanarev, F.M., "Model of the Electron," APERON, Vol. 7. Nr. 3-4, July-October, 2000.

Kanarev, F.M., *The Foundation of Physchemistry of Microworld*, Kuban State Agrarian University (KSAU), Krasnodar, Russia, 2002.

Kane, G., *The Particle Garden - Our Universe as Understood by Particle Physicists*, Addison-Wesley Publishing Company, Reading, MA, 1995.

Kanigel, R., *The Man Who Knew Infinity - A life of the Genius Ramanujan,* Washington Square Press, Published by Pocket Books, New York, London, 1991.

Kaplan, R., *The Nothing That Is - A Natural History of Zero*, Oxford University Press, Oxford, New York, 1999.

Kaplan, R. and Kaplan, E., *The Art of The Infinite*, Oxford University Press, Oxford, New York, 2003.

Kaufmann, W.J., *Black Holes and Warped Spacetime*, W.H. Freeman and Company, San Francisco, CA.

Kimura, Y.G., *The Book of Balance*, (Translation), The University of Science and Philosophy, Contact Printing, North Vancouver, B.C., Canada, 2002.

Kimura, Y.G., "The Transcendent Unity of Science and Spirituality," *VIA – Vision in Action*, Vol. 2, No. 1 & 2, 2004.

King, M.B., "Vortex Filaments, Torsional Fields and the Zero-Point Energy," *Journal of New Energy*, Vol. 3, No. 2/3, 1998.

King, M.B., "Dual Vortex Forms: The Key to a Large Zero-Point Energy Coherence," *Journal of New Energy*, Vol. 5, No. 2, 2000.

King, M.B., *Quest for Zero Point Energy*, Adventures Unlimited Press, Kempton, IL, 2001.

King, M.B., *Tapping the Zero-Point Energy,* Paraclete Publishing, Provo, Utah, 1989.

Knight, D.C., *The Science Book of Meteorology*, Franklin Watts, Inc., New York, 1964.

Krauss, L.M., *Quintessence - The Mystery of Missing Mass in the Universe*, Basic Books, New York, NY, 2000.

Kumar, S., "A Spiral Structure for Elementary Particles," Int. J. Res. Vol. 1, Issue 6, July 2014.

Kumar, S., "Journey of the Universe from Birth to Rebirth with Insight into Unified Interaction of Elementary Particles with Spiral Structure," Int. J. Res. Vol. 1, Issue 9, October 2014.

Kumar, S., "Quantum Spiral Theory," Int. J. Res. Vol. 2, Issue 1, January 2015.

Kumar, S., "Spiral Hashed Information Vessel," International Journal of

Scientific & Engineering Research, Vol. 4, Issue 6, April 2015.

Kumar, S., "Spiral Structure of Elementary Particles Analogous to Sea Shells: A Mathematical Description," International Journal of Current Research, Vol. 7, Issue 02, Feb. 2015, p. 12814.

Kumar, S., "Mass-Energy Equivalence in Spiral Structure for Elementary Particles and Balance of Potentials," International Journal of Scientific & Engineering Research. Vol. 6, Issue 6, July 2015.

L

Lamb, G.L., "Solutions and the Motion of Helical Curves," *Physical Review Letters*, Vol. 37, No. 5, August 1976.

Lakhtakia, A. and Weiglhofer, W.S., "Time-Dependent Beltrami Fields in Free Space: Dyadic Green Functions and Radiation Potentials," *Physical Review E*, Vol. 49, Number 6, June 1994.

Lakhtakia, A. and Weiglhofer, W.S., "Covariances and Invariances of the Beltrami-Maxwell Postulates," *IEE Proc. - Sci. Meas. Technol.*, Vol. 142, No. 3, May 1995.

Lang, T.G., "Proposed Unified Field Theory – Part II: Protons, Neutrons and Fields," *Galilean Electrodynamics*, Vol. 12, No. 6, Nov./Dec. 2001.

Larsen, R., et al, *Emanuel Swedenborg - A Continuing Vision*, Swedenborg Foundation, Inc., New York, 1988.

Laugwitz, D., *Differential and Riemannian Geometry*, Academic Press, New York, 1965.

Lauwerier, H., *Fractals - Endlessly Repeated Geometrical Figures*, Princeton University Press, Princeton, New Jersey, 1991.

Lederman, L.M. and Teresi, D., *The God Particle*, Bantam Doubleday Dell Publishing Group, Inc., New York, 1993.

Lederman, L. and Hill, C.T., *Symmetry and the Beautiful Universe*, Prometheus Books, New York, 2004.

Lederman, L.M. and Hill, C.T., *Quantum Physics for Poets*, Prometheus Books, New York, 2013.

Lederman, L.M. and Hill, C.T., *Beyond the God Particle*, Prometheus Books, New York, 2011.

Lerner, E.J., *The Big Bang Never Happened*, Vintage Books, Random House, Inc., New York, 1992.

Lewis, H., *Geometry – A Contemporary Course,* Third Edition, McCormick-Mathers Publishing Company, Cincinnati, Ohio, 1973.

Lewis, J.R., *Scientology*, Cary, NC, Oxford University Press, 2009.

Lindgren, C.E., *Four-Dimensional Descriptive Geometry*, McGraw-Hill Book Company, New York, 1968.

Lindley, D., *The End of Physics - The Myth of a United Theory*, HarperCollins Publishers, Inc., 1993.

Lipschultz, M.M, *Differential Geometry,* Schaum's Outline Series, McGraw-Hill, New York, 1969.

Livio, M., *The Equation That Couldn't Be Solved – How Mathematical Genius Discovered the Language of Symmetry*, Simon and Schuster, New York, 2005.

Livio, M., *The Golden Ratio*, Broadway Books, New York, 2002.

Lockwood, E.H., *A Book of Curves*, Cambridge University Press, New York, 1961.

Lomberg, J., *Unified Force Theory, Dark Matter and Consciousness,* The Aenor Trust, PO Box 4706, Salem, Oregon, 2004.

Lorentz, H.A., *Problems of Modern Physics - A Course of Lectures Delivered in the CA Institute of Technology*, Ginn and Company, Boston, 1927.

Lucas, C.W., "A Classical Electromagnetic Theory of Elementary Particles," *Journal of New Energy*, Vol. 6, No. 4, 2002.

Lucas, C.W. "A Classical Electromagnetic Theory of Elementary Particles Part 2, Interwining Charge-Fibers," The Journal of Common Sense Science, Foundation of Science, May 2005, Vol. 8 No. 2.

Ludwig, C., *Michael Faraday - Father of Electronics*, Herald Press, Scottdale, PA, 1978.

Lugt, H.J., *Vortex Flow in Nature and Technology*, Krieger Publishing Company, Malabar, Florida, 1995.

Lykken, J. and Spiropulu, M., "Supersymmetry and the Crisis in Physics," *Scientific American*, May 2014.

M

Maldacena, J., "The Illusion of Gravity," *Scientific American*, pp. 57-63, November 2005.

Magueijo, J., *Faster Than the Speed of Light - The Story of A Scientific Speculation,* Penguin Books, New York, 2003.

Manning, J., *The Coming Energy Revolution - The Search for Free Energy*, Avery Publishing Group, Garden City Park, New York, 1996.

Manning, H.P., *The Fourth Dimension Simply Explained*, Dover Publications, Inc., New York, 1960.

Maor, E., *e: The Story of a Number*, Princeton University Press, Princeton, NJ, 1994.

Maor, E., *Trigonometric Delights*, Princeton University Press, Princeton, NJ, 1998.

Marsden, J.E. and McCracken, M., *The Hopf Bifurcation and Its Applications*, Springer-Verlag, New York, Berlin, 1976.

Mazur, B., *Imagining Numbers (particularly the square root of minus fifteen)*, Picador, New York, 2003.

McCrea, W.H., "Arthur Stanley Eddington," *Scientific American*, June 1991.

McCutcheon, M., *The Final Theory – Rethinking Our Scientific Legacy*, Universal Publishers, Boca Raton, Florida, 2004.

McLeish, J., *The Story of Numbers*, Fawcett Columbine, New York, 1991.

Meacher, M., *Destination of the Species – The Riddle of Human Existence*, Books, Winchester, UK, Wasington USA, 2009.

Melker, A.A. and Krupina, M.A., "Designing Muni-Fullerences and Their Relatives on Graph Basis," *Materials Physics and Mechanics 20*, 18-24 2014.

Messent, J., *Embroidery & Architecture,* B.T Batsford Ltd., London, 1985.

Millar, D., et al, The Cambridge Dictionary of Scientists, Cambridge University Press, 1996.

Miller, A.I., *137 – Jung, Pauli and the Pursuit of the Scientific Obsession*, W.W

Norton & Company, Inc., New York, 2009.

Miller, A.I., *Albert Einstein's Special Theory of Relativity,* Springer-Verlag, New York, Berlin, 1997.

Milton, R., *Alternative Science - Challenging the Myths of the Scientific Establishment*, Park Street Press, Rochester, Vermont, 1996.

Mitchell, W.C., *Bye Bye Bing Bang – Hello Reality,* Cosmic Sense Books, Carson City, Nevada, 2002.

Mitsopoulos, T.D., "Similarity Between Elementary Particles and Electric Circuits," *Galilean Electrodynamics*, Vol. 12, No. 6, Nov./Dec. 2001.

Moore, W., *Schrodinger - Life and Thought*, Cambridge University Press, UK, 1992.

Mortimer, S., *Techniques of Spiral Work – A Practical Guide to the Craft of Making Twists by Hand,* Linden Publishing, Fresno, California, 1995.

Moyer, M., "Is the Space Digital," *Scientific American*, February 2012.

Mugnai, D., et al, "Observation of Superluminal Behaviors in Wave Propagation," *Physical Review Letters*, Vol. 84, Number 21, May 2000.

Murchie, G., *The Seven Mysteries of Life - An Exploration in Science and Philosophy*, Houghton Mifflin Company, Boston, 1978.

N

Nahin, P.J., *An Imaginary Tale – The Story of* $\sqrt{-1}$, Princeton University Press, Princeton, New Jersey, 1998.

Nakahara, M., *Geometry, Topology and Physics,* Second Edition, Taylor & Francis, Taylor & Francis Group, New York, London, 2003.

Nash, C. and Sen, S., *Topology and Geometry for Physicists*, IBI Global, London, UK, 1983.

Nernst, W., Verh. Dtsch. Phys. Ges., 18, 83, 1916.

Newton, I., *The Principia*, Prometheus Books, Amherst, New York, 1995.

Nierengarten, J-F., Editor, *Fullerenes and Other Carbon-Rich Nanostructures (Structure and Bonding)*, Springer, 2014.

Niven, W.D., *The Scientific Papers of James Clerk Maxwell*, Dover Publications, Inc., New York, 1890.

Novikov, I.D., *The River of Time*, Cambridge University Press, Cambridge, UK, 1998.

O

Okun, L.B., "The Concept of Mass," *Physics Today*, Vol. 42, June 1989.

Oliwensrein, L., "Bent out of Shape," *Discover*, July 1993.

Oros di Bartini, R., "Relations Between Physical Constants," *Progress in Physics*, v. 3, pp. 34-40, October 2005.

Oros di Bartini, R., "Some Relations Between Physical Constants," *Doklady* Acad. Nauk USSR, v. 163, No. 4, pp. 861-864, 1965.

Oschman, J.L. and Schman N.H., "Vortical Structure of Light and Space: Biological Implications," *J Vortex Sci Technol.* 2:1, 2015.

Oschman, J.L. and Schman N.H., "The Heart as a Bi-Directional Scalar Field Antena," *J Vortex Sci Technol.* 2:2, 2015.

Oschman, J.L., *Energy Medicine – The Scientific Basis,* Second Edition,

Elsevier, New York, 2016.

P

Pagels, H.R., *The Cosmic Code – Quantum Physics as the Language of Nature*, Bantam Books, New York, 1982.

Panek, R., *The 4% Universe – Dark Matter, Dark Energy, and the Race to Discover the Rest of Reality*, Houghton Mifflin Harcourt, Boston, New York, 2011.

Pappas, T., *The Joy of Mathematics – Discovering Mathematics All Around You*, Wide World Publishing, Tetra, 1989.

Parry, A., *The Russian Scientist*, The Macmillan Company, New York, 1973.

Parson, A.L., "A Magneton Theory of the Structure of the Atom," Smithsonian Miscellaneous Collections, Vol. 65, No. 11, Publication 2371, Nov. 29, 1915.

Pauli, W., *Theory of Relativity*, Pergamon Press, London, UK, 1958.

Peat, F.D., *Superstrings and the Search for The Theory of Everything*, Contemporary Books, Lincolnwood (Chicago), ILL,1988.

Pedoe, D., *Geometry - A Comprehensive Course*, Dover Publications, Inc., New York, 1988.

Peebles, P.J.E., *Principles of Physical Cosmology*, Princeton University Press, Princeton, New Jersey, 1993.

Penrose, R., "Twistor Quantization and Curved Space-time." *International Journal of Theoretical Physics* (Springer Netherlands), **1**: 61-99, 1968.

Penrose, R., *The Road to Reality - A Complete Guide to the Laws of the Universe*, Alfred A. Knopf, New York, 2005.

Penrose, R., *Shadows of the Mind – A Search for the Missing Science of Consciousness,* Oxford University Press, 1994.

Peratt, A.L., "Birkeland and the Electromagnetic Cosmology," *Sky & Telescope*, May 1985.

Physical Review D: Particles and Fields, Vol. 54, The American Physical Society, 1996.

Pierson, H.O., *Handbook of Carbon, Graphite, Diamond and Fullerenes: Properties and Applications (Material Science and Process Technology)*, Noyes Publications, 1994.

Pickover, C.A., *Mathematics and Beauty II; Spirals and "Strange" Spirals in Civilization, Nature, Science, and Art*, IBM Thomas J. Watson Research Center, Yorktown Heights, NY, 1987.

Polyakov, A., "Gauge Fields and Strings," Harwood Academic Publishers 1987, *Nucl. Phys.* **B396**, 367, 1993.

Ponomarev, C.D. and Andreeva, L.E., *The Calculation of Elastic Elements of Machines and Sensors,* Machinostroenie, Moscow, 1980.

Porter, R., *The Biographical Dictionary of Scientists*, Oxford University Press, New York, 1994.

Posamentier, A.S. and Lehmann, I., *A Biography of the World's Most Mysterious Number*, Prometheus Books, Amherst, New York, 2004.

Potemra, T. A., "Hannes Alfven, Father of Space Plasma Physics," Geomagnetism and Aeronomy with Special Historical Case Studies, IAGA Newsletters 29/1997, Published by IAGA, Germany, p.101, 1997.

Price, W.C., et al, *Wave Mechanics; The First Fifty Years - A Tribute to Professor Louis De Broglie*, John Wiley & Sons, New York-Toronto, 1973.

Price, H., *Time's Arrow and Archimedes' Point*, Oxford University Press, New York, 1996.

Prikhodko, I.P., etal, "Microscale Glass-Blown Three-Dimensional Spherical Shell Resonators," *IEEE Journal of Microelectromechanical Systems*, 2011.

Purce, J., *The Mystical Spiral- Journey of the Soul,* Thames and Hudson, 1974.

Purdy, S., and Sandak, C.R., *Ancient Greece*, Franklin Watts, New York, 1982.

Puthoff, H.E., et al, "Engineering the Zero-Point Field and Polarizable Vacuum for Interstellar Flight," *Journal of New Energy*, Vol. 6, No. 1, 2001.

R

Randless, J. *Breaking the Time Barrier*, Paraview Pocket Books, New York, London, 2005,

Reed, D., "Excitation and Extraction of Vacuum Energy Via EM-Torsional Field Coupling - Theoretical Model," *Journal of New Energy*, Vol. 3, No. 2/3, 1998.

Reed, D., "A New Paradigm for Time – Evidence From Empirical and Esoteric Sources," *Journal of New Energy*, Vol. 6, No. 2, 2001.

Resnick, R., *Introduction to Special Relativity,* Jon Wiley & Sons, Inc., New York, London, 1968.

Ridley, B.K., *Time, Space and Things*, Cambridge University Press, Cambridge, UK, 1994.

Riordan, M., *The Hunting of the Quark - A True Story of Modern Physics*, Simon and Schuster/Touchstone, New York, 1987.

Riordan, M. and Schramm, D.N., *The Shadows of Creation - Dark Matter and the Structure of the Universe*, W.H. Freeman and Company, New York, 1991.

Rucker, R., *The Fourth Dimension*, Houghton Mifflin Company, Boston, 1984.

Russell, P., *The White Hole in Time*, Harper San Francisco, 1992.

Russell, W. *The Universal One*, University of Science and Philosophy, Swannanoa, Waynesboro, Virginia, 1974.

Russell, W., *A New Concept of the Universe*, The University of Science and Philosophy, Swannanoa, Waynesboro, VA, 1989.

Russell, W., *The Secret of Light*, The University of Science and Philosophy, Swannanoa, Waynesboro, VA, 1994.

Ryu, C., *The Grand Unified Theory – A Scientific Theory of Everything*, PublishAmerica, Baltimore, 2004.

S

Saito, R., Author, Editor, *Physical Properties of Carbon Nanotubes*, Imperial College Press, London, 1998.

Salem, K.G., *The New Gravity - A New Force - A New Mass - A New Acceleration - Unifying Gravity with Light*, Salem Books, Johnstown, PA, 1994.

Sagan, C., *Cosmos,* Ballantine Books, New York, 1980.

Sanders, P.A. Jr., *Scientific Vortex Information,* Free Soul Publishing, Sedona, AZ, 1992.

Sano, C., "Twisting & Untwisting of Spirals of Aether and Fractal Vortices Connecting Dynamic Aethers," *Journal of New Energy,* Vol. 6, No. 2, 2001.

Sarg, S., "A Physical Model of the Electron According to the Basic Structure of Matter Hypothesis," *Physics Essays,* Vol. 16, No.2, 180-195, 2003.

Sarg, S, "Basic Structure of Matter – Supergravitation Unified Theory Based on an Alternative Concept of the Physical Vacuum," Proceedings of the 17th Annual Conference of the NPA at Long Beach, CA, Vol. 7, pp. 479-484, 23-26 June, 2010.

Savov, E., *Theory of Interaction - The Simplest Explanation of Everything,* Geones Books, Sofia, Bulgaria, 2002.

Schneider, M.S., *A Beginner's Guide to Constructing the Universe,* HarperPerennial, New York, 1995.

Schweighauser, C.A., *Astronomy from A to Z – A Dictionary of Celestial Objects and Ideas,* Sangamon State University, Springfield, Illinois, 1991.

Schwenk, T., *Sensitive Chaos – The Creation of Flowing Forms in Water and Air,* Rudolf Steiner Press, Hillside House, East Sussex, 2008.

Schwerdtfeger, H., *Geometry of Complex Numbers – Circle Geometry, Moebius Transformation, Non-Euclidean Geometry,* Dover Publication, Inc., New York, 1979.

Scientific American, *The Enigma of Weather,* Scientific American, New York, NY, Oct. 2004.

Segal, V.M., "Materials Processing by Simple Shear," *Mat. Sci & Eng.,* vol. 197, 157-164, 1995.

Segal, V.M., "Equal Channel Angular Extrusion: From Macro Mechanics to Structure Formation," *Mat. Sci & Eng.,* vol. 271, 322-333, 1999.

Segal, V.M., "Severe Plastic Deformation: Simple Shear versus Pure Shear," vol. 338, pp. 331-344, 2002.

Seggern, D.H. von, *CRC Handbook of Mathematical Curves and Surfaces,* CRC Press, Boca Raton, Florida, 1990.

Seggern, D.H. von, *CRC Standard Curves and Surfaces,* CRC Press, Boca Raton, Florida, 1993.

Segre, E., *Nuclei and Particles - An Introduction to Nuclear and Subnuclear Physics,* W.A. Benjamin, Inc., New York, 1965.

Seife, C., *Zero - The Biography of a Dangerous Idea,* Viking Penquin, New York, 2000.

Semat, H., *Introduction to Atomic and Nuclear Physics,* Rinehart & Company, Inc., New York, 1958.

Series, G.W., *Advances - The Spectrum of Atomic Hydrogen,* World Scientific, New Jersey, 1988.

Serway, R.A., *Physics for Scientist & Engineers with Modern Physics,* 3rd Edition, Saunders Golden Sunburst Series, Saunders College Publishing, Philadelphia, PA, 1990.

Seward, C., "Ball Lightning Events Explained as Self-Stable Spinning High-Density Plasma Toroids or Atmospheric Spheromaks," IEEE *Access*

Practical Innovations, Volume 2, 2014, 153-59.

Sharlin, H.I., *Lord Kelvin - The Dynamic Victorian*, The Pennsylvania State University Press, PA, 1979.

Sheka, E., *Fullerences: Nanochemistry. Nanomagnetism, Nanomedicine, Nanophotonics,* CRC Press, 2011.

Siegfried, T., *Strange Matters – Undiscovered Ideas at the Frontiers of Space and Time,* The Berkley Publishing Group, A division of Penguin Group, New York, 2004.

Siegfried, T., *The Bit and the Pendulum*, John Wiley & Sons, Inc., New York, 2000.

Simhony, M., *Invitation to the Natural Physics of Matter, Space, Radiation*, World Scientific, New Jersey, London, 1994.

Smolin, L., *The Trouble With Physics: The Rise of String Theory, The Fall of a Science, and What Comes Next,* Houghton Mifflin, 2006.

Sprott, J.C., *Strange Attractions - Creating Patterns in Chaos*, M&T Books, New York, 1993.

Sproull, R.L., *Modern Physics – A Textbook for Engineers,* John Wiley & Sons, New York, 1956.

Stenger, V.J., *God and the Atom – From Democritus to the Higgs Boson: The Story of a Triumphant Idea*, Prometheus Books, New York, 2013.

Sternberg, S., *Curvature in Mathematics and Physics*, Dover Publications, Inc., New York, 2012.

Sternglass, E.J., *Before the Big Bang - The Origins of the Universe*, Four Walls Eight Windows, New York, NY, 1997.

Strogatz, S., *Sync - The Emerging Science of Spontaneous Order*, Hyperion, New York, 2003.

Sunden, O., "Time-Space-Oscillation: Hidden Mechanism Behind Physics," *Galilean Electrodynamics*, Vol. 12, Special Issue 2, Fall 2001.

Swedenborg, E., *The Principia*, Swedenborg Society, London, 1912.

Synge, J.L., "The Electrodynamic Double Helix," In *Magic Without Magic: John Archibald Wheeler* - A Collection of Essays in Honor of His Sixtieth Birthday, edited by John R. Klauder, W. H. and Company, San Francisco, 1972.

T

Tanaka, K., Editor, Iijima, S., *Carbon Nanotubes and Graphene, Second Edition,* Nanotube Research Center, Tsukuba, Japan, 2014.

Talbot, M., *The Holographic Universe*, Harper Perennial, 1992.

Tewari, P., *Universal Principles of Spacetime and Matter - A Call for Conceptual Reorientation,* Crest Publishing House, New Deli, 2002.

Tewari, P., "On the Space-Vortex Structure of the Electron," www.tewari.org/ Theory_Papers/Tewari-Final%20Proof.pdf. 2005.

Thomson, J.J., *A Treatise on the Motion of Vortex Rings*, MacMillan and Co., London, 1883.

Thomson, J.J., *Electricity and Matter*, Charles Scribner's Sons, New York, 1904.

Thomson, D.W. and Bourassa, J.D., *Secrets of Aether*, Published by The Aenor Trust, Salem, OR, 2004.

Time-Life Books, *A Soaring Spirit - Time Frame BC 600-400*, The Time Inc.

Book Company, Alexandria, VA, 1987.

Time-Life Books, *Empires Ascendant - Time Frame 600 BC - AD 200*, The Time Inc. Book Company, Alexandria, VA, 1987.

Tricker, R.A., *The Contributions of Faraday and Maxwell to Electrical Science*, Pergamon Press Ltd., London, UK, New York, 1987.

Thorne, K.S., *Black Holes & Time Warps - Einstein's Outrageous Legacy*, W.W. Norton & Company, New York, 1994.

Treasures of Early Irish Art: 1500 B.C. to 1500 A.D., From the Collections of the National Museum of Ireland, Royal Irish Academy, Trinity College, Dublin, 1977.

U

Unger, R.M. and Smolin, L., *The Singular Universe and the Reality of Time*, Cambridge University Press, Cambridge, UK, 2015.

V

Von Stade, S., "Owner". *Flowtoys*. Flowtoys (Toroflux).

Valens, E.G., *The Attractive Universe: Gravity and Shape of Space*, Motion, Magnet, 1969.

Van der Laan, C., "The Vortex Theory of Atoms," Thesis for the Master's Degree in History and Philosophy of Science, Institute for History and Foundation of Science, Utrecht University, Dec. 2012.

Van Eenwik, J.R., *Archetypes & Strange Attractors - The Chaotic World of Symbols*, Inner City Books, Toronto, Canada, 1997.

Van Flandern, T., *Dark Matter, Missing Planets & New Comets - Paradoxes Resolved, Origins Illuminated*, North Atlantic Books, Berkeley, CA, 1993.

Valone, T., "Inside Zero Point Energy," *Journal of New Energy*, Vol. 5, No. 4, 2001.

Van Nostrand's Scientific Encyclopedia, D. Van Nostrand Company, Inc., 1958.

Veltman, M., *Facts and Mysteries in Elementary Particle Physics*, World Scientific, New Jersey, 2003.

Venable, W.M., *The Interpretation of Spectra*, Reinhold Publishing Corporation, New York, 1948.

Volk, G., "Toroids, Vortices, Knots, Topology and Quanta," Proceedings of of the 18th Annual Conference the NPA, 6-9 at the University Maryland College Park, MD, Vol. 8, July 2011.

Vrooman, J.R., *Rene Descartes - A Biography*, G.P. Putnam's Sons, New York, 1970.

W

Wagner, O.E., "Structure in the Vacuum," *Frontier Perspectives*, Vol. 10, No. 2, Fall 2001.

Walker, F.L., "The Fluid Space Vortex: Universal Prime Mover," *Physics Essays*, Vol. 15, No. 2, 2002.

Wallace, D.F., *Everything and More - A Compact History of ∞*, W.W. Norton & Company, New York, London, 2003.

Watson, J.D., *The Double Helix*, W.W. Norton & Company, New York, 1980.

Weber, C.S., "VRML Gallery of Fullerenes," Fullerene Library, JSV1.08, 1999.

Weinberg, S., *Dreams of a Final Theory*, Pantheon Books, New York, 1992.

Weir, S.T., Mitchell, A.C. and Nellis, W.J., "Metallization of Fluid Molecular Hydrogen," *Physics Review Letters* 76, 1860, 1996.

Westfall, R., *The Life of Isaac Newton*, Cambridge University Press, New York, NY, 1994.

Wheeler, J.A., *Geons, Black Holes & Quantum Foam – A Life in Physics,* W.W. Norton & Company, Inc., New York, 1998.

Wheeler, J.A., *Geometrodynamics, Topics of Modern Physics, Vol.1*, Academic Press Inc., New York, NY, 1962.

White, H.E., *Introduction to Atomic Spectra*, McGraw-Hill Book Company, Inc., New York, 1934.

Whitney, S.K., "9 Editor's Essays," *Galilean Electrodynamics*, Vol. 16, Special Issue 3, Winter 2005.

Whitney, S.K., *Algebraic Chemistry – Applications and Origins,* Nova Science Publishers, Inc., New York, 2013.

Wiener, N., *Cybernetics or Control and Communication in the Animal and the Machine*, 2nd edition, The MIT Press and John Wiley & Sons, Inc., New York, 1961.

Wigner, E. and Huntington, H.B., "On the Possibility of a Metallic Modification of Hydrogen, *Journal of Chemical Physics* 3 (12): 764, 1935

Wilczek, F., *Longing for the Harmonies – Themes and Variations from Modern Physics*, W.W. Norton & Company, New York, 1987.

Wilczek, F., *The Lightness of Being – Mass, Ether, and the Unification of Forces*, Basic Books, New York, 2008.

Wilczek, F., *A Beautiful Question – Finding Nature's Deep Design*, Penguin Press, New York, 2015.

Williamson, J.G, and van der Mark, M.B., "Is the electron a Photon with Toroidal Topology?," Annales de la Fondation Louis de Broglie, Volume 22, No. 2, 133, (1997).

Witten, E., "Perturbative Gauge Theory as a String Theory in Twistor Space," (2004) (http://arxiv.org/abs/hep-th/0312171)" *Commun Math. Phys.* 252: 189-258.

Woit, P., *Not Even Wrong: The Failure of String Theory and the Search for Unity in Physical Law*, Basic Books, 2006.

Wolff, M., "Origin of the Mysterious Instantaneous Transmission of Events in Science," *The Cosmic Light*, Vol. 4, No. 2, Spring 2002.

Wolfram, S., *A New Kind of Science*, Wolfram Media, LLC, Champaign, IL, 2002.

Wong, H.-S. P., *Carbon Nanotubes and Graphene Device,* Cambridge University Press, Cambridge, UK, 2011.

Z

Zeman, R.K. et al., *Helical/Spiral CT - A Practical Approach*, McGraw-Hill, Inc., New York, 1995.

Zombeck, M.V., *Handbook of Space Astronomy and Astrophysics*, Cambridge

University Press, Cambridge, UK, 1990.

Zwikker, C., The Advanced Geometry of Plane Curves and Their Applications, Dover Publications, Inc., New York, 1994.

PUBLICATIONS BY V.B. GINZBURG

1. Theory & Application of Magnetoelastic Effect

Ginzburg, V.B., "Magnetoelastic Self-Compensating Device for Measurement of Drilling Load MKN-1," *Machinery and Oil Equipment*, Moscow (USSR), Vol. 6, pp. 23-6, 1967.

Ginzburg, V.B., "The Calculation of Design Parameters of Magnetoelastic Force Transducers," *Machinery and Oil Equipment*, Moscow (USSR), Vol. 10, pp. 26-8, 1967.

Ginzburg, V.B. and Airapetov, V.A., "Magnetoelastic Self-Compensating Pressure Measuring Device," *Machinery and Oil Equipment*, Moscow (USSR), Vol. 2, pp. 33-35, 1968.

Ginzburg, V.B., "Depth Magnetoelastic Transducers," *Machinery and Oil Equipment*, Moscow (USSR), Vol. 8, pp. 24-27, 1968.

Ginzburg, V.B. and Ginzburg, P.B. "Non-Contact Magnetoelastic Torque Transducer," *Machinery and Oil Equipment*, Moscow (USSR), Vol. 4, pp. 25-28, 1969.

Ginzburg, V.B., "Magnetoelastic Pressure Transducer," *Machinery and Oil Equipment*, Moscow (USSR), Vol. 6, pp. 30-33, 1969.

Ginzburg, V.B., *Magnetoelastic Transducers*, Energy, Moscow (USSR), 1970.

Ginzburg, V.B., "New Magnetoelastic Devices for Measurement of Drilling Load MKN2 and Washing Fluid Pressure MID1," *Machinery and Oil Equipment*, Moscow (USSR), Vol. 11, pp. 24-28, 1970.

Ginzburg, V.B., "Increasing Accuracy of Magnetoelastic Devices for Measurement of Drilling Load," *Machinery and Oil Equipment*, Moscow (USSR), Vol. 6, pp. 29-32, 1971.

Ginzburg, P.B. and Ginzburg, V.B., "Improvement of Maintenance and Metrological Characteristics of Magnetoelastic Torque Measuring Devices," *Automation and Telemechanization of Oil Industry*, Vol.5, pp. 18-22, 1973.

Ginzburg, V.B., "The Calculation of Magnetization Curves and Magnetic Hysteresis Loops for a Simplified Model of a Ferromagnetic Body," IEEE Transaction on Magnetics, Vol. MAG-12, No. 2, March 1976.

Ginzburg, V.B., "The Magnetoelastic Properties of a Simplified Model of a Ferromagnetic Body in Low Magnetic Field," IEEE Transaction on Magnetics, Vol. MAG-13, No. 5, March 1977.

2. Mathematical Modeling of Metal Production Processes

Ginzburg, V.B., "Dynamic Characteristics of Automatic Gage Control System with Hydraulic Actuators," *Iron and Steel Engineer*, pp. 57-65, January

1984.

Ginzburg, V.B., "Basic Principles of Customized Computer Models for Cold and Hot Strip Mills," *Iron and Steel Engineer,* pp. 21-35, September 1985.

Ginzburg, V.B. and Schmiedberg W.F., "Heat Conservation Between Roughing and Finishing Mills of Hot Strip Mills," *Iron and Steel Engineer,* pp. 29-39, April 1986.

Ginzburg, V.B., *Reradiating Heat Shield,* U.S. Patent No. 4,595,358, Jun. 17, 1986.

Kelk, G.F., Ellis, R.H. and Ginzburg, V.B. "New Developments Improve Hot Strip Shape: Shapemeter-Looper and Shape Actimeter," *Iron and Steel Engineer,* pp. 48-56, August 1986.

Ginzburg, V.B., "Strip Profile Control with Flexible Edge Backup Rolls," *Iron and Steel Engineer,* pp. 23-33, July 1987.

Ginzburg, V.B., Kaplan, N.M., James, K.L. and Zickefoose, W.F., "Application of Off-Line Computer Model MILLMAX at Weirton Steel's Hot Strip Mill," *Iron and Steel Engineer,* pp. 24-33, June 1988.

Ginzburg, V.B, *Steel-Rolling Technology: Theory and Practice,* Marcel Dekker, New York, 1989.

Ginzburg, V.B., Bakhtar, F. and Dittmar, R.W., "Theory and Design of Reheating Type Heat Retention Panels," *Iron and Steel Engineer,* pp. 17-25, December 1989.

Ginzburg, V.B., Kaplan, N.M., Bakhtar, F. and Tabone, C.J., "Width Control in Hot Strip Mills," *Iron and Steel Engineer,* pp. 25-39, June 1991.

Ginzburg, V.B., and Di Giusto, B., "Self-Compensating Back-up Rolls for the Improvement of Strip Profile," *Metallurgical Plants and Technology International,* pp. 98-100, Vol 2, 1993.

Ginzburg, V.B, *High-Quality Steel Rolling: Theory and Practice,,* Marcel Dekker, New York, 1993.

Ginzburg, V.B., Fanchini, R., Bakhtar, F.A., and Azzam, M., "Selection of Optimum Mill Configurations for Cold Mills," *AISE Steel Technology,* November 1999.

Ginzburg, V.B., *Continuous Spiral Motion System for Rolling Mills,* U.S. Patent No. 5,970,771, Oct. 26, 1999.

Ginzburg, V.B, and Ballas, R., *Flat Rolling Fundamentals,* Marcel Dekker, New York, 2000.

Ginzburg, V.B., *Superlarge Coil Handling System,* U.S. Patent No. 6,009,736, Jan. 4, 2000.

Ginzburg, V.B., *Continuous Spiral Motion System and Roll Bending System for Rolling Mills,* U.S. Patent No. 6,029,491, Feb. 29, 2000.

Ginzburg, V.B., *Metallurgical Design of Flat Rolled Steels,* Marcel Dekker, New York, 2005.

Ginzburg, V.B., Author and Editor, *Flat Rolled Steel Processes – Advanced Technologies,* Taylor & Francis Group LLC, Boca Raton, Florida, 2009.

Ginzburg, V.B., Author and Editor, *The Making, Shaping and Treating of Steels, Flat Products Volume,* Association of Iron & Steel Technology (AIST), Warrendale, PA, 2014.

3. Universal Spacetime Theory

Ginzburg, V.B., "Toroidal Spiral Field Theory," *Speculations in Science and Technology*, Vol. 19, 1996.

Ginzburg, V.B., *Spiral Grain of the Universe - In Search of the Archimedes File*, University Editions, Inc., Huntington, WV, 1996.

Ginzburg, V.B., "Structure of Atoms and Fields," *Speculations in Science and Technology*, Vol. 20, 1997.

Ginzburg, V.B., "Double Helical and Double Toroidal Spiral Fields," *Speculations in Science and Technology*, Vol. 22, 1998.

Ginzburg, V.B., *Unified Spiral Field and Matter - A Story of a Great Discovery*, Helicola Press, Pittsburgh, PA, 1999.

Ginzburg, V.B., "Nuclear Implosion," *Journal of New Energy*, Vol. 3, No. 4, 1999.

Ginzburg, V.B., "Dynamic Aether," *Journal of New Energy*, Vol. 6, No. 1, 2001.

Ginzburg, V.B., *The Unification of Strong, Gravitational & Electric Forces*, Helicola Press, Pittsburgh, PA, 2003.

Ginzburg, V.B., "Electric Nature of Strong Interactions," *Journal of New Energy*, Vol. 7, No. 1, 2003.

Ginzburg, V.B., "Unified Spiral Field Theory – A Quiet Revolution in Physics," *VIA-Vision in Action*, Vol. 2, No. 1 & 2, 2004.

Ginzburg, V.B., "The Relativistic Torus and Helix as the Prime Elements of Nature," *Proceedings of the Natural Philosophy Alliance*, Vol. 1, No. 1, Spring 2004.

Ginzburg, V.B., *Prime Elements of Ordinary Matter, Dark Matter & Dark Energy*, Helicola Press, Pittsburgh, PA, 2006.

Ginzburg, V.B., *Prime Elements of Ordinary Matter, Dark Matter & Dark Energy – Beyond Standard Model & String Theory*, The second revised edition, Universal Publishers, Boca Raton, Florida, 2007.

Ginzburg, V.B., "The Unification of Forces," *Proceedings of the Natural Philosophy Alliance*, Vol. 4, No. 1, 2007.

Ginzburg, V.B., "The Origin of the Universe, Part 1: Toryces," *Proceedings of the Natural Philosophy Alliance*, The 17th Annual Conference of the NPA 23-26 June 2010 at California State University Long Beach, Vol. 7, 2010.

Ginzburg, V.B., "Basic Concept of 3-Dimensional Spiral String Theory (3D-SST)," *Proceedings of the Natural Philosophy Alliance*, The 18th Annual Conference of the NPA, 6-9 July 2011 at the University of Maryland, College Park, USA, Vol. 8, 2011.

Ginzburg, V.B., *The Spacetime Origin of the Universe*, Helicola Press, Pittsburgh, PA, 2013.

Ginzburg, V.B., *The Spacetime Origin of the Universe with Visible Dark Matter & Energy*, Third edition, Helicola Press, Pittsburgh, PA, 2016.

Ginzburg, V.B., "A Novel Method of Modeling of Fundamental Properties of.

Ginzburg, V.B., "A Novel Method of Modeling of Fundamental Properties of Materials," *Proceedings of MS&T17 Materials Science & Technology*, Lawrence L. Convention Center, Pittsburgh, Pennsylvania, USA, October 8-12, 2017.

Index

www.ingramcontent.com/pod-product-compliance
Lightning Source LLC
Chambersburg PA
CBHW022052210326
41519CB00054B/323